Österreichische Akademie der Wissenschaften
Schriftenreihe der Erdwissenschaftlichen Kommissionen
Band 8

Geotectonic Evolution and Metallogeny of the Mediterranean Area and Western Asia

Proceedings of the Final Symposium of IGCP Project 169
Leoben, October 1984

Edited by

W. E. Petrascheck and S. Janković

In Kommission bei

Springer-Verlag Wien New York

1986

ISBN-13: 978-3-211-86527-9 e-ISBN-13: 978-3-7091-5822-7
DOI: 10.1007/978-3-7091-5822-7

Fotosatz und Druck: R. Spies & Co., 1050 Wien

ISSN 0171-2225

Inhalt/Contents

Vorwort

IGCP-Projekt 169 „Tectonic Evolution and Metallogeny in the Eastern Mediterranean and Western Asia" erwuchs aus dem schon mehrere Jahrzehnte zurückliegenden Bestreben der beiden Projektleiter, Prof. Dr. SLOBODAN JANKOVIĆ (Belgrad) und des Herausgebers dieses Bandes, die Bildung der Erzlagerstätten SE-Europas und der Alpen mit dem tektonischen Geschehen in diesem Abschnitt des alpin-mediterranen Orogens zu korrelieren. Schon vor 30 Jahren, im Oktober 1955, hatte der Verfasser in Leoben eine südostgeologische Tagung veranstaltet, bei der trotz der Behinderungen durch die Nachkriegsverhältnisse Geologen aus Ost und West zusammenkamen. Den tektonischen Rahmen hatte ein Vortrag des Altmeisters der Tektonik, HANS STILLE, gesteckt. (Vorträge der südostgeologischen Tagung in Leoben 1954, Berg- u. Hüttenmänn. Monatshefte, 101, 2, Wien 1956). Etwa gleichzeitig hatte S. JANKOVIĆ, aufbauend auf die Synthese von A. CISSARZ, die metallogenetische Stellung verschiedener Gruppen von Erzlagerstätten in Jugoslawien analysiert und anschließend eine Zusammenstellung der erzbildenden Epochen und der Erzbezirke des Landes veröffentlicht. (S. JANKOVIĆ, Metallogenetska epoki rudonosna podrucje Jugoslawije, Beograd 1967.)

Der Einzug der Begriffe der Plattentektonik in die geologische Betrachtungsweise brachte neue Vorstellungen in das Bild von der tektonisch-magmatisch-metallogenetischen Entwicklung des östlichen Mittelmeergebietes. Zusammengefaßt wurden diese Vorstellungen in einem von Janković herausgegebenen Band, der dem IGCP-Projekt 3 gewidmet war (Metallogeny and Plate Tectonics in the Northeastern Mediterranean, Fac. Min. & Geol., Belgrade 1977) und in einem Artikel von W. E. PETRASCHECK „Mineral Zoning and Plate Tectonics in the Alpine-Mediterranean orogen", GeolAssoc. Canada, Spec. paper 14, 1976.

Bei der Verfassung dieser Beiträge ergaben sich zwei fundamentale Probleme: erstens, daß die plattentektonischen Vorgänge im alpin-südosteuropäischen Raum nicht mit den plattentektonischen Modellen der pazifischen Kontinentalränder erklärbar sind, und zweitens, daß die ursprüngliche geotektonische Lage der geologischen Einheiten zur Eurasischen und zur Afrikanischen Platte noch wenig geklärt sind. Aus diesem Grund beschlossen die beiden Projektleiter, eine räumliche Erweiterung der Betrachtung des Tethys-Orogens nach Osten bis zum Pamirgebirge und nach Westen bis in den Atlas. Die neue und bisher erste metallogenetische Synthese des östlichen Abschnittes, also von Mittel-Asien, wurde von S. JANKOVIĆ zusammengestellt, der während mehrerer Jahre als Experte der Vereinten Nationen in diesem Gebiet arbeitete. Der zentral-mediterrane Abschnitt wurde von einer italienischen Arbeitsgruppe unter der Leitung von P. ZUFFARDI bearbeitet, das W-mediterrane Gebiet von Y. FUCHS, zusammen mit Kollegen aus Algerien und Tunesien. Aus dem Vergleich der einzelnen Regionen ergibt sich ein überraschend ähnlicher tektonisch-metallogenetischer Entwicklungsprozeß im ganzen Orogengürtel von Afghanistan bis Algerien: eine triassische Krustendehnung mit Magmen und Lagerstätten des Rifttyps, eine jurassisch-unterkretazische Bildung von Mikroozeanen mit Lagerstätten der ophiolithischen Assoziation, anschließend senone bis paleogene Subduktionen, gekennzeichnet durch den Porphyrie-Kupfererztyp und endlich Kontinent-Kontinent-Kollision mit anatektischer Granitoidbildung mit einer z. T. bis in das Jungtertiär wirksamen hydrothermal-aszendenten Erzbildung.

Die Sonderstellung von Alpen und Westkarpaten ist nach der Meinung des Verfassers durch deren tektonischen Stil, also durch deren komplexen Krustenbau zu erklären. Im Atlas ist ein gewisser metallogenetischer Einfluß der Afrikanischen Plattform erkennbar, desgleichen auch Beziehungen zur Verbreitung salinarer Ablagerungen.

Hypothetisch bleiben weiterhin der Ursprung und die Rotation der Mikrokontinente. Hier fehlen noch palaeogeographische und palaeomagnetische Korrelationen, durch die der Zusammenhang von Erzbezirken festgestellt werden kann, woraus sich auch für die regionale Prospektion nützliche Folgerungen ergeben würden.

Somit sind die 23 Beiträge dieses Bandes wertvolle und vielfach erstmalige Beschreibungen der Mosaiksteine des Tethysorogens im Sinne einer plattentektonischen Metallogenese. Die Zusammensetzung der Mosaik-Elemente ist eine interessante Aufgaben für die nähere Zukunft.

Eine vereinfachte kartographische Darstellung der Ergebnisse im Maßstab 1 : 5 Millionen ist für das Mediterrangebiet unter der Redaktion von Herrn A. EMBERGER (Orléans) in einem Atlas vorgesehen. Mehrere Autoren dieses Bandes haben die Entwürfe der Karten ihrer Länder für diesen Atlas verfaßt. Kurze Zusammenfassungen in deutscher Sprache wurden vom Herausgeber den einzelnen Artikeln vorangestellt.

Ich möchte allen Kollegen, die aus dem weiten Raum zwischen Dushambe (USSR) und Alger nach Leoben gekommen sind und hier vorgetragen haben, auf das herzlichste für ihre Beiträge zu einer interessanten wissenschaftlichen Zusammenarbeit danken. Für wertvolle redaktionelle und organisatorische Mitarbeit beim Final Meeting in Leoben sei Frau A. PÄRTAN mein aufrichtiger Dank gesagt.

Wien, Mai 1985 W. E. PETRASCHECK

Preface

IGCP-Project 169 "Geotectonic Evolution and Metallogeny in the Eastern Mediterranean and Western Asia" emanated from several earlier attempts of both Project Leaders, Prof. S. JANKOVIĆ (Belgrade) and the editor of this volume, to correlate the formation of ore deposits of Eastern Europe and the Alps with the tectonic events in this section of the alpine-mediterranean orogen. Already 30 years ago, in October 1955, the author had organized a Southeast-Geological Symposium in Leoben, at which, inspite of the impediment by the post-war situation, geologists from the East and West came together. The tectonic frame was defined by the famous master of tectonics, HANS STILLE. (Vorträge der Südostgeologischen Tagung in Leoben, 1954, Berg- u. Hüttenmänn. Monatshefte, 101, Wien 1956.)

About the same time S. JANKOVIĆ, setting up on the synthesis by A. CISSARZ, analysed the metallogenic position of different groups of ore deposits in Yugoslavia and, adjoining to this, published a summary on the ore forming epochs and ore districts of the country. (S. JANKOVIĆ, Metallogenetske epoki i rudonosna podrucja Jugoslawije, Beograd 1967.)

The entry of plate tectonic terms into geological considerations brought new concepts for the view on the tectonic-magmatic-metallogenic development. These concepts were summarized in a volume, edited by S. JANKOVIĆ and dedicated to IGCP-Project 3 (Metallogeny and Plate Tectonics in the Northeastern Mediterranean, Fac. Min. & Geol., Belgrade 1977), as well as in an article by W. E. PETRASCHECK "Mineral zoning and plate tectonics in the Alpine-Mediterranean orogen" (Geol. Assoc. Canada, Spec. paper 14, 1976.)

When compiling these contributions, two fundamental problems arised: first, plate tectonic events in the alpine and Southeast European area are not explainable by the plate tectonic models of the Pacific continental margins, and secondly, the former geotectonic positions of the geological units in relation to the Eurasian and African Plate are still not well defined. Furthermore the two project leaders decided on a spatial extension of the investigations as far as the Pamir Mts. towards the East and as far as the Atlas towards the West.

The new and so far first metallogenic synthesis of the Eastern section, i. e. of Central Asia, was compiled by S. JANKOVIĆ, who during several years has been working as an expert for UNDP in this area; the Central Mediterranean section was analysed by an Italian working group under the leadership of P. ZUFFARDI (Milano). The

West mediterranean area was composed by Y. Fuchs together with colleagues from Algeria and Tunisia.

When comparing these different regions, a surprisingly similar tectonic-metallogenic developing process is showing in the entire orogenic belt from Afghanistan to Algeria: a crustal dilation with magmas and rift-type deposits during the Triassic, a Jurassic-Lower Cretaceous formation of micro-oceans with deposits of the ophiolitic association, followed by Senone to Paleogene subductions, characterized by the Porphyrie-Copper type, and finally continent-continent-collision with anatectic granitoid formation with a rise of hydrothermal-ascendent ores, partly effective until the Neogene. The exceptional metallogenic character of the Alps and Western Carpathians, according to the author's opinion can be explained by their tectonic feature, i. e. their complex crustal structure. In the Atlas a certain metallogenic influence of the African platform is manifest, combined with the distribution of saline formations.

Still hypothetical remain the origin and the rotations of the microcontinents. Here the palaeogeographic and palaeomagnetic correlations are still missing, by which the connection of the ore districts could be established. Also useful results for regional prospection would emanate from a palaeomagnetic analysis.

In this way the **23** articles of this volume are a most valuable contribution and in many cases deliver descriptions of the Tethyan orogene in connection with plate tectonic and metallogeny for the very first time. The composition of the mosaic-like elements is an interesting task for the near future.

A simplified cartographic presentation of the results on a scale of 1 : 5 Mill. in an atlas is intended to be edited by A. Emberger (Orléans) for the mediterranean area. Several authors of this volume have composed drafts of the maps of their respective countries, to be incorporated into this atlas. Short abstracts in german have been put ahead of each article by the editor.

I would like to express my cordial thanks to all colleagues having come to Leoben as far as from Dushambe (USSR) to Alger to hold their lectures and for their contribution in achieving an international scientific co-operation. My thanks also go to Mrs. A. Pärtan who has been most helpful in organizing the Symposium and in editing this volume.

Vienna, May 1985 — W. E. Petrascheck

Correlation of the Alpidic geotectonic features in the Eastern Mediterranean

By A. TOLLMANN, Vienna*)

Abstract

A survey on the main Alpidic geotectonic structures, their genesis and correlation in the Eastern Mediterranean is given, useful for the understanding of the principal types and phases of ore mineralization.

Zusammenfassung

Ein Überblick über die Hauptzüge der alpidischen tektonischen Strukturen des östlichen Mediterran-Gebietes, über deren Genese und Korrelation wird vorgelegt, mit dem Zweck, das Verständnis für die grundlegenden Typen und Phasen der alpinen Erzlagerstättenbildung dieser Region zu fördern.

Content

1. The significance of the tectonic events for metallogeny

After more than hundred years of intensive geological investigations in the Eastern Mediterranean realm the structure of this region is now known in its princi-

*) Address of the author: Prof. Dr. Alexander TOLLMANN, Institut für Geologie Universität Wien, Universitätsstraße 7, A-1010 Wien, Austria.

pal features. Of course still a lot of questions concerning the correlation of the different units of the individual sectors of this mountain chain remains to be clarified.

During the last years new methods like paleomagnetic investigations, radiometric dating etc. and new concepts like the plate tectonics brought light into some genetic problems.

Important phases of geotectonics in the Alpidic era, connected with metallization, are the following events:

1. The attenuation of the earth crust and somewhat later the rifting and opening of first rift-systems: This first step began in the region mentioned during the Permian (attenuation) and the lower Middle Triassic (first rifting) by a progressive division of Pangea into two supercontinents on both sides of the scissorlike opening Tethys. During this phase particulary the stratiform ore deposits of sulphides of Pb, Zn, Fe, Cu, Sb and Hg have been formed.

2. The formation of wide true ocean-floors with thick ophiolite sequences followed during Jurassic and (Lower) Cretaceous time, combined with island-arc volcanism. Magmatic ultramafic to mafic and volcano sedimentary types of ore deposits prevailed in this phase with metals like Fe, Cu, Ni, Cr, Co, Zn, Mn.

3. By the Upper Triassic to Liassic Cimmerian (-Indosinian) phase in the eastern part of the northern branch of Tethys and by a sequence of Alpidic orogenic phases since the Lower Cretaceous (since the Austroalpine phase in the Hauterivian-Barrémian) in the entire Tethys began the subduction of the margins of the great plates, began the orogenesis with nappe formation and an calcalcaline volcanism with andesitic to dacitic magmas. It was connected with the formation of porphyry copper ores, and with deposits of Mo, Au, Pb, Zn etc. This type of volcanism and also a granodioritic plutomism lasted still until the younger phases of the orogenetic development in Miocene time, accompanied or followed by hydrothermal ore deposits (Pb, Zn, Sb, Mo, Hg, As etc.).

2. The main features of the Alpidic paleogeography of the Eastern Mediterranean

The attenuation of the crust in the Eastern Mediterranean realm began with the birth of the Tethys during Permian time, when the Mediterranean region was invaded by the sea, coming from the east and advancing towards west. The formation of this Permo-Mesozoic Tethys was totally independent of older Mediterranean oceans, like the extremely broad Hercynic "Prototethys" in Central Europe. This Alpidic Tethys was formed completely discordant above the eroded Hercynian mountain system, as well as above the northernmost part of the old socle of the Afro-Arabian block.

The advance toward west of the Tethys in Permian time reached the Carnian Alps, Sicily and Tunisia, in Triassic time (with the Sephardic facies as marginal sea) the Betic Cordillera and the Rif, in Jurassic time by opening of the Tethyan Atlantic, the region of Middle America.

Starting already in Permian time, a characteristic subdivision of the Tethys area into three longitudinal zones is recognizable, remaining with somewhat changing details also during the Mesozoic era:

a) A northern trough of the Tethys sea above an attenuated crust received a thick sequence of neritic sediments during Permian time. Here the rifting of narrow channels with oceanic floor with ophiolites, pillow lavas and radiolarites had started since the earlier Middle Triassic: in the Western Carpathians since the Middle Triassic (Meliata series), in the Eastern Carpathians (Transsylvanian nappe) and in the Dobrodgea since the Lower Anisian. The eastern prolongation of this northern channel from Crimea to Caucasus and a parallel secondary rift system in Northern Anatolia (Karakaya Marginal Sea Belt) have been closed during Upper Triassic time by the Cimmerian orogenesis. But under a somewhat changed shape the northern branch of the Tethys sea had regenerated since the Lower Jurassic.

b) A median rise, formed by a long, later divided microcontinent with thick crust, named Kreios. This broad geoanticline sometimes was flooded by shallow sea, bearing typical alpine-mediterranean macro- and microfaunas, or fell dry, forming gaps of different extent in the corresponding series or showing masses of detritic material. The flora of this microcontinent formed – e. g. during the Ladinian/Carnian time – a separate Mediterranean phytogeographic region with mixed northern (Laurasia) and southern (Gondwana) types of flora (N. Pantić et al., 1983, p. 6). The position of this median rise as a middle ridge yet in Permian time was pointed out by I. Argyriadis (1975, p. 60, p. 95 ff.; 1980), and in Triassic time and later by A. Tollmann (1978, p. 341: "Kreios").

c) The southern trough also was marked by a thick series of neritic sediments during the Permian. In Triassic time the rifting of the crust had begun here since the Middle or Upper Anisian by the opening of a series of parallel narrow troughs: The longest external deepwater basin with pelagic sediments, flysch and lavas in Middle Triassic time represents the Budva-Cukali-Pindus zone in the Dinarides and Hellenides, Cyprus and an equivalent in the Taurus of Antalya. Further internal rifting zones in Triassic time were mentioned by G. Kauffmann (1976) from Greece. During the Jurassic and the Cretaceous a broad ocean formed in the eastern prolongation of the southern Mediterranean branch of the Tethys in Zagros and in Oman.

3. Some remarks on regional tectonics

a) The northern branch of the Alpine orogenetic system

The northern branch of the Eastern Mediterranean comprises the Eastern Alps, the Carpathians, the Balkan chain and the Pontides in Anatolia, moreover also a spur running from the Dobrodgea to Crimea and Caucasus.

The connection of the nappe system of the Eastern Alps and Western Carpathians was established in a satisfactory manner by D. Andrusov (1960, 1968), A. Tollmann (1960, 1969, 1975), M. Mahel (1979) et al. For details see their publications: As equivalents of Flysch, Helvetides, Penninic, Lower-, Middle-and

Upper-Austroalpine nappe systems of the Alps one can recognize Flysch, Pienides, Tatrides, Vysoká and Krížna, Veporides (with Föderata- and Velký Bok-series) and Choč-Strážov-Gemerides in the Carpathians. The Meliata unit and the Südbükk-schiefer unit represent a new Carpathian element above the oceanic crust (H. KOZUR & R. MOCK, 1973; K. BALOGH et al., 1984). In general one can not agree with the ideas of B. LEŠKO & I. VARGA (1980) about the tectonic correlation based on some geophysical data: The Penninic zone of the Alps shurely does not reappear in the Klippenbelt, the Veporides and Gemerides.

The great break between the Transdanubian Mts. and Western Carpathians on the one hand and Tisia and Eastern Carpathians on the other hand is explained by H. P. LAUBSCHER (1971, Fig. 2), B. GECZY (1973, p. 426), F. HORVATH et al. (1977, p. 210, Fig. 1), S. KOVÁCS (1980, Fig. 5; 1982, Fig. 1) et al. as a considerable transverse strike-slip fault along the Tisia block, including the Bihor autochthonous in the east, which was transported from the outer zone of the Carpathians (marginal Gresten facies today in the south in Mecsek, Villany and Bihor!) to an inner position. The opinions about the timing of this shift vary from Lower Cretaceous to Lower Tertiary. The north facing nappe system of the Apuseni Mts. therefore is considered as the displaced prolongation of the nappe system of the Western Carpathians, supported also by the facies sequence within this nappe pile (M. BLEAHU, 1976).

The correlation of the tectonic units of the Eastern and Southern Carpathians has been given in 1969 by A. TOLLMANN (map Fig. 1) and in excellent manner in recent time – including the Kraistides and the Balkan chain – by M. SANDULESCU (1975, Figs. 2, 3, 16, 17, 26, 30) and M. BLEAHU et al. (1974), so that it is sufficient to draw the attention to these publications.

The nappe structure in the area of the Paring window with a sequence of Danubian (below), Severin-, Getic- and Supragetic nappe can be traced to the Balkanides (tectonic division cf. E. BONČEV et al., 1974): The Pre-Balkan corresponds with the external Danubian, the Stara Planina nappe with the internal Danubian, the Srednogora nappe with the vast Getic nappe (M. SANDULESCU, 1975, Fig. 16).

The problematic structure of the (Sakar-) Strandsha zone in the eastern part of the Srednogorie unit is either explained as a tectonic window with metamorphic Triassic and Jurassic (A. TOLLMANN, 1965; 1968, p. 59; P. BECK-MANNAGETTA et al., 1978, map: "Thrakian window") or as a fault trough (graben) with a metamorphic Lower Paleozoic and also Mesozoic series (I. NATSCHEW et al., 1979) or as an anticlinorium with metamorphic Triassic-Jurassic in local synclines (S. SAVOV, 1962; S. SAVOV et al., 1971).

The movements of the Balkanide nappes doubtlessly are directed towards north. A nappe structure is evident. Therefore we can not agree with the hypothesis of a slightly folding without any southward dipping subduction zone in the sense of E. AIELLO et al. (1977, Fig. 3) or K. HSÜ et al. (1977, Fig. 5).

Recently even the idea arised that the vast Srednogorie nappe has been thrusted as a whole above the Rhodope massif, which was interpreted as a tectonic window (P. BECK-MANNAGETTA & W. MEDWENITSCH, 1978, Nebenkarte 1 : 9,000.000).

The prolongation of the main northern branch of the Mediterranean orogen runs eastwards to the Pontides in northern Anatolia. A new synthesis of A. ŞENGÖR & Y. YILMAZ (1981) gives a pretty survey including a lot of details on the Tethyan evolution of Turkey in space and time. In Permotriassic time two ocean channels of the northern branch of the Tethys acted as mobil zones: The Crimea-Caucasus-Mashad trough in the north and the Karakaya Marginal Sea as a narrow channel south of the Pontides. By the Cimmerian orogenesis in the Upper Triassic both areas have been reduced, the southern one was totally closed. After A. ŞENGÖR et al. (1981, p. 210 ff. and Fig. 6B) two new ocean channels were formed in northern Anatolia during Early Jurassic time: the Intrapontic branch and – divided by the Sakarya microcontinent – the Izmir-Ankara channel. Since the Middle Jurassic the sea has again invaded a broad area of northern Turkey ("Vardar ocean") north of the central Anatolian platform. The alpine orogenesis in this region began with the subduction north of the Pontides island arc during Mid Cretaceous time, continued with ophiolite obduction (Bozkir ophiolite nappe) in Late Cretaceous and was finished mainly in Mid-Eocene – much sooner than in the southern branch of the Tethys in this country. A. ŞENGÖR et al. (1981) think about north dipping subduction zones beneath the European continent and Sakarya island and therefore a southward facing of nappe structures in the Pontides, still has not been proved sufficiently (cf. I. KETIN, 1956, Fig. 1).

b) The southern branch of the Alpine orogen

Concerning the tectonic correlation of the units in the southern branch of the eastern Mediterranean realm we can be brief with regard to some recent papers, summerizing our knowledge from the region of Dinarides-Hellenides. After the "classic" papers of L. KOBER (1952) and K. PETKOVIĆ (1958) about the great tectonic lines of the Dinarides we mention at least the articles by J. AUBOUIN (1963, 1973), J. AUBOUIN et al. (1970), M. HERAK (1986), W. MEDWENITSCH (1956), W. MEDWENITSCH & B. SIKOŠEK (1965), J. PAMIĆ (1984 a, b), B. SIKOŠEK (1974), B. SIKOŠEK & W. MEDWENITSCH (1965) as some reviews about the structure of the Dinarides and the papers by J. AUBOUIN et al. (1970, 1976), D. BERNOULLI et al. (1972, 1974), V. JACOBSHAGEN et al. (1976, 1978), J. CADET et al. (1980), M. BONNEAU (1984), R. HALL et al. (1984), A. POISSON (1984) about the Hellenides and their correlation to the Taurides. The article by R. BLANCHET & J. MERCIER (1978) and the volume by J. DIXON & A. ROBERTSON (1984) summarizes all important facts about the whole region.

As summary of all these papers results: From the external side in the west, from the foreland Apulia, with thick neritic sediments of Triassic and Jurassic we meet prograding towards East, towards the internal region, however, the following facies zones (respectively nappes in later time):

1. The Adriatic-Ionien Zone, a miogeosynclinal trough zone, with neritic series up to the Middle Liassic, covered by a pelagic sequence. This zone is relatively autochthonous, with Schuppen structure, detached in the Triassic gipsum level.

2. The Dalmato-Gavrovo-Tripolitza Zone, a miogeosynclinal shallow platform, rich in reef-buildups, only in Crete allochthonous.

3. The Budva-Cukali-Pindus Zone, an eugeosynclinal deep water trough with abundant pelagic series including radiolarits and basic lavas from Triassic to Eocene time, finally transported as a cover nappe far towards the West.

4. The Karst-Parnass Zone, a platform with thick neritic series mainly from Triassic to Upper Cretaceous, namely in the High- and Pre-Karst. Nappe structure.

5. The Bosnian-Grammos-Beotian Zone, a trough with pelagic sediments from Triassic to Jurassic and a flysch of the Cretaceous. The facies corresponds in best manner with that of the Pindus Zone.

6. The Serbo-Subpelagonian Zone, a deep trough during Upper Jurassic and Lower Cretaceous with masses of ophiolites and radiolarites, covered by neritic series of Higher Cretaceous. Transported as a nappe system (Merdita-, Durmitor-nappe etc.) with rests of a sole of Paleozoic (in Yougoslavia) and Triassic formations.

7. The Golija-Korab-Pelagonian Zone, a ridge with a neritic Triassic and Jurassic series covered by a sheet of ophiolites and a gap up to the Upper Cretaceous. This unit was transported as a nappe system (Korab-Perister nappe etc.) with a socle of crystalline basement and granites. In the eastern part of this zone and its eastern prolongation we find the windows of Olympus, Attica, Eubea, the Cyclades and Menderes. The units 5–7 have been summarized by L. Kober (1952) as "Extern Radiolarite-Ophiolit Zone" = "ERO".

8. The Vardar Zone ("IRO-Zone" L. Kober), the innermost trough zone in front of the Serbo-Macedonian hinterland (9), is characterized by masses of ophiolites and radiolarites (diabase-hornfels formation) in the Upper Jurassic, deposited on neritic Triassic-Jurassic limestones. Three phases of flysch(oid) formations (boundary Jurassic/Cretaceous, Middle and Upper Cretaceous) follow.

In a comprehensive manner we can group the units 1–2 into the relative autochthonous miogeosynclinal external area, the unit 3–8 into the allochthonous eugeosynclinal internal area. The distances of the nappe transport of these inner units exceed about hundred, probably in some examples also two hundred kilometers. The formation of nappes took place essentially during the Lower Cretaceous and the Late Eocene (D. Bernoulli & H. Laubscher, 1972). As an indicator for the wandering of movements from the inner to outer zones we can – as usually – utilize the beginning of the flysch formation. According to J. Aubouin (1964) and particularly D. Richter (1976, 1978a, b) we find the oldest flysch (1) of Dogger age in the inner Vardar Zone (Svoula flysch), then (2) the Eohellenic flysch of Upper Jurassic to Lower Cretaceous in the Vardar- and Pelagonian zone, followed (3) by the main flysch stage of the inner zones (Pelagonides, Parnass, Pindus) from the end of the Cretaceous to the Upper Eocene, and finally (4) the main flysch stage of the outer zones (Gavrovo, Ionian zone) from Upper Eocene to Lower Miocene, terminated by a L./M. Miocene flysch of the outer fold zone.

The metamorphosis began in the inner region, the Vardar zone, after a first precursor of the Upper Jurassic ophiolite phase in the Midcretaceous before the sedimentation of the Albian flysch.

The correlation from the Aegean region to Anatolia and to the Taurides was stated by papers of W. E. PETRASCHECK (1960), J. AUBOUIN et al. (1976), D. BERNOULLI et al. (1974), R. BRINKMANN (1976), J. BRUNN et al. (1976), J. DIXON & A. ROBERTSON (1984), L. MOUSSOULOS et al. (1977), O. MONOD (1977) etc.

In the Taurides of southern Anatolia the rifting of the crust started in Upper Anisian to Ladinian time with production of radiolarites, tuffites, pelagic limestones, turbidites, slumpings and olistholits. A second phase in Carnian time brought up to 600 m of pillow lavas, of ophiolites, accompanied by radiolarites, pelagic and turbiditic sediments (TH. JUTEAU, 1970, 1979; J. MARCOUX, 1974; O. MONOD et al., 1974). This deep ocean channel can be traced since the Triassic from the Antalya nappes regions across Bitlis to the Zagros, with the maximum of tectonic activity in Upper Cretaceous (L. RICOU, 1971, Fig. 1). The maximum of the oceanic stage in the Taurides was attained in Early Cretaceous. Since Midcretaceous time here the orogenesis started along a north dipping subduction zone, forming the pile of Lycian nappes in the region of Antalya (I. KETIN, 1960; E. BINGÖL, 1974; J. BRUNN et al., 1972, 1974; J. MARCOUX & A. POISSON, 1972; L. RICOU et al., 1974, 1975, 1979; J. DIXON et al., 1984; R. MALL et al., 1984; A. ŞENGÖR et al., 1984).

4. Review about the great steps of Alpidic evolution

The formation of the Mediterranean Tethys began by the—above mentioned—attenuation of the Hercynian crust during Permo-Skythian time as the result of the start of the scissorlike meridional drifting of the two supercontinents. Since the Lower Anisian the first oceanic rift channels opened in the northern and southern branch of the Tethys. The attenuation of the crust continued. The facies of the sediments is also dictated by eustatic see level movements (Lower Carnian event etc.). The Jurassic evolution is controlled by the transversal drifting of Africa from Northern America: By this rifting broad oceans opened in the Penninic area in échelon pattern. The attenuation and rifting of the crust rise up to its maximum during the Upper Jurassic with its ophiolite-radiolarite formations in the northern and southern branch of this sector of the Tethys and with formation of grand gravity sliding nappes of the Hallstatt zone in the northern branch. The main subduction by a convergent drifting of the supracontinents started in both branches of the Tethys by the Austroalpine phase of Hauterivian-Barrémian (130—125 million years ago), followed by the first metamorphic processes. The nappe formation realized during Midcretaceous and Lower Tertiary, followed the advance of the subduction front from the inner to the outer zone in both branches. In younger time great block movements along transverse faults interferred with the overthrust tectonics. The ore mineralization during the whole Alpidic era was corresponding with these tectonic events, as outlined in the first chapter.

Literature

1. Synoptic geological maps

Beck-Mannagetta, P. & Medwenitsch, W.: Atlas der Donauländer, Blatt Geologie mit Tektonik, 1 : 2,000.000. – Wien (Österr. Ost- u. Südosteuropa-Inst.) 1978.

Bogdanoff, A.: Carte géologique de l'Europe 1 : 10,000.000. – Moscou (Comm. Cartes tectoniques internat.) 1970.

Choubert, G., & Faure-Muret, A. [Coord.]: Europe. Feuille 9 de l'Atlas Géologique du Monde, 1 : 10,000.000. – Paris (Unesco, 26e congr. géol. intern.) 1980.

Dercourt, J., Zonenshain, L. P. et al.: Présentation de 9 cartes paléogéographiques au 1/20.000.000e s'étendant de l'Atlantique au Pamir pour la période du Lias à l'Actuel. – Bull. Soc. géol. France, (8) *1,* 637–652, 1 fig., 10 maps, Paris 1985.

Gaertner, H.-R. v., & Walther, H. W.: Carte Géologique Internationale de l'Europe, 3. Ed., 1 : 1,500.000, 47 Bl., D 5: Budapest; Hannover (Bundesanst. Bodenforsch.) 1969.

– International Geological Map of Europe and the Mediterranean Region, 1 : 5,000.000, 2 Bl., Hannover (Bundesanst. Bodenforsch.) 1971.

Laffitte, P. [Coord.]: Carte métallogénique de l'Europe, 1 : 2,500.000, 9 feuilles, Paris, (UNESCO) 1970.

Mahel, M. [Ed.]: Tectonic Map of the Carpathian-Balkan Mountain System etc. – 1 : 1,000.000, 8 sh., Paris (UNESCO/Geol. ústav D. Stúra) 1973.

– Note explicative à la Carte tectonique du système Carpates-Balkanes, 1 : 1,000.000. – 44 p., Paris (UNESCO) 1975.

Shiw, D. I. [Red.]: Carte tectonique de l'Europe et des régions avoisinantes, 1 : 10,000.000. – Moscou (Comm. Carte géol. monde) (1975) 1979.

Schatzsky, N. [Dir.]: Carte tectonique international de l'Europe, 1 : 2,500.000, 16 feuilles, Moscou (Comm. Carte géol. monde) 1962.

Szadeczky-Kardoss, E. [Ed.]: Map of metamorphites of the Carpatho-Balkan-Dinaride Area, 1 : 1,000.000; 4 sheets, Budapest (Lab. Geochem. Res. Hungar. Acad. Sci.) 1976.

Zwart, H. [Coord.]: Metamorphic Map of Europe, 1 : 2,500.000; 16 sheets, Paris (UNESCO) 1973.

2. Review-papers and -books

Ager, D.: The Geology of Europe. – XIX, 535 p., num. fig., tab., Maidenhead (Mc Graw-Hill) 1980.

Alpan, S., Bingöl, E. et al. [Ed.]: Papers Congr. Earth Sci. Occas. 50th Anniv. Turkish Rep., 608 p., num. fig., tab., pl., Ankara (Miner. Res. Explor. Inst.) 1975.

Aubouin, J., Debelmas, J., & Latreille, M.: Géologie des chaînes alpines issues de la Téthys. – Mém. BRGM., *115,* 354 p., num. fig., Orléans 1980.

Biju-Duval, B.: De la Téthys aux mers intra-alpines actuelles: Introduction. – Mém. BRGM., *115,* (Coll. C 5, 26° congr. géol. int.), p. 232–236, 2 fig., Paris 1980.

Biju-Duval, B., & Montadert, L. [Ed.]: Structural history of the Mediterranean Basins. – Symp. int. 25 congr. comm. int. explor. sci. Méditerranée, Split 1976. 448 p., num. fig. and tab., pl. 1–8, Paris (Édit. Technip) 1977.

Bogdanoff, A. et al. [Ed.]: Note explicative pour la carte tectonique internationale de l'Europe au 1 : 2,500.000. – Congr. géol. internat., Comm. Carte géol. Monde, 360 p., 121 fig., Moskau (Nedra) 1964.

Closs, H., Roeder, D., & Schmidt, K.: Alps, Apennines, Hellenides. – Sci. Rep. Inter Union Comm. Geodyn., *38,* 620 p., 279 fig., 14 tab., 5 pl., Stuttgart (Schweizerbart) 1978.

Dercourt, J., Zonenshain, L. P., et al.: Présentation de 9 cartes paléogéographiques au 1/20.000.000e s'étendant de l'Atlantique au Pamir pour la période du Lias à l'Actuel. – Bull. Soc. géol. France, (8) *1*, 637–652, 1 fig., 10 maps, Paris 1985.

Dewey, J., Pitman, W., Ryan, K., & Bonnin, J.: Plate Tectonics and Evolution of the Alpine System. – Bull. geol. Soc. Amer., *84*, 3137–3180, 19 fig., Tulsa 1973.

Dixon, J. E. & Robertson, A. [Ed.]: The Geological Evolution of the Eastern Mediterranean. – Spec. Publ. geol. Soc., *17*, 824 p., num. fig. and tab., Oxford etc. (Blackwell) 1984.

Hsü, K.: Mediterranean Sea. – Mc. Graw Hill Yearbook Science and Technology, 4 p., 3 fig., Maidenhead/Berkshire (Mc Graw Hill Book Comp.) 1978.

Hsü, K., & Bernoulli, D.: Genesis of the Tethys and the Mediterranean. – Initial Rep. Deep Sea Drilling Proj., *42* (1), 943–949, 3 fig., Washington 1978.

Ilavský, J. [Ed.]: Metallogénese de l'Europe alpine centrale et du sud-est. – 415 p., 126 fig., sev. tab., Bratislava (Inst. Geol. D. Štúr) 1979.

Janković, S. [Ed.]: Metallogeny and plate Tectonics in the Northeastern Mediterranean. – 559 p., 121 fig., Belgrade (Fac. Mining Geol. Univ.) 1977.

Laubscher, H., & Bernoulli, D.: Mediterranean and Tethys. – [In:] A. Nairn et al. [Ed.]: The Ocean basis and margins, *4A*, 1–28, 9 fig., New York (Plenum Publ. Comp.) 1977.

Lemoine, M. [Ed.]: Geological Atlas of Alpine Europe and Adjoinig Alpine Areas. – XV, 584 p., 655 fig., 2 maps, Amsterdam (Elsevier) 1978.

Mahel, M. [Ed.]: Tectonics of the Carpathian Balkan Regions. – 456 p., num. fig., tab., Bratislava (Geol. Inst. D. Štúr) 1974.

– [Ed.]: Tectonic problems of the Alpine System. (D. Andrusov-Festschr.). – 259 p., num. fig., tab., pl., Bratislava (VEDA) 1975.

– Geotectonic position of magmatites in the Carpathians, Balkan and Dinarides. – Západné Karpaty, ser. geol., *4*, 173 p., 17 fig., Bratislava 1978.

Petrascheck, W. E.: Magmatismus und Metallogenese in Südosteuropa. – Geol. Rdsch., *42*, 128–143, 1 fig., pl. 3, Stuttgart 1953.

– Mineral zoning and plate tectonics in the Alpine-Mediterranean Area. – Spec. Pap. geol. Assoc. Canada, *14*, 353–359, 3 fig., Ottawa 1976.

Tollmann, A.: Entstehung und früher Werdegang der Tethys etc. – Mitt. österr. geol. Ges., *77* (1984), 93–113, 4 fig., Wien 1984.

Tollmann, A., & Kristan-Tollmann, E.: Paleogeography of the European Tethys from Paleozoic to Mesozoic etc. – [In:] K. Nakazawa [Ed.]: The Tethys – her paleogeography and paleobiofacies from Paleozoic to Mesozoic Eras. – 25 p., 5 fig., Tokyo (Tokai Univ. Press) 1985.

3. Regional literature

Aiello, E., Bartolini, C. et al.: Sedimentary features of the Srednogorie zone (Bulgaria): An Upper Cretaceous intra-arc basin. – Sediment. Geol., *19*, 39–68, 20 fig., 2 tab., Amsterdam 1977.

Andrusov, D.: Neues über die Epirogenese und Orogenese in den Westkarpaten. – Geol. Rdsch., *50*, 404–415, pl. 9, Stuttgart 1960.

– Grundriß der Tektonik der Nördlichen Karpaten. – 188 p., 70 fig., Bratislava (Slovak. Akad. Wiss.) 1968.

Argyriadis, I.: Mésogée permienne, chaîne hercynienne et cassure téthysienne. – Bull. Soc. géol. France, (7) *17*, 56–67, 6 fig., Paris 1975.

Argyriadis, I., Graciansky, P. C. de et al.: The opening of the Mesozoic Tethys between Eurasia and Arabia-Africa. – Mém. BRGM., *115*, 199–214, 7 fig., Paris 1980.

AUBOUIN, J.: Esquisse paléogéographique et structurale des chaînes alpines de la Méditerranée moyenne. – Geol. Rdsch., *53*, 480–534, 4 fig., Stuttgart 1963.
– Réflexions sur le problème des Flyschs et des Molasses: Son aspect dans les Hellénides (Grèce). – Eclogae geol. Helv., *57*, 451–496, 12 fig., Basel 1964.
– L'exemple des Dinarides etc. – Bull. Soc. géol. France, (7) *15*, 401–644, 8 fig., Paris 1973.
– Des tectoniques superposées et de leur signification par rapport aux modèles geóphysiques: L'exemple des Dinarides. – Bull. Soc. géol. France, (7) *15*, 426–458, 8 fig., Paris 1973.
AUBOIN, J., BLANCHET, R. et al.: Essai sur la géologie des Dinarides. – Bull. Soc. géol. France, (7) *12*, 1970, 1060–1095, 3 fig., 1 tab., Paris 1972.
AUBOUIN, J., BONNEAU, M. et al.: Contribution à la géologie des Hellénides etc. – Annales Soc. géol. Nord, *90*, 277–306, 14 fig., Lille 1970.
AUBOUIN, J., BONNEAU, M., DAVIDSON, J. ET AL.: Esquisse structurale de l'Arc égéen externe: des Dinarides aux Taurides. – Bull. Soc. géol. France, (7) *18*, 327–336, 3 fig., Paris 1976.
BALOGH, K., KOZUR, H., & PELIKAN, P.: Die Deckenstruktur des Bükkgebirges. – Geol. paläont. Mitt. Innsbruck, *13* (3), 89–96, 2 fig., 1 tab., Innsbruck 1984.
BERNOULLI, D. GRACIANSKY, P. CH. DE, & MONOD, O.: The Extension of the Lycian Nappes (SW Turkey) into the Southeastern Aegean Islands. – Eclogae geol. Helv., *67*, 39–90, 31 fig., pl. 1, Basel 1974.
BERNOULLI, D., & LAUBSCHER, H.: The Palinspastic Problem of the Hellenides. – Eclogae geol. Helv., *65*, 107–118, 6 fig., Basel 1972.
BINGÖL, E.: Discussions on the metamorphic map of Turkey in a scale of 1 : 2,500.000 and geotectonic evolution of some metamorphic belts. – Bull. Min. Res. Expl. Inst. Turkey, *83*, 132–138, 5 fig., 1 map, Ankara 1974.
BLANCHET, R., & MERCIER, J.: The Dinarides and the Hellenides. – [In:] M. LEMOINE [Ed.]: Geological Atlas of Alpine Europe etc., 461–508, 49 fig., Amsterdam (Elsevier) 1978.
BLEAHU, M.: Structural position of the Apuseni Mountains in the Alpine System. – Rev. roum. géol. géoph. géogr. géol., *20* (1), 7–19, 2 fig., 1 tab., Bucureşti 1976.
BLEAHU, M. et al.: Rumanian Carpathians. – [In:] M. MAHEL [Ed.]: Tectonics of the Carpathian Balkan Regions, 221–283, 12 fig., Bratislava (Geol. Inst. D. Štúr) 1974.
BONČEV, E. et al.: The Balkanides. [In:] M. MAHEL [Ed.]: Tectonics of the Carpathian Balkan Regions, 307–345, fig. 4–23, Bratislava (Geol. Inst. D. Štúr) 1974.
BONNEAU, M.: Correlation of the Hellenides nappes in the south-east Aegean and their tectonic reconstruction. – Spec. Publ. geol. Soc., *17*, 517–527, 5 fig., Oxford 1984.
BREZSNYÁNSZKY, K. [Ed.]: Hungary. Excursion 104. – Int. Geol. Congr. XXVIIth Sess. Moscow, Exc.–Guide, *104*, 93 p., 43 fig., Budapest 1984.
BRINKMANN, R.: Geology of Turkey. – IX, 158 p., 68 fig., Stuttgart (Enke) 1976.
BRUNN, J. H., ARGYRIADIS, I. et al.: Eléments majeurs de liaison entre Taurides et Hellénides. – Bull. Soc. géol. France, (7) *18*, 481–497, 2 fig., Paris 1976.
BRUNN, J. H., GRACIANSKY, P.-CH. DE et al.: Structures majeurs et corrélations stratigraphiques dans les Taurides occidentales. – Bull. Soc. géol. France, (7) *12* (1970), 515–556, 9 fig., pl. 1–4, Paris 1972.
BRUNN, J. H., RICOU, L. et al.: Le problème de l'origine des nappes et de leurs translations dans les Taurides occidentales. – Bull. Soc. géol. France, (7) *16*, 101–106, 2 fig., Paris 1974.
CADET, J.-P., BONNEAU, M. et al.: Les chaînes de la Méditerranée moyenne et orientale. – Mém. BRGM., *115*, p. 98–118, 10 fig., Paris 1980.
DERCOURT, J.: L'expansion océanique actuelle et fossile: ses implications géotectoniques. – Bull. Soc. géol. France, (7) *12*, 261–317, 17 fig., Paris 1970.
GECZY, B.: Plate Tectonics and Paleogeography in the East-Mediterranean Mesozoic. – Acta geol. Acad. Sci. Hungar., *17*, 421–428, Budapest 1973.

GRACIANSKY, P.-CH.: Existence d'une nappe ophiolitique à l'extrémite occidentale de la chaîne sud-anatolienne. – C. R. Acad. Sc. Paris, *264*, sér. D., 2876–2879, 1 fig., Paris 1967.
HALL, R., AUDLEY-CHARLES, M., & CARTER, D.: The significance of Crete for the evolution of the Eastern Mediterranean. – Spec. Publ. geol. Soc., *17*, 499–516, 11 fig., Oxford 1984.
HERAK, M.: A new concept of geotectonics of the Dinarides. – Prirodoslovna istraž., *53*, Acta geol. jugosl. akad. znan. umjet., *16* (1), 1–42, 10 fig., Zagreb 1986.
HORVATH, F., VÖRÖS, A., & ONUOHA, K.: Plate-tectonics of the western Carpatho-Pannonian Region: A working hypothesis. – Acta geol. Akad. Sci. Hung., *21*, 207–221, 3 fig., Budapest 1977.
HSÜ, K., NACHEV, I., & VUCHEV, V.: Geologic evolution of Bulgaria in light of plate tectonics. – Tectonophysics, *40*, 245–256, 5 fig., Amsterdam 1977.
JACOBSHAGEN, V., DÜRR., ST. et al.: Structure and Geodynamic Evolution of the Aegean Region. – [In:] H. CLOSS et al. [Ed.]: Alps, Apennines, Hellenides; 537–564, 8 fig., Stuttgart (Schweizerbart) 1978.
JACOBSHAGEN, V., MAKRIS, J. et al.: Alpidischer Gebirgsbau und Krustenstruktur des Peloponnes. – Z. dt. geol. Ges., *127*, 337–363, 7 fig., Hannover 1976.
JUTEAU, TH.: Pétrogénèse des ophiolites des nappes d'Antalya (Taurus lycien oriental, Turquie). – Sciences de la terre, *15*, 265–288, 2 fig., 2 pl., Nancy 1970.
– Ophiolites des Taurides: Essai sur leur histoire océanique. – Rev. Géol. dyn. Géogr. phys., *21*, 191–214, 9 fig., Paris 1979.
KAUFFMANN, G.: Perm und Trias im östlichen Mittelgriechenland und auf einigen ägäischen Inseln. – Z. dt. geol. Ges., *127*, 387–398, 4 fig., Hannover 1976.
KETIN, I.: Über einige meßbare Überschiebungen in Anatolien. – Berg- u. hüttenm. Mh., *101*, 22–24, 1 fig., Wien 1956.
– Note explicative de la carte tectonique de Turquie an 1 : 2,500.000. – Bull. miner. res. explor. Inst. Turkey, for. ed., *54*, 1–7, 1 map, Ankara 1960.
KOBER, L.: Leitlinien der Tektonik Jugoslawiens. – Geol. Inst. Serb. Akad. Wiss., vol. spec. *189*, No. 3, 81 p., 2 pl., Beograd 1952.
KOVÁCS, S.: Paleogeographical significance of the Triassic Hallstatt limestone facies in the North Alpine facies region. – Földtani Közlöny, *110*, 360–381, 5 fig., Budapest 1980.
– Problems of the "Pannonian Median Massif" and the plate tectonic concept. Constributions based on the distribution of Late Paleozoic-Early Mesozoic isopic zones. – Geol. Rdsch., *71*, 617–639, 4 fig., Budapest 1982.
KOZUR, H., & MOCK, R.: Zum Alter und zur tektonischen Stellung der Meliata-Serie des Slowakischen Karstes. – Geol. zborn. geol. Carpathica, *24* (3), 265–274, Bratislava 1973.
LAUBSCHER, H. P.: Das Alpen-Dinariden-Problem und die Palinspastik der südlichen Tethys. – Geol. Rdsch., *60*, 813–833, 6 fig., Stuttgart 1971.
LESKO, B., & VARGA, I.: Alpine elements in the West Carpathian structure and their significance. – Mineralia slovaca, *12*, 97–130, 2 fig., Bratislava 1980.
MAHEL, M.: Palinspastic picture of the West Carpathians in the basic evolutionary stages. – Geodyn. Invest. Czechoslov., Final Report, 179–186, 3 fig., Bratislava 1979.
MARCOUX, J.: "Alpine type" Triassic of the Upper Antalya Nappe (Western Taurids-Turkey). – Schriftenr. erdwiss. Kommiss. österr. Akad. Wiss., *2*, 145–155, 3 fig., pl. 1–2, Wien 1974.
MARCOUX, J., & POISSON, A.: Une nouvelle unité structurale majeure dans les nappes d'Antalya etc. – Comptes r. séanc. Acad. Sci., *275*, Sér. D, 655–658, 1 fig., Paris 1972.
MÁRTON, E.: Tectonic implications of paleomagnetic results for the Carpatho-Balkan and adjacent areas. – [In:] J. DIXON & A. ROBERTSON [Ed.]: The geological Evolution of the Eastern Mediterranean. – Spec. Publ. geol. Soc., *17*, 645–654, 11 fig., Oxford 1985.

Márton, E.: Paleomagnetism and tectonics in the Mediterranean region. – J. of Geodynamics, in press, 1986.
Medwenitsch, W.: Zur Geologie Vardarisch-Makedoniens (Jugoslawien) etc. – Sitzber. österr. Akad. Wiss., math.-natw. Kl., Abt. I, *165,* 397–473, 11 fig., 2 pl., Wien 1956.
Medwenitsch, W., & Sikošek, B.: Abriß der Geologie und Tektonik Jugoslawiens. – Rep. VII Congr. Carpatho-Balkan geol. Assoc., *1,* 293–303, 2 fig., Sofia 1965.
Monod, O.: Toward a tectonic connection between Hellenides and Taurides. – Note VIth Colloqu. Geol. Aegean Reg. Izmir, 9 p., 1 fig., Orsay 1977.
Monod, O., Marcoux, J., Poisson, A., & Dumont, J.-F.: Le domaine d'Antalya, témoin de la fracturation de la plate-forme africaine au cours du Trias. – Bull. Soc. Géol. France, (7) *16,* 116–127, 3 fig., Paris 1974.
Moussoulos, L. et al. [Ed.]: Coll. Abstracts VI Colloqu. on the Geology of the Aegean Region. – 185 p., sev. fig., Athens (Inst. Geol. Mining Res.) 1977.
Natschew, I., & Maljakow, I.: Raswitije na posnanijata sa geologijata na Strandsha. – Rev. bulgar. geol. Soc., *40,* 1–9, 2 fig., Sofia 1979.
Pamić, J.: Permo-Triassic rift faulting and magmatism of the Dinarides. – Annales Soc. Géol. Nord, *103* (1983), 133–141, 4 fig., Lille 1984 a.
– Triassic magmatism of the Dinarides in Yugoslavia. – Tectonophysics, *109,* 273–307, 11 fig., 4 tab., Amsterdam 1984 b.
Pantić, N., Grubič, A., & Sladič-T., M.: The importance of Mesozoic floras and faunas from intraoceanic carbonate platforms. – Boll. Soc. Paleont. Ital., *22,* 5–14, 6 fig., Modena 1983.
Petković, K.: Neue Erkenntnisse über den Bau der Dinariden. – Jb. geol. B.-A., *101,* 1–24, 2 fig., pl. 1–2, Wien 1958.
Petrascheck, W. E.: Magmatismus und Metallogenese in Südosteuropa. – Geol. Rdsch., *42,* 128–143, 1 fig., 1 pl., Stuttgart 1953.
– Über ostmediterrane Gebirgszusammenhänge. – Abh. dt. Akad. Wiss. Berlin, Kl. III, *1960* (1), 9–18, 2 fig., Berlin 1960.
Poisson, A.: The extension of the Ionian trough into southwestern Turkey. – Sec. Publ. geol. Soc., *17,* 241–249, 6 fig., Oxford 1984.
Richter, D.: Das Flysch-Stadium der Helleniden – Ein Überblick. – Z. dt. geol. Ges., *127,* 467–483, 2 fig., pl. 1, Hannover 1976.
– Die geodynamische Entwicklung der Helleniden im Spiegel ihrer Flysch- und Molasse-Bildungen. – Z. geol. Wiss., *6,* 1439–1459, 12 fig., Berlin 1978.
– The Main Flysch Stages of the Hellenides. – [In:] H. Closs et al. [Ed.]: Alps, Apennines, Hellenides. – 434–438, 1 fig., Stuttgart (Schweizerbart) 1978.
Ricou, L.: Le croissant ophiolithique peri-Arabe etc. – Rev. Géogr. phys. Géol. dyn., *13,* 327–349, 11 fig., Paris 1971.
Ricou, L.-E., Argyriadis, I., & Lefevre, R.: Proposition d'une origine interne pour les nappes d'Antalya et le massif d'Alanya (Taurides occidentales, Turquie). – Bull. Soc. géol. France, (7) *16,* 107–111, 2 fig., Paris 1974.
Ricou, L., Argyriadis, I., & Marcoux, J.: L'Axe Calcaire du Taurus, un alignement de fenêtres arabo-africaines sous des nappes radiolaritiques, ophiolitiques et métamorphiques. – Bull. Soc. géol. France, (7) *17,* 1024–1044, 8 fig., Paris 1975.
Ricou, L., Marcoux, J., & Poisson, A.: L'allochtonie des Bey Dağlari orientaux. Reconstruction palinspastique des Taurides occidentales. – Bull. Soc. géol. France, (7) *21,* 125–133, 5 fig., Paris 1979.
Sandulescu, M.: Essai de synthèse structurale des Carpathes. – Bull. Soc. géol. France, (7) *17,* 299–358, 30 fig., Paris 1975.

– The Moesic Platform and the North Dobrodgean Orogene. – [In:] M. LEMOINE [Ed.]: Geological Atlas of Alpine Europe etc., 427–442, 28 fig., Amsterdam (Elsevier) 1980.
– The Balkans. – [In:] M. LEMOINE [Ed.]: Geological Atlas of the Alpine Europe etc., 443–460, 25 fig., Amsterdam (Elsevier) 1980.
– Analyse géotectonique des chaînes alpines situées autour de la Mer Noire occidentale. – Anuarul Inst. Géol. Géofiz., *56,* 5–54, 8 fig., 1 map, Bucureşti 1980.
SAVOV, S.: Tectonics of South Strandza. – Contrib. geol. Bulgarie, *1,* 254–298, 11 fig., 3 pl., Sofia 1962.
SAVOV, S., ČATALOV, G., & BUDUROV, K.: Über die Petrographie und das Alter der Karbonatgesteine im Tal des Rezovska-Flusses etc. – C. r. Acad. bulg. Sci., *24* (11), 1529–1532, 4 fig., Sofia 1971.
ŞENGÖR, A. M. C., & YILMAZ, Y.: Tethyan evolution of Turkey: a plate tectonic approach. – Tectonophysics, *75,* 181–241, 7 fig., Amsterdam 1981.
ŞENGÖR, A., YILMAZ, Y., & SUNGURLU, O.: Tectonics of the Mediterranean Cimmerides etc. – Spec. Publ. geol. Soc., *17,* 77–112, 7 fig., 1 tab., Oxford 1984.
SIKOŠEK, B., KARAMATA, S., & STEVANOVIĆ, P. M.: The Jugoslavian Dinarides. – [In:] MAHEL, M. [Ed.]: Tectonics of the Carpathian Balkan Regions. – 347–364, 12 fig., 2 tab., Bratislava (Geol. Inst.) 1974.
SIKOŠEK, B., & MEDWENITSCH, W.: Neue Daten zur Fazies und Tektonik der Dinariden. – Z. dt. geol. Ges., *116* (1964), 342–358, 7 fig., 1 pl., Hannover 1965.
TOLLMANN, A.: Neue Ergebnisse über den Deckenbau der Ostalpen auf Grund fazieller und tektonischer Untersuchungen. – Geol. Rdsch., *50,* 506–516, pl. 15, Stuttgart 1960.
– Das Strandscha-Fenster, ein neues Fenster der Metamorphiden im alpinen Nordstamm des Balkans. – N. Jb. Geol. Paläont. Mh., *1965,* 234–248, 1 fig., Stuttgart 1965.
– Bemerkungen zu faziellen und tektonischen Problemen des Alpen-Karpaten-Orogens. – Mitt. Ges. Geol. Bergbaustud. Wien, *18* (1967), 207–248, pl. 8, Wien 1968 a.
– Der Deckenbau im mediterranen Orogen mit besonderer Berücksichtigung des Balkangebirges. – Iswestija Geol. Inst. Blgarska Akad. Nauk., *17,* 53–60, Sofia 1968 b.
– Karpatische Züge in Fazies und Tektonik der Ostalpen sowie Anmerkungen zur Großgliederung des Subtatrikums. – [In:] MAHEL, M. [Ed.]: Tectonic Problems of the Alpine System, 109–120, 2 fig., 1 tab., Bratislava (VEDA) 1975.
– Plattentektonische Fragen in den Ostalpen und der plattentektonische Mechanismus des mediterranen Orogens. – Mitt. österr. geol. Ges., *69* (1976), 291–351, 11 fig., Wien 1978.
– Geologie von Österreich. – Vol. *1*: XIV, 766 p., 200 fig., 25 tab., Wien (Deuticke) 1977; vol. *2*: XV, 710 p., 286 fig., 27 tab., 1985; vol. 3: in press, 1986.
– The Alpidic evolution of the Eastern Alps. – [In:] P. FAUPL & H. FLÜGEL [Ed.]: Geodynamics of the Eastern Alps, 15 p., 8 fig., Vienna (Österr. Bundesverlag) 1986.

Genetic Types of Alpine ore Deposits and Tectonic Settings in the Northeastern Mediterranean and Southwest Asia

By S. JANKOVIĆ*)

With 1 Figure

Zusammenfassung

Die metallogenetischen Bezirke des „Tethyan-Eurasian Belt" zwischen SE-Europa und Mittel-Asien werden in regional-temporaler Folge beschrieben: 1. Bezirke des triassischen Rifting, 2. Bezirke des Ophiolithgürtels, 3. Bezirke der Subduktionszonen, 4. Bezirke der kontinentalen Plattenkollision, 5. Bezirke einer jüngeren endogenen Mineralisation. Bei allen Erzbezirken werden die Lagerstättentypen in genetischen Klassen aufgezählt und durch Beispiele belegt. Gegenüber den altbekannten südosteuropäischen Beispielen bringen die Ausführungen über die westasiatischen und zentralasiatischen Lagerstättenbezirke grundsätzlich Neues. Das gilt auch für den bisher unbekannten vulkano-sedimentären Sb-Tl-Lagerstättentyp jungtertiären Alters in der Türkei.

Summary

The Alpine ore deposits of the northeastern Mediterranean and southwest Asia are associated with a part of the Tethyan Eurasian metallogenic belt (JANKOVIĆ, 1976), and they are situated along the southern margin of the Eurasian plate in the domain between the Afro-Arabian and the Indian plates in the south, and the Eurasian plate in the north. The tectonic evolution of this area is in close connection with the development of the Tethys oceans – the closure of the Palaeo-Tethys and opening and evolution of the Neo-Tethys, its margins and island arcs, convergence, closure, subduction of oceanic crust(s), continent-continent collisions, and development of magmatic complexes and volcano-intrusive belts and zones.

*) Faculty of Mining & Geology, University of Belgrade, Djusina 7, YU-11000 Belgrade.

The emplacement of ore deposits in the area under consideration is related to a definite time interval, and a specific tectonic setting:

1. Intracontinental rifting

The intracontinental rifting along the northern margin of Gondwanaland and/or within its already separated fragments is particularly widespread during the late Permian-Middle Triassic and the late Triassic-early Jurassic, but intracontinental rifting took place locally even later during the Eocene-Miocene interval; the deep penetrating fractures in the Sistan block in Afghanistan in the Quaternary and the carbonatite deposit may belong to this type of tectonic settings.

The lateral spreading of continental crust and commencement of drift produced its thinning, forming the graben floor, by deep crustal flowage and tensional faulting. These processes in the area under consideration are largely of a short life span and failed to achieve the stage of development of oceanic floor features. But locally, the sea-floor spreading along rift system continues right to the stage of oceanization, i. e. the development of a floor of oceanic crust such as in the area of Mirdita in Albania. Intracontinental spreading axes are closely similar to mid-ocean ridges, but there are differences because the geothermal field linked with the spreading is not dissipated by great volumes of oceanic water.

Apart from thinning of the underlying continental crust and high heat flow caused by rising mantle current, the intracontinental rifting is often accompanied by basic and intermediate volcanism derived from contaminated basic (mafic) magma (the spilite keratophyre complexes), in places by the alkaline complexes, as well.

The ore metals originated from the intermediate, mafic or alkaline magmatic complexes, and/or from hydrothermal mobilization from the surrounding rocks. In some areas volcanic sources at depth supplied only heat which drives hydrothermal systems. Apart from submarine exhalations, hot solutions and metal-rich muds in some basins, there exist intracontinental mineral-bearing basins without volcanic activity, the mineralization of which is associated with shallow-water environment, while the sources of metals are most probably non-volcanic.

The following main genetic types and metallogenic environments can be distinguished, both in the Northeastern Mediterranean and Southwest Asia:

1. Volcano-sedimentary and volcanogenic hydrothermal deposits, both syngenetic and epigenetic are related to volcanic activity, close and/or at the floor of a epicontinental sea. Locally, ore deposits are related to subvolcanic intrusives (diorite, even gabbro).

Among ore deposits the most significant are the following:

1.1. The proximal and distal hydrothermal volcano-sedimentary lead-zinc deposits ("massive sulphide type"), locally accompanied by barite (the deposits: Brskovo and Bobija in Yugoslavia).

1.2. Bedded ferromanganese oxide and iron carbonate (the Vares deposit in Yugoslavia).

Scarcity of major copper deposits is a specific feature of such metallogenic environments.

1.3. Hydrothermal veins and disseminated lead-zinc sulphides in the volcanics (the Suplja Stijena deposit in Yugoslavia).

2. Low temperature hydrothermal deposits related to the marginal seas are usually located close to continental margins, forming elongated metallogenic zones near the continental margins.

2.1. Low temperature Pb–Zn–Ba±F association in the carbonate rocks are usually distributed along the continental margin (the lead-zinc deposits in Central Iran, the lead-zinc-barite deposits in Baluchistan, Pakistan – JANKOVIĆ, 1983; the Borovitsa-Olovo-Zavlaka zone in the Dinarides – JANKOVIĆ, 1967).

Hydrothermal siderite mineralization (stratiform and vein types) occurs often in the vicinity of Pb–Zn mineralization and/or as subordinate mineral constituent of the low temperature Pb–Zn±Ba association in the carbonate rocks.

2.2. The Triassic lead-zinc deposits in the carbonate rocks (Bleiberg, Raible and Mezica in the Alps; the Silesian ore deposits in Poland).

3. Skarn deposits are scare (the magnetite deposit at the gabbro/limestone contact – Tovarnitsa in Yugoslavia – JANKOVIĆ, 1982).

4. Hydrothermal syngenetic and epigenetic mercury mineralization is largely confined to active plate margins where geothermal gradient was sufficient to mobilize it from the mantle (in the Dinarides in Yugoslavia: the Idrija deposit – MLAKAR and DROVENIK, 1971; the Great Caucasus).

5. Carbonatite deposits have been found so far in North Pakistan (the Loe Shilman deposits) and in South Afghanistan (the Quaternary Hanneshyne deposit – ABDULAH and CHMYRIEV, 1980). They are associated with alkaline rocks.

2. Mineralization associated with opholites

When the lateral spreading of continental crust continues from the stage of intracontinental rifting, new oceanic crust begins to be formed and a mid-oceanic ridge develops. The floor of the ocean has many tectonic elements that are considered to be significant metallotects of ore deposits (the active spreading axes, hot-spots, convection currents etc.).

Within an ophiolite assemblage, three groups can be distinguished, each being characterized by specific metallogenic features:

a) a plutonic group of ultramafic and mafic (gabbroic) rocks,

b) a silicic plutonic group that ranges from quartz diorite to plagiogranite; whereas the ultramafics and gabbro are allogenic rocks (JACKSON and THAYER, 1972), the plagiogranites were intruded as fluid magma into the mafic plutonic rocks after their principal deformation, ofter overlapping the early stages of sheeted dyke intrusion*), and

c) a volcanic and volcano-sedimentary rocks (swarms of basalts dykes, overlying flows, and deep-sea sediments); the volcanic rocks display major structural discordance with gabbro and peridotite.

Among the ore deposits associated with ophiolite suites in the NE Mediterranean and SW Asia, endogenous as well as exogenous ore deposits can be distinguished as two principal types:

2.1. Endogenous ore deposits

2.1.1. Magmatic deposits are related to both ultramafics and mafic complexes:

Chromite deposits: Numerous podiform chromite deposits are known in the area under consideration. The dunite is usually the country rock; the harzburgite seldom host chromite deposits, and pyroxenite – exceptionally only.

Apart from typical podiform, there are deposits clearly exhibiting features of stratiform type.

Platinum mineralization is rarely found (Veluce in Yugoslavia – JANKOVIĆ, 1967; traces in the Troodos complex).

Ni–Cu–Co sulphides (pyrrhotite – chalcopyrite – pentlandite ± magnetite association accompanied by gold and silver) are locally found in serpentinites, mainly of small size. They often exhibit transition into hydrothermal stage (Petkovic in Yugoslavia – JANKOVIĆ, 1967).

The scarcity or even lack of sulphides in the Alpine peridotites, both residual and cumulate peridotites and gabbro, in the NE Mediterranean is due to the removal of S in the melt formed by partial fusion, as suggested by NALDRETT (1973) for residual Alpine peridotites. It is not quite certain whether a large-scale desulphurization of the upper mantle really tooks place as argued by NALDRETT (1973).

Magnetite deposits occur occasionally in some serpentinized ultramafics. Mineral association consists of magnetite and subordinate amounts of Ni–Cu–Fe sulphides (Lipovac in Yugoslavia – JANKOVIĆ, 1967).

Titanoferrous magnetite veins and disseminated mineralization occur occasionally in the gabbro, accompanied by traces of chalcopyrite and pyrite. They are mainly of small size (Rastishte, Tara mountain in Yugoslavia – JANKOVIĆ, 1967; Szarvasko in Hungary – LENGYEL, 1957).

2.1.2. Volcano-sedimentary deposits include two principal types, occasionally lying in the same ophiolite complex:

---

*) In some regions, the ultramafic and gabbro group occurs without silicic plutonic and volcanic rocks.

Bedded ferromanganese deposits are usually associated with pillow lavas and with tuffaceous beds between flows or with overlying marine sediments. These Fe—Mn oxides interbedded by chert display often colloform textures. They are numerous in the NE Mediterranean, but small in size.

Sea-floor sulphides (pyritic copper sulphides – the Cyprus type) were probably formed by interreaction between sea water and hydrothermal ore-bearing solutions from deep-seated volcanic and meteoritic water sources; the ore metals originated from the volcanics from which they were mobilized by hydrothermal solutions.

These deposits are numerous in the NE Mediterranean and SW Asia, but they are mostly of small size (up to a few million tons of ore, exceptionally 20 to 30 million tons).

2.2. Exogenous deposits

These deposits were mainly formed by agencies unrelated to the original ophiolite long after crystallization and emplacement. The ore metals originating from the ophiolites were mobilized and transported to the precipitation site by hydrothermal solutions and/or cold water. Hydrothermal solutions are mainly related to younger magmatic activity as the source of heat, while the water is mainly of meteoric origin.

Table 1 shows the principal exogenous deposits in the area of the NE Mediterranean.

Table 1. Principal types and mineral associations of the exogenous deposits in the NE Mediterranean and SW Asia

Type	Mineral composition	Deposit
	Magnesite in serpentinite	Vavdos and Eubea, Greece; Petrascheck 1961 Goles, Yugoslavia – Janković, 1982
Hydrothermal/ Metamorphic:	Asbestos: Chrysotile ± Amphibole	Korlača, Yugoslavia–Vakanjac, 1964 Zidiani, Greece
Vein and/or stockwork	Talk	
	Magnetite	Mineralization only. See: Paraskevopoulus and Economou, 1980
Hydrothermal-sedimentary	Magnesite	Bela Stena and Beli Kamen, Ilić in Yugoslavia
Sedimentary	Magnesite – Huntite – Hydromagnesite	near Servia, North-Greece
Residual	Lateritic nickel deposits	Goles-Mulina in Yugoslavia, Maksimović, 1966

Table 1. (Continued)

Type	Mineral composition	Deposit
Redeposited	Lateritic nickeliferous iron deposits	Kukës a. a., Albania (Pumo et al., 1982); Larymna, Greece (Albandakis, 1974); Rzhanovo (Ivanov, 1962), Mokra Gora (Janković, 1967, 1982) in Yugoslavia
Marine placer	Chromite, Ilmenite, magnetite	Ulcinj, Yugoslavia

3. Mineralization in subduction-related settings

Along the southern margin of the Eurasian plate, above the subducted Afro-Arabian and Indian plates, very significant non-ferrous hydrothermal deposits have been formed. These highly mineralized sectors of the Tethyan Eurasian metallogenic belt are mainly associated with the volcano-intrusive complexes of calc-alkaline suites (Janković, 1977, 1984), locally alkaline.

The principal genetic types of ore deposits are:

3.1. Porphyry copper deposits

The main features of porphyry copper deposits in the area under consideration can be briefly summerized as following:

Host intrusive rocks: The most common associated igneous rock types are quartz diorite, granodiorite and monzonite, and their porphyritic equivalents. The host intrusives are mostly composite, multistage and multiperiod complexes.

The differentiation sequences are calc-alkaline or calcic, locally slightly alkaline.

Age: The age of certain porphyry copper occurrences and host intrusions shows a general correspondance with the age of relevant tectonic events in development of the Tethyan domain.

Porphyry copper mineralization in the Tethyan Eurasian metallogenic belt was formed during several intervals: a) Upper Cretaceous-Lower Tertiary, b) Eocene-Oligocene, and c) Miocene. The Upper Cretaceous-Lower Tertiary deposits are dominant in Southeastern Europe, while the Tertiary deposits are mostly developed in Southwest Asia.

Alteration: Hydrothermal alteration of wall-rocks ranges from skarn in carbonate wall-rocks through potassic, sericitic to propylitic alteration of intrusive and extrusive igneous rocks. Biotitization, in some deposits, was found to be a good measure of intensity of chalcopyrite concentration (Sar Cheshmeh in Iran, Saindak in Pakistan). Intense silicification is a common characteristic of alteration, sometimes coupled with advanced argillic alteration in the uppermost parts of porphyry copper systems (Bor in Yugoslavia).

Alteration zoning is a characteristic feature of the alteration patterns. In some deposits argillic and advanced argillic alterations, as well as telescoping, were found. Pervasive pyritization often accompanies alteration.

Wall rocks: Copper mineralization is exceptionally developed entirely, or nearly so, in intrusive rock progenitors. Ore grade mineralization occurs both in the intrusive host rocks and in the sourrounding rocks.

Copper-bearing intrusions mostly invaded volcanic rocks (andesite prevails) and/or volcanic-sedimentary sequences; crystalline schists are seldom penetrated by the hypabissal intrusive host rock (the Medet deposit in Bulgaria – BOGDANOV, 1982).

Supergene effects: Oxidation and enrichment of some of the porphyry copper deposits (Sar Cheshmeh in Iran, Yulong in Tibet, China, etc.) have given rise to formation of thick enrichment blankets, but the enrichment zones have seldom little more than 20 m in thickness, and they are not of high grade; they are entirely lacking, particularly in some major porphyry copper deposits in SE Europe.

Related massive sulphide deposits: In some porphyry copper districts such as Recsk in Hungary (BAKSA et al., 1980), Bor in Yugoslavia (JANKOVIĆ ET AL., 1980) massive (pyrite) sulphide deposits, of volcanogenic replacement type in andesite, are in close genetic and spatial association with porphyry copper systems. These massive sulphide deposits were emplaced under advanced argillic conditions in a volcanic sequence, above the porphyry copper systems; they most probably represent the uppermost parts of the porphyry copper systems. SILLITOE (1973) considers advanced argillic alteration coupled with pervasive silicification as the common alteration type in the uppermost parts of a porphyry copper system, where volcanic rocks, comagmatic with the copper-bearing stocks, tend to predominate.

3.2. Skarn deposits

These deposits are mostly associated with calcic skarn, very seldom with magnesian skarn. They are usually of small size.

The skarn deposits most frequently involve the mineral associations shown at Table 2.

Table 2. Skarn deposits: Major mineral associations

Mineral associations	Deposit
Magnetite ± Fe–Cu sulphides; with magnesian skarn: Ludwigite – magnetite	Ocna de Fier a. a. in western Romania – Ianovici and Borcos, 1982
Scheelite ± cassiterite	SE Afghanistan (Abdulah and Chmyriov, 1980)

Table 2. (Continued)

Mineral associations	Deposit
Molibdenite – Bi-minerals	Baita Bihorului in western Romania (Ianovici and Borcos, 1982)
Chalopyrite ± pyrite, magnetite, galena, spalerite, locally scheelite as well	
Galena – sphalerite ± chalcopyrite	Rushita, western Romania (Kräutner, 1972*))
Boron mineralization with magnesian skarns	Western Romania (Ivanovici and Borcos, 1982)

*) from: IANOVICI and BORCOS, 1982.

3.3. Greisen deposits

The greisen mineralization is widespread in the ore fields of southeastern Afghanistan, where cassiterite is often accompanied by wolframite mineralization (ABDULAH and CHMYRIOV, 1980).

3.4. Volcanogenic hydrothermal deposits

These deposits are mainly associated with volcano-intrusive complexes of calc-alkaline suites. They are commonly located in subvolcanic level, genetically associated with deep-seated sources (hypabyssal intrusions). The country rocks are both volcanics and surrounding rocks. Mineralization is epigenetic with respect to volcanics, replacement type of mineralization occurs in places as well.

The spatial distribution of mineralization is often controlled by the volcanic structures.

These deposits are usually characterized by a relatively short vertical interval of mineralization. The vertical zonation of mineral associations and ore metals is often found in some of these deposits (from top downward: gold → lead/zinc sulphides → copper sulphides). Telescoping of mineralization occurs occasionally, as well as grading into skarn mineralization.

Hydrothermal alteration is commonly widespread. The composition of the alteration facies depends on the composition of the country rock. Potassic, argillic alteration, as well as silicification, chloritization are the most widespread facies.

Among major deposits, belonging to this group, the following will be reviewed:

Copper massive sulphides: These deposits are associated with andesite-dacite volcanics. They are of replacement type and they are known mainly in SE Europe (Bor in Yugoslavia – JANKOVIĆ, 1982, JANKOVIĆ et al., 1980; Radtka, Krassen and Chelopech in Bulgaria – BOGDANOV, 1982). Gold is often a significant constituents of ore. Pyrite is usually the most abundant mineral, accompanied by many copper sulphides (chalcopyrite, enargite, covellite, chalcosite). Locally, native sulphur with copper sulphides was found.

Copper vein deposits in volcanics are rare, particularly those of economic importance (the Rossen ore field in Bulgaria – BOGDANOV, 1982).

Lead-zinc deposits: They are located along fractured zones in volcanics (andesite-dacite), less frequently in surrounding rocks, often in close connection with caldera structures.

The vein and replacement types are the most common, but stockwork and disseminated types occur as well.

Gold-silver deposits, vein and stockwork types, occur in subvolcanic level of andesite-dacite suites in the Carpathians (the Neogene Transylvania deposits such as Brad and Baia Mare in Romania – IANOVICI and BORCOS, 1982; Banska Stiavnica in Czechoslovakia – ILAVSKY et al., 1979).

Cassiterite deposits in subvolcanic level of andesite-dacite suites occur sporadically only. Some cassiterite stockwork mineralization in the Shindan-Kyshmaran district in Afghanistan (ABDULAH and CHMYRIOV, 1980) resembles porphyry tin type in the Andes.

It may be of interest to mention that cassiterite occurs in many sectors of the Tethyan Eurasian metallogenic belt, occasionally in or near copper deposits related to calc-alkaline complexes (the Bor district in Yugoslavia, the Chagai district in Pakistan, then around the Arghandab batholit in SE Afghanistan).

3.5. Volcano-sedimentary deposits

These deposits are associated with island arcs and andesite–dacite (± rhyolite) suites. Such deposits, both distal and proximal, contain copper and copper–lead–zinc sulphides; pyrite is as a rule the most abundant mineral. Barite occurs occasionally, as well.

In many metallogenic districts and/or within individual ore deposit such as those in the Eastern Pontids in Turkey, a lateral zonal distribution of ore metals was observed; in the Madneul copper deposit, Lesser Caucasus however a vertical one: zinc–lead sulphides ± barite dominate in the upper part of deposit, copper sulphides in the lower section of ore deposit.

4. Mineralization in continent-continent collision setting

Some of ore deposits in the Tethyan Eurasian metallogenic belt are associated with tectonic events after subduction of oceanic crust and the beginning of continent-continent collision.

The endogenous mineralization is commonly associated with granitoids emplaced in the overriding continental plate following continent-continent collision. Such granitoids derived predominantly from continental crust (S-type). Li, Sn and W are the dominant metals in mineralization generated in such a tectonic setting.

The Himalayas are probably the best example for a continent-continent collision in the Tethyan Eurasian metallogenic belt. Collision between the Indian and Eurasian plates, and renewed post-Eocene northwards motion of the Indian plate, resulted in crustal telescoping both north and south of the Indus-Tsangpo suture zone by means of deformation creep at depth and thrusting at higher crustal level (DEWEY and BURKE, 1973). This tectonic shortening and intracrustal thrusting in Oligocene and Miocene times can be considered as a mechanism for the Barrovian-type metamorphism and partial melting of deeper parts of the continental crust followed by intrusion of the leucogranites into higher levels (DIETRICH and GANSER, 1981).

So far known mineralization linked with anatectic granites in the Himalayas is not numerous. Tin, tungsten, locally fluorine, copper, antimony, lead-zinc are the most common, but large deposits have not been found. Greisen mineralization is rather frequent within marginal albitized, silicified, greisenised and/or tourmalinized facies of the leucogranites. Tin mineralization was reported by TALALOV (1976) for the margins of the Dandeldhura granite in the Lower Himalayas of West Nepal. The granite locally displays extensive greisen and cassiterite, scheelite and tantalo-columbite occur near the margin.

5. Post-collision*) magmatic activity and mineralization

Some regional metallogenic units such as ore districts and provinces in the SE Europe and the Asia Minor are associated with the Oligocene-Miocene calc-alkaline complexes. The origin of these magmatic complexes cannot be related to subduction of an oceanic crust and its partial melting, although they are situated in the vicinity of a suture zone, formed after the closure of the Vardar-Izmir-Ankara ocean. It is more likely, but still as a tentative model, that the widespread calc-alkaline igneous suites resulted from an anatectic partial melting of the lowermost part of continental crust and that locally even some ophiolites were involved. These processes took place during the late Palaeogene through early Neogene along the suture Vardar-Izmir-Ankara zone, preceeded by uplifting of the central parts of suture zone owing lateral compression.

The ore deposits were emplaced in hypabyssal and volcanic levels, later often associated with caldera structures. The ore deposits were formed from submarine brines, syngenetic and/or epigenetic with respect to country rocks; they represent a specific group of ore deposits developed in this tectonic setting. Some deposits were formed above ophiolites and they contain some elements which were mobilized by hydrothermal solution passing through ophiolites.

Lead-zinc and antimony are the dominant metals in this tectonic setting. The major ore deposits and the most significant associations of metals and minerals are the following:

---

*) Collision of fragments of the former Afro-Arabian and the Eurasian plates.

5.1. Lead-zinc deposits: Skarn, volcanogenic vein and replacement deposits are the most common, characterized in places by significant vertical extent of mineralization (up to 2000 m such as in the Trepča deposit in Yugoslavia – JANKOVIĆ, 1967), Rodopi in Bulgaria, Laurion in Greece (Petrascheck and Marinos 1953).

Galena and sphalerite accompanied by pyrite, occasionally pyrrhotite, are the main minerals. Sulphosalts are often abundant. These deposits contain silver, particularly skarn type, gold and bismuth, as well as Cd, Ga, Sb as minor constituents of ore.

5.2. Antimony deposits: Numerous antimony deposits occur in the Serbian-Macedonian metallogenic province in SE Europe (JANKOVIĆ, 1960). Apart from almost monomineral stibnite deposits, there are other associations of metals and minerals such as Sb–As–Tl (Alshar in Yugoslavia; Gümüsköy in Turkey – JANKOVIĆ, 1982), Sb–W (stibnite-wolframite deposit at Philadelphion in Greece – JANKOVIĆ, 1979), Sb–Pb/Zn–As (Rujevats and Rajiceva Gora in Yugoslavia – JANKOVIĆ, 1979; JANKOVIĆ and ZARIC, 1980c).

A distinct zonation of Pb–Zn and Sb distribution occurs in some ore fields in Yugoslavia and Western Anatolia in Turkey.

5.3. Porphyry copper deposits occur occasionally mainly along a narrow fractured zone in the souteastern part of the SE Europe (the Buchim-Skouries zone – JANKOVIĆ, et al., 1980b).

5.4. Molybdenum mineralization is sparce. Some of them, such as the stockwork-disseminated low grade deposit at Machkatitsa in Yugoslavia (JANKOVIĆ, 1967; MILOVANOVIĆ and ILIC, 1953) can be very large.

The quartz-molybdenite veins are rather frequent, but commonly of small size (the veins in granite near Axiopolis in Greece – MARINOS, 1982), and the occurrences at Biga Penninsula in Turkey.

5.5. Hydrothermal sedimentary: These deposits were formed in the Neogene basins, originating from ascending hydrothermal springs and migrating through thin fractures into the basin of sedimentation. They supplied metal ions and caused a high-metal content in the water. This led to the depositional enrichment and formation of ore bodies. Such processes are commonly multi-cyclic.

The most significant mineral associations will be reviewed here:

- The large bedded magnesite such as bela Stena in Yugoslavia; some of these deposits contain increased concentrations of boron and boron minerals.
- The large bedded boron deposits, occasionally accompanied by As, are known in Western Turkey (EMET a. a.).
- the Sb–As–Tl deposit at Gümüsköy near Kütahya belongs to this type (JANKOVIĆ, 1982, JANKOVIĆ and BELL, 1976). The epigenetic quartz-stibnite veins occur in the basement of the sedimentary basin where Sb–As–Tl association was deposited.

- The complex Ag–Ba–Sb–As–Pb/Zn deposits at Gümüsköy was formed in the same Neogene basin as the Sb–As–Tl association.

Fig. 1 shows the distribution of major Alpine deposits and their relation with specific tectonic settings in the NE Mediterranean.

References

ABDULLAH, SH., and CHMYRIEV, V. M. (1980): Geology and mineral deposits of Afghanistan. Vol. 2. Mineral deposits. – Moscow, Nedra, 384 p.

ALBANDAKIS, N. (1974): Nickeliferous iron ores in the Locris and Euboea districts. Min. Metall. Chron., no. 18, 9–28, no. 19, 17–41.

BAKSA, CS., CSEH-NEMETH, J., and CSILLAG, J. (1980): The Recsk porphyry copper and skarn copper deposit, Hungary. – *In:* European Copper Deposits - S. JANKOVIĆ and R. SILLITOE, eds., Mining-Geol. Fac., Belgrade p. 73–76.

BOGDANOV, B. (1982): Bulgaria. – *In:* Mineral Deposits of Europe, Vol. 2. Southeast Europe. – Mineral. Soc. and IMM, London, p. 214–232.

DEWEY, J. F., and BURKE, K. (1973): Tibetan, Variscan and Precambrian basement reactivation products of continental collision – J. Geol., Vol. 81, p. 683–692.

DIETRICH, V., and GANSSER, A. (1981): The leucogranites in the Bhutan Himalaya. – Schweiz. Miner. und Petr. Mitt., Bd. 61, S. 177–202.

IANOVICI, V., and BORCOS, M. (1982): Romania. – *In:* Mineral Deposits of Europe, Vol. 2. Southeast Europe. – Mineral. Soc. and IMM, London, p. 55–142.

ILAVSKY, J., et al. (1979): Metallogenese de l'Europe Alpine Central et du Sud-est. – Inst. Geol. Dionyz Štur, Bratislava, 414 p.

IVANOV, T. (1962): Nickeliferous iron ore – Ržanovo on the mountain Kožuf, NR Macedonia. – III-ème Congrès des Géologues de Yougoslavie, Budva 1959: Rapports, Vol. 2, p. 249–264 (English summary).

JACKSON, E. D., and THAYER, T. P. (1972): Some criteria for distinguishing between stratiform, concentric and alpine peridotite-gabbro complexes. – 24th Int. Geol. Congr., Montreal, Proc. sect. 2, p. 289–296.

JANKOVIĆ, S. (1960): Allgemeine Characteristika der Antimon-Erzlagerstätten Jugoslawiens. – Neues Jb. Miner. Abh., *94,* S. 506–538.

JANKOVIĆ, S. (1967): Metalogenetske epohe i rudonosna područja Jugoslavije. – Rud.-geol. fak. i Rud. Inst., Belgrade, 215 p. (German summaries).

JANKOVIĆ, S. (1976): The copper deposits and geotectonic setting of the Tethyan Eurasian metallogenic belt. – 25th Int. Geol. Congr., Sydney, 1976. Miner. Deposita, *12,* 1977, p. 37–47.

JANKOVIĆ, S. (1979): Antimony deposits in Southeast Europe. – Zav. za geol. i geofiz. istr., Ser. A, Vol. 37. Belgrade, p. 25–48.

JANKOVIĆ, S. (1982): Yugoslavia. – *In:* Mineral Deposits of Europe, Vol. 2. Southeast Europe. Mineral. Soc. and IMM, London, p. 143–202.

JANKOVIĆ, S. (1983): The strata-bound low temperature Pb–Zn–Ba–F deposits in carbonate rocks in western Asia: Geotectonic setting and main metallogenic features. – Festband Prof. C. G. Amstutz, Heidelberg University.

JANKOVIĆ, S. (1984): Metallogeny of the Alpine granitoids in the Tethyan Eurasian metallogenic belt. – 27th Int. Geol. Congr., Moscow, Sect. S. 12, Vol. 12, p. 136–148.

JANKOVIĆ, S., et LE BEL, L. (1976): Le thallium dans le minerai de la Bözcukur pres Kytahya, Turkuie. – Bull. suiss Mineral. Petr., *56,* p. 69–77.

Janković, et al. (1980): Metallogenic features of copper deposits in the volcano-intrusive complexes of the Bor district, Yugoslavia. – *In:* European Copper Deposits – S. Janković and R. Sillitoe, eds., Mining-Geol. Fac., Belgrade, p. 42–49.

Janković, S., Petkovic, M., Tomson, I. N., and Kravcov, V. (1980b): Porphyry copper deposits in the Serbo-Macedonian province, southeastern Europe. – *In:* European Copper Deposits. – S. Janković and R. Sillitoe, eds., Mining-Geol. Fac., Belgrade, p. 96–102.

Janković, S., and Zaric, P. (1980c): Metallogenic features of the antimony mineralization in the southeastern Kopaonik, Yugoslavia (Rajićeva Gora ore field). – Zbor. rad. Rud.-geol. fak., Vol. 22, Belgrade, p. 43–56.

Lengyel, E. (1957): New results of titanium-vanadium and iron ore prospecting around Szarvasko. – Ann. Hung. Geol. Inst., Vol. XLVI, p. 251–368, Budapest.

Maksimović, Z. (1966): Mineralni i hemijski sastav fosilne kore raspadanja harcburgita na Golešu. – VI Sav. Jug. Geol., Ohrid, Vol. 2, 530–546.

Marinos, G. (1982): Greece. – *In:* Mineral Deposits of Europe. Vol. 2. Southeast Europe. – Mineral. Soc. and IMM, London, p. 233–253.

Milovanovic, B., and Ilic, M. (1953): Die Molybdenlagerstätten des Eruptivmassivs von Surdulica. – Zbor. rad. Rud.-geol. fak., Beograd, S. 113–136.

Mlakar, I., and Drovenik, M. (1971): Structural and genetic particularities of the Idrija mercury ore deposit. – Geol. razpr. in poroč., Ljubljana, *14,* p. 67–126.

Mulina, Č. (1973): Hidrotermalno-egzogena ležišta silikatnih ruda nikla i kobalta Glavice i Čikatova. – Rud. Metal., Tehn., *28,* no. 10, 1919–1925.

Naldrett, A. J. (1973): Nickel sulphide deposits – their classification and genesis, with special emphasis on deposits of volcanic association. – Canad. Min. Metall. Bull., Vol. 76, p. 183–201.

Paraskevopoulos, G. M., and Economou, M. I. (1980): Genesis of magnetite ore occurrences by metasomatism of chromite ores in Greece. – N. Jb. Miner. Abh., *140,* 1, p. 29–53.

Petrascheck, W. E. (1954): Die Eisenerz- und Nickellagerstätten von Lokris in Ostgriechenland. Bull. Inst. Geol. and Subseil Res. Vol. III Athens.

Petrascheck, W. E., and Marinos, G. (1956): Laurion. Bull. Geol. and Subseil Res. Vol. IV Athens.

Petrascheck, W. E. (1961): Zur Bildung griechischer Magnesitlagerstätten. Radex Rundschau 4.

Pumo, E., Melo, V., and Ostrosi, B. (1982): Albania. – *In:* Mineral Deposits of Europe. Vol. 2. Southeast Europe. – Mineral. Soc. and IMM, London, p. 203–214.

Sillitoe, R. H. (1973): The tops and bottoms of porphyry copper deposits. – Econ. Geol., Vol. 68, p. 799–815.

Talalov, V. A. (1976): Main features of magmatism and metallogeny of the Nepalese Himalayas. – *In:* Coll. ecologie et de la geologie de l'Himalaya. P. Centr. Nat. Res. Sci., N 268, p. 409–430.

Vakanjac, B. (1964): Geneza hrizotil-azbestnih ležišta u rejonu Bosanskog Petrovog Sela. – Geol. gl., Sarajevo, *10,* 197–225.

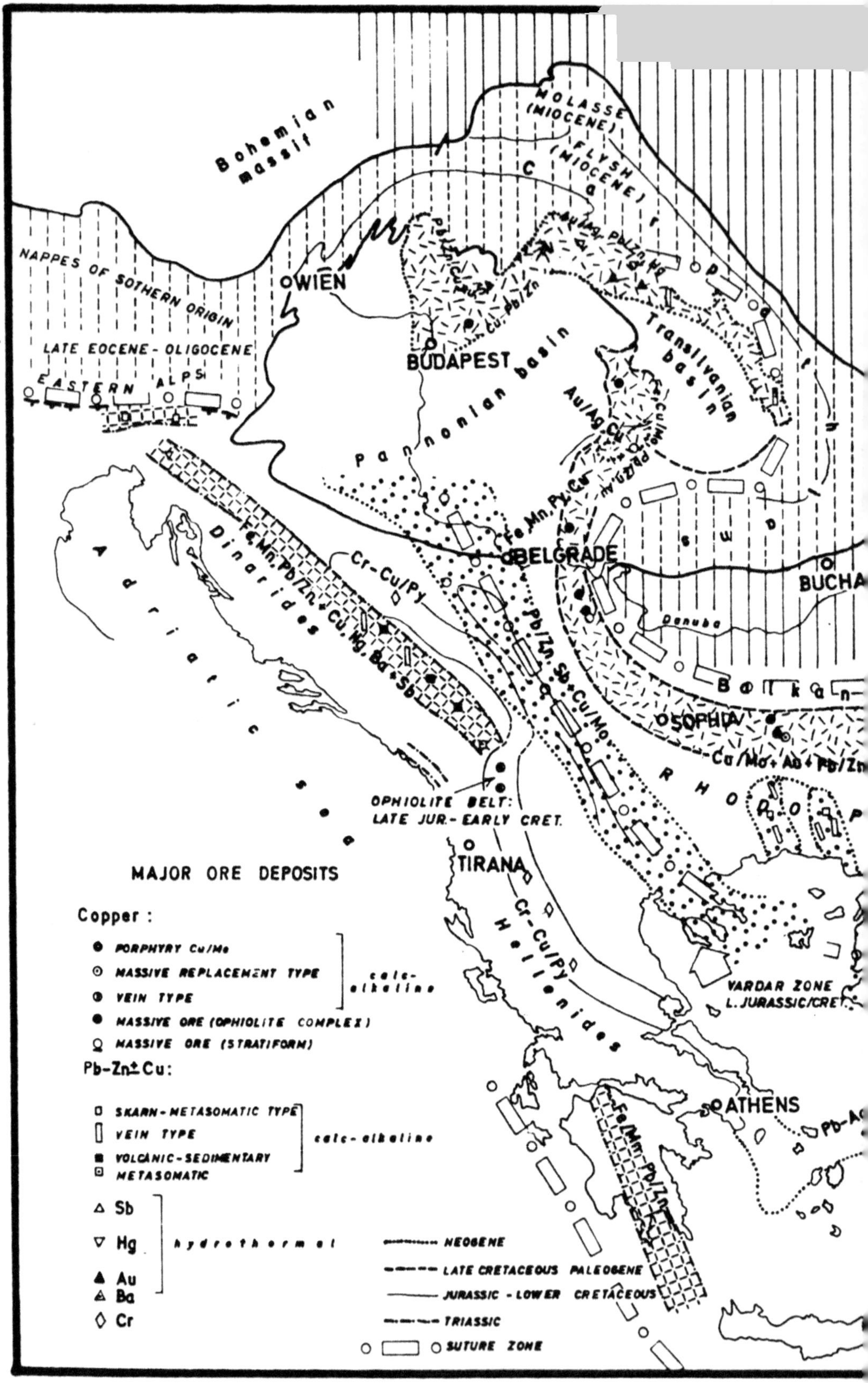
Bohemian massif
MOLASSE (MIOCENE)
FLYSH (MIOCENE)
C a r p a t h i a n s
NAPPES OF SOTHERN ORIGIN
LATE EOCENE - OLIGOCENE
EASTERN ALPS
WIEN
BUDAPEST
Pannonian basin
Transilvanian basin
Au/Ag Cu
Fe,Mn,Py,Cu
BELGRADE
BUCHA
Danube
Balkan
SOPHIA
Cu/Mo+Au+Pb/Zn
RHODOPE
Dinarides
Adriatic sea
Fe,Mn,Pb/Zn+Cu,Hg,Ba+Sb
Cr-Cu/Py
Pb/Zn,Sb+Cu/Mo
OPHIOLITE BELT: LATE JUR.-EARLY CRET.
TIRANA
Cr-Cu/Py
Hellenides
VARDAR ZONE L. JURASSIC/CRET.
ATHENS
Fe/Mn,Pb/Zn
MAJOR ORE DEPOSITS
Copper:
PORPHYRY Cu/Mo
MASSIVE REPLACEMENT TYPE
VEIN TYPE
calc-alkaline
MASSIVE ORE (OPHIOLITE COMPLEX)
MASSIVE ORE (STRATIFORM)
Pb-Zn±Cu:
SKARN-METASOMATIC TYPE
VEIN TYPE
VOLCANIC-SEDIMENTARY
METASOMATIC
calc-alkaline
Sb
Hg
Au
Ba
Cr
hydrothermal
NEOGENE
LATE CRETACEOUS PALEOGENE
JURASSIC - LOWER CRETACEOUS
TRIASSIC
SUTURE ZONE

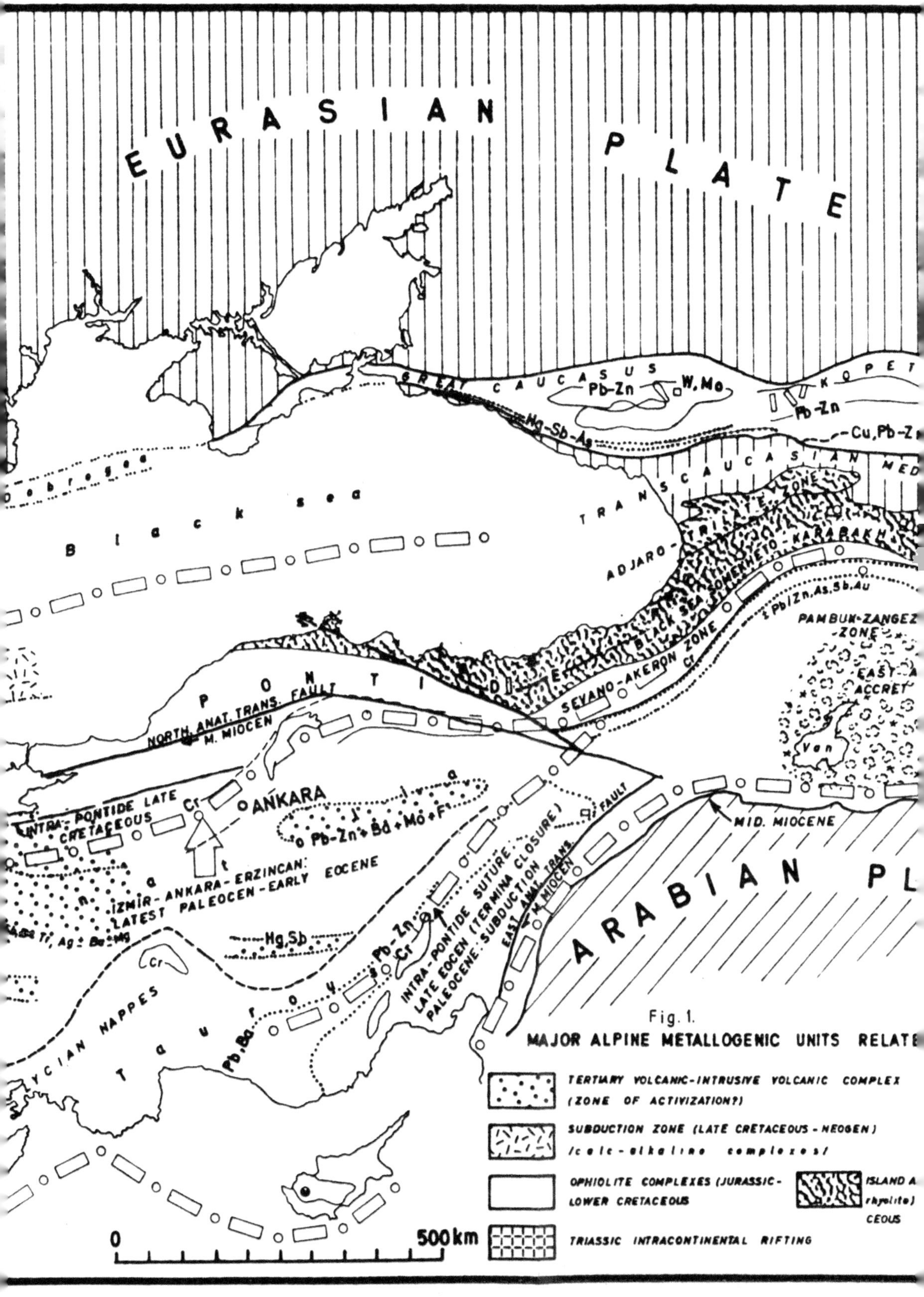

Fig. 1.
MAJOR ALPINE METALLOGENIC UNITS RELATE

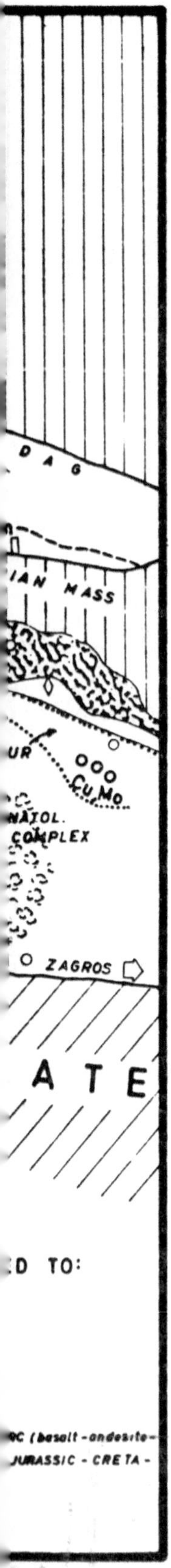
D A G
IAN MASS
UR
Cu Mo
NATOL.
COMPLEX
ZAGROS
A T E
D TO:
C (basalt - andesite -
JURASSIC - CRETA -

The Subduction-related Magmatism in Southeastern Europe and Southwestern Asia

By St. Karamata*)

With 4 Figures

Zusammenfassung

Ein im Mittel granodioritischer Magmatismus kretazisch-tertiären Alters, teils intrusiv, teils extrusiv, begleitet den Südrand der Euroasiatischen Platte in wechselndem Abstand vom Ophiolithgürtel. Der stark wechselnde Chemismus der Magmatite zwischen den Süd-Karpaten, der Srednegorje, den Pontiden, dem Kaukasus, den Iraniden und den Chagai Hills wird verglichen; es ist ein typischer Magmatismus der Subduktionszonen, vergleichbar dem der pazifischen Küsten Amerikas, wobei eine Kontaminierung durch kontinentale Kruste im K-Gehalt und Pb-Gehalt erkennbar ist. Für das Alter dieses gegen Vorderasien immer jünger werdenden Magmatismus sowie seine Dauer ist das Ausmaß der Subduktionen, besonders der Breite des vorgelagerten ozeanischen Beckens, bestimmend gewesen. Der Magmatismus ist der Träger des Tethys-Eurasischen Kupfergürtels im Sinne von Janković.

Introduction

The subduction-related magmatism of the southern margin of the Euroasian plate, occuring in marginal parts of the plate or along its margin, is interesting from many aspects. It is continuous from SW Carpathians (SW Romania), passing through the Timok magmatic area (Eastern Serbia, Yugoslavia), Srednegorie (Bulgaria), Northeastern Pontides (Turkey), Central Iranian belt, to the Chagai-Hills (Baluchistan, Pakistan). All along the belt it exhibits the same geochemical features and metallogeny, despite of variations in geological setting and age.

---

*) Address: Faculty of Mining and Geology, University of Belgrade, Djušina 7, YU-11000 Belgrade.

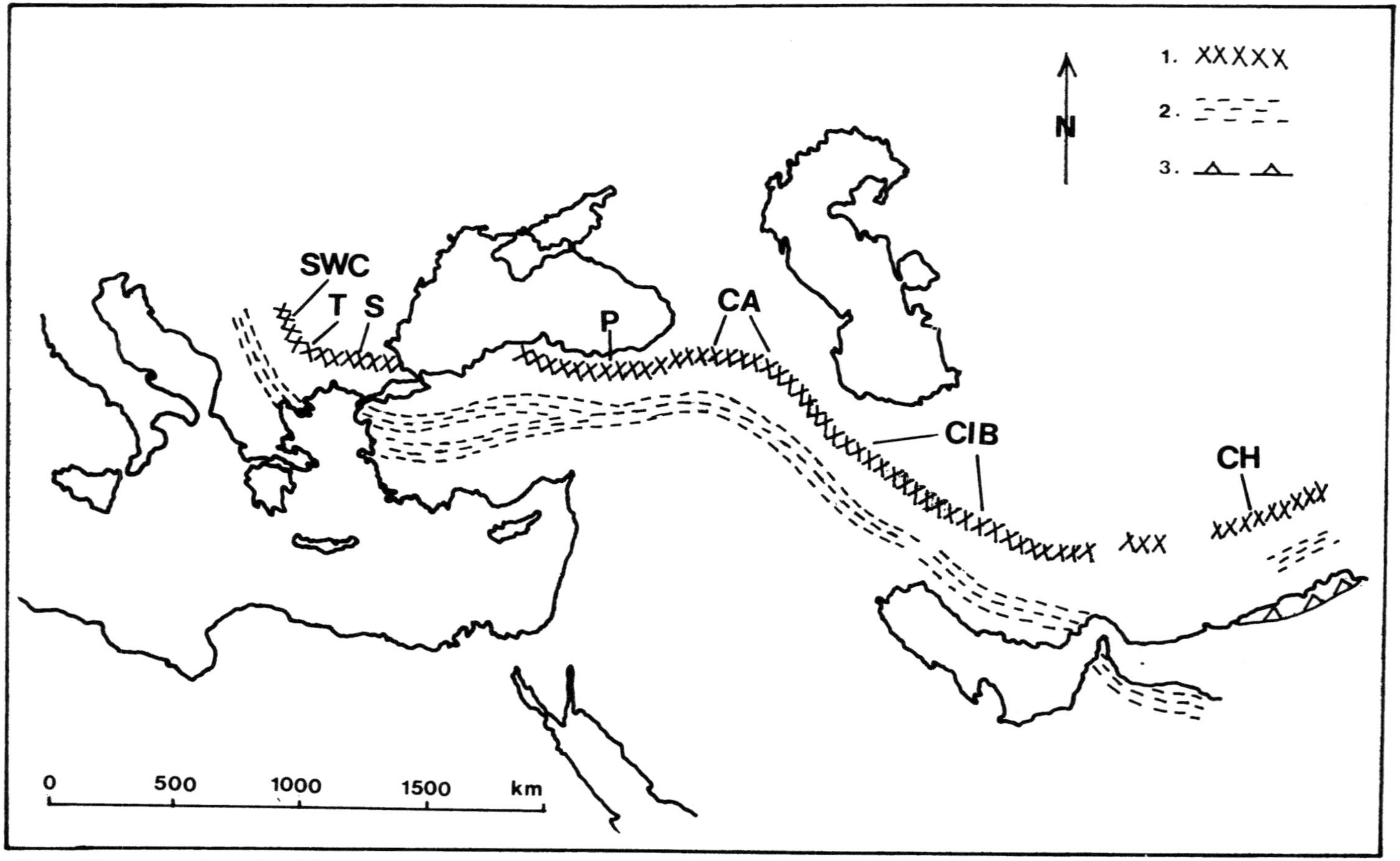

Fig. 1. The extension of subduction-related magmatism in southeastern Europe and southwestern Asia. 1 = the subduction-related magmatism; 2 = the ophiolite belts; 3 = active subduction zone. SWC = Southwestern Carpathians, Romania; T = Timok magmatic complex; Eastern Yugoslavia; S = Srednegorie, Bulgaria; P = Pontides, Northeastern Turkey; CA = Caucasus, USSR; CIB = Central Iranian belt; CH = Chagai-Hills, Baluchistan, Pakistan.

It is parallel to the ophiolite belts, and in the western parts to the young granodioritic-granitic/dacitic-rhyodacitic magmatism belt of crustal origin.

The distribution of the subduction-related magmatic rocks and of the ophiolite belts is presented in figure 1.

Petrology of igneous rocks

All along the belt volcanic and intrusive rocks of this association are occuring. In deep eroded parts the intrusive rocks are more exposed, in very young volcanic areas intrusives are absent. The uplift after the formation of these rocks associated with erosion exposes deeper seated intrusive rocks.

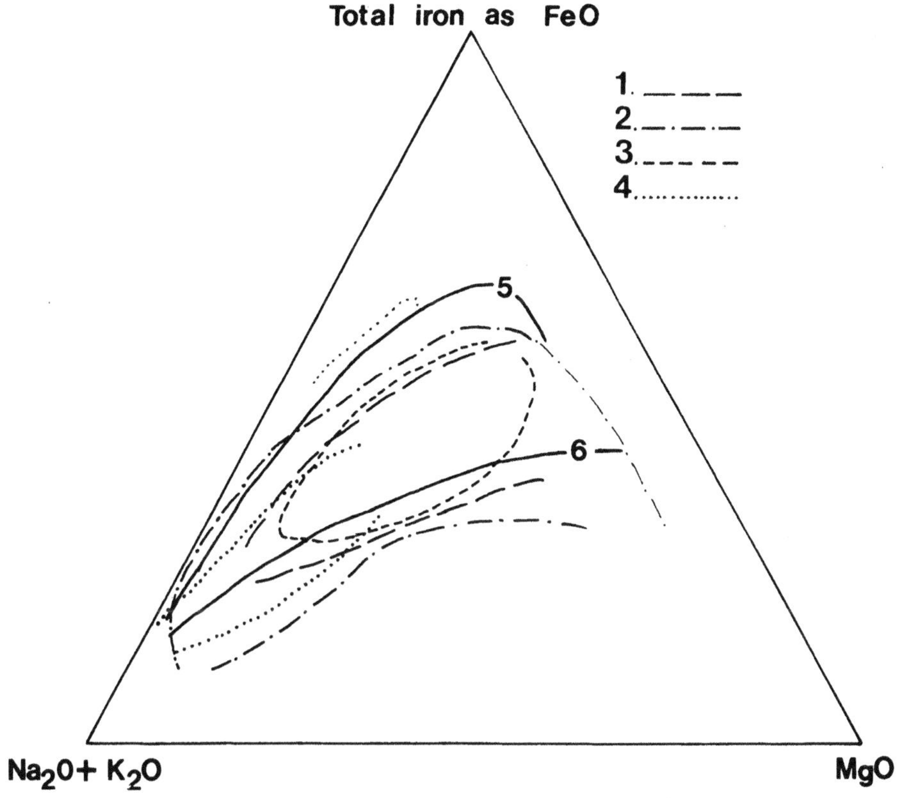

Fig. 2. Comparative AFM diagram for rocks of some areas of the SW Carpathians–Chagai-Hills zone. 1 = Timok magmatic complex (Karamata et al., 1967, supplemented by new data); 2 = Pontides (Akinci 1980, supplement from Akin, 1978, and Moore et al., 1980); 3 = Central Iranian belt (Förster et al., 1972); 4 = Chagai-Hills; 5 = differentiation trend for igneous rocks of the Tonga-Marianas-S. Sandwich island arcs and 6 = for the Cascades-N. Chile-New Guinea areas (after Brown, 1982). For 1–4 the signs embrace the main concentration of the data, the same signs but thinner indicate the field of scattered data.

The volcanic rocks range in composition from basalt, through basaltic andesite, andesite and dacite to rhyolite, but the intermediate and acidic rocks may be sodic or enriched in potassium (K-poor andesite or trachyandesite and even latite, K-poor dacite or rhyodacite, and Na-rhyolite or K-Na rhyolite, which is very rare). The main group of this series are andesites. The intrusive rocks corresponding to these volcanics, are ranging from gabbro through diorite, quartz diorite to granodiorite and even to granite (sodic), and because of enrichment in potassium to monzodiorite, monzonite, and locally to syenite and granite (K-Na). The quartz diorites are the most abundant rocks among the intrusives.

Both groups of igneous rocks, the volcanic and the intrusive ones, are corresponding, and the rocks of the series differ only in the fabric, i. e. in the depth and time of consolidation.

The intrusive rock association corresponds to the I-type granitoids (CHAPELL and WHITE, 1974), but if the intrusive rocks are treated as members of an I-type granitoid association then the associated analogous volcanics have to be considered as members of the same association. It means that the I-type granitoid association has to be treated as an intrusive-volcanic association.

The chemical characteristics of rocks from this association are ploted at the AFM diagram (Fig. 2). For comparison at the same diagram are indicated the differentiation trends of magmatic rocks from primitive island arcs (i. e. Tonga, Mariana, S. Sandwich) and of mature island arcs or marginal parts of continents (i. e. New Guinea, Cascades, N. Chile) given by BROWN (1982).

The igneous rocks of different parts of the SW-Carpathians – the Chagai-Hills belt are situated in the AFM diagram between these two extreme differentiation trends. All these rocks have calc-alkaline differentiation trends characteristic for igneous activity related to subduction zones.

A similar feature is noticeable when the data are plotted at the alkalies versus silica diagram (Fig. 3). Some differences among the rocks from separate parts of the belt may be explained by the thickness and heating of the crust or by the island arcs maturity at the time when magmas passed and consolidated. The data are concentrated in the field of high-alumina basalts and their differentiates or at both sides of it. Only the igneous rocks of the Central Iranian belt are generally poorer in alkalies and therefore situated in the field of tholeiitic basalts and their differentiates.

The silica contents of these magmatic rocks range from about 45 to 75% (in Pontides and Srednegorie between 40 and 78%), but the rocks with about 60% of silica (± 2 to 4%) are most frequent. This most common silica content is analogous to the mean silica content of andesites from Pacific Ocean island arcs (TAYLOR, 1969).

Althoug the most frequent silica contents of subduction-related igneous rocks from the SW-Carpathians – Chagai-Hills zone and the andesites from Pacific Ocean island arcs are identical, the rocks of these two regions differ clearly in K_2O contents. The mean K_2O content of island arcs andesites of Pacific Ocean is 1.60% (TAYLOR, 1969) and much lower than the most common K_2O contents of subduction-related igneous rocks of the SW-Carpathian – Chagai-Hillszone. The K_2O contents of subduction-related magmatics ranges from 0.2 to about 8%, with concentration

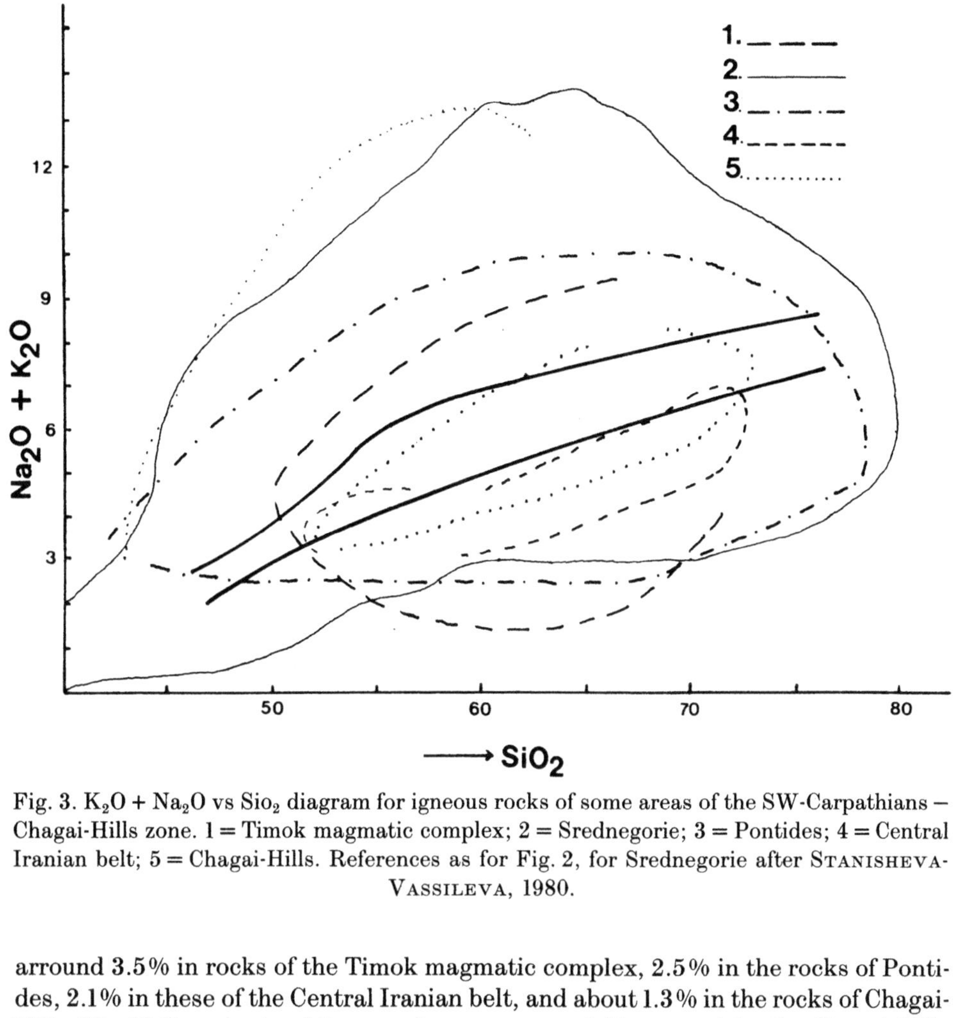

Fig. 3. $K_2O + Na_2O$ vs Sio_2 diagram for igneous rocks of some areas of the SW-Carpathians – Chagai-Hills zone. 1 = Timok magmatic complex; 2 = Srednegorie; 3 = Pontides; 4 = Central Iranian belt; 5 = Chagai-Hills. References as for Fig. 2, for Srednegorie after STANISHEVA-VASSILEVA, 1980.

arround 3.5% in rocks of the Timok magmatic complex, 2.5% in the rocks of Pontides, 2.1% in these of the Central Iranian belt, and about 1.3% in the rocks of Chagai-Hills. The K_2O contents of these rocks are very variable, except for the Chagai-Hills area, wherefrom analytical data are scarce, and the K_2O content may be very different even for the same silica content (Fig. 4), i. e. sodic, sodic-potassic and potassic rocks occur among rocks of all acidities. Only a weak increase of K_2O with SiO_2 is noticeable. Here should be noted that K_2O content is not dependent on the distance from the suture zone, it shows only a tendency of increase with evolution of igneous activity in some areas.

Geochemistry and metallogeny

The data on trace element contents of the subduction-related igneous rocks of the SW-Carpathians – Chagai-Hills zone are very scarce. Therefore only selected elements occuring or almost missing in the ore deposits are considered here.

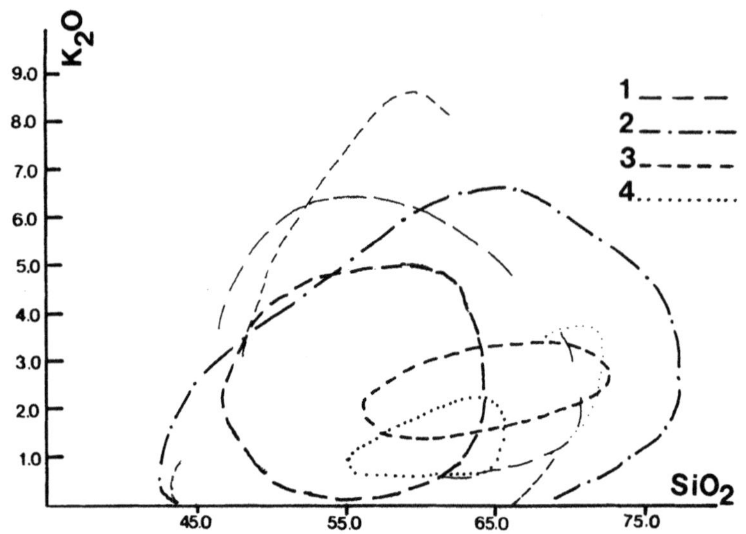

Fig. 4. K_2O vs Sio_2 diagram for igneous rocks of some areas of the SW Carpathians–Chagai-Hills zone. Legend as for Fig. 2.

With this igneous activity are connected copper deposits of different type (JANKOVIĆ, 1977, 1980). Molybdenum, gold, silver, arsenic, selenium, germanium, and locally platinum group elements are associated with copper in these deposits although often in minimal contents. Zinc and lead occur but are very subordinated to copper. From all these elements some analytical data exist only for copper and lead. These data are summarised in Table 1.

It is clear from these data that the igneous rocks of the SW-Carpathians – Chagai-Hills zone (if neglected some anomalous values belonging to weakly mineralised rocks) have copper contents arround 40 to 60 ppm. There is no essential difference between intrusive and volcanic rocks, as well as among rocks from different areas. These mean copper contents are similar to the mean copper content of Pacific Ocean island arc andesites (54 and 55 ppm Cu according to TAYLOR, 1968). However between these two groups of rocks exists a clear difference in lead contents. The Western Pacific island arc andesites are poor in lead (4 to 6.7 ppm mean content according to TAYLOR, 1968) compared to the igneous rocks of the western part of the SW-Carpathians – Chagai-Hills zone (lead contents: 10 to 55 ppm). Only in the easternmost part of the zone, in the Chagai-Hills, the mean lead content is already 12 ppm, approaching the low lead contents of the Western Pacific island arcs andesites.

The relatively high copper contents and low lead contents of the igneous rocks of the studied zone may be in relation with the metallogenic character of this magmatism. The enrichment of lead in comparison to Western Pacific island arcs andesites is probably connected with a contamination of magmas by continental crust material. This is an explanation for the enrichment of potassium in these rocks too.

Table 1. Cu and Pb concentrations in some magmatic complexes related to Mesozoic-Tertiary subduction in Eastern Mediterranean and Southwestern Asia (Romania-Pakistan)

Area (in brackets the number of analysis)	Cu range	x̄	σ	Pb range	x̄	σ	Metallogeny	Remark	Reference
Banatites, SW Romania									
Monzonitic-dioritic-granodioritic association	57– 163			11– 22			Cu		Ianovici et al., 1977
Granodioritic-granitic association	12– 40			13– 32			Cu, Zn, Pb	hybridised	Ianovici et al., 1977
Timok magmatic area, Eastern Serbia									
Volcanics (40)	3– 97	39	28	2– 92	22	18	Cu		Karamata, 1974 a
Intrusives (3)	38	38	–	43– 67	55	10	Cu (Pb-Zn)	hybridised	Karamata, 1974 a
Pontides, NE Turkey									
Volcanics (27/26)	15–1000	243	240	5–100	30	21	Cu	Mostly mineralised samples	Karamata et al., 1979
Volcanics (15)	15– 150	63	36					Unmineralised samples	Karamata et al., 1979
Intrusives (16)	4– 650	288	228	7–105	40	26	Cu (Pb-Zn)	Mostly Cu-mineralised samples	Karamata et al., 1979
Intrusives (7)	4– 125	57	44					Unmineralised samples	Karamata et al., 1979
Volcanics (15)	8– 52			10– 60	10		Cu		Akin, 1978
Intrusives (6)	10– 57			10– 66	10		Cu (Pb-Zn)		Akin, 1978
Intrusives (12)	3– 330	51	87	4– 31	15	9	Cu (Pb-Zn)	1 sample high in Cu	Moore et al., 1980
Intrusives (11)	3– 65	25	25						Moore et al., 1980
Chagai Hills, Baluchistan, Pakistan									
Volcanics (6)	24– 77	50	21	5– 17	12	4		Quarternary volcanics	
Calc-alkaline andesites (18)	25– 150	54		2– 11	6, 7			Ngauruhoe (New Zealand), Asama (Japan), Bouganville (Solomon Islands), Fiji	Taylor, 1968
Saipan andesites, average		54			4				Taylor, 1968

x̄ = mean content; σ = standard deviation.

Geological setting and age of the igneous rocks

All these igneous rocks are situated in the southern marginal parts of the Euroasian plate, or they represent island arcs added later to the Euroasian plate. The belt of these igneous rocks occur at a distance of 100 to 250 km from the ophiolite belt. The igneous rocks originated during differend time intervals depending on the time and other conditions of the subduction which caused this magmatism.

In southwestern Banat these magmatic rocks originated during Senonian until Paleocene (Cioflica and Vlad, 1984), in Eastern Serbia the igneous activity began at Albian and lasted until the beginning of Tertiary (Karamata et al., 1967, Janković et al., 1980), in the Srednegorie area the magmatic activity lasted from Turonian to Paleogene (Bogdanov et al., 1983). In all these areas the magmatism occured on or in the continental crust. In the Pontides (Akin, 1978, Akinci, 1980) and along the Caucasus (Azisbekov and Dzodzenidze, 1970, Adamia et al., 1977) the igneous activity was partly in island arcs, partly in a continental crust environment, beginning at Jurassic and continuing to Oligocene or Miocene respectively. The igneous activity in the Central Iranian belt (Förster et al., 1972, Förster, 1974) began during Paleocene and is still continuing. It is situated in the marginal part of the Euroasian plate. The igneous rocks of Chagai-Hills originated during two intervals, the first was Upper Cretaceous to Lower Miocene, the second from Pliocene to recent time. This is related to a displacement of the subduction zone.

This magmatism occurs, as was mentioned before, along the southern margin of the Euroasian plate, inside the plate or along its margin at added islands arcs. This depends mainly on the type and width of the oceanic area which was closed, i. e. subducted. If the oceanic area was narrow, this type of magmatism does not occur; if it was wider, a short lasting magmatism at the marginal part of the Euroasian plate (of Andean type) developed, but if the oceanic area was very wide, a long lasting magmatism of Andean type or island arcs later added to the continental margin, originated. From SW-Carpathians to Chagai-Hills all these cases exist.

Summary

The Upper Cretaceous to Tertiary, and in the East to Quaternary magmatism on the southern margin of the Euroasian plate or next to this margin is, according to all its features, typical for a subduction related environment. Its differences from one region to the next depend on the type of the former oceanic basin, on the time of the subduction and its continuance and the development or absence of island arcs. However, such differences are logical for an Tethys-type area with microblocks, their displacements and rotations during the approach of the African plate towards the Euroasian.

This magmatism is of calc-alkaline character, embracing very different igneous rocks. The differentiation processes associated with hybridisation by compounds of the continental crust especially in later phases gave a I-type granitoid association together with its volcanic equivalents. All these processes influenced the geochemistry of these igneous rocks too, particularly in late phases.

However the fundamental geochemical and metallogenetic features of this magmatism are preserved and clearly expressed. The rocks are enriched in copper and are initially poor in lead. Finally, the copper wealth of the Euroasian metallogenic belt (JANKOVIĆ, 1977) is genetically related to this igneous activity.

Acknowledgement

My thanks go to the many colleagues with whom I discussed the characteristics of this magmatism.

The work was supported by the funds of the Serbian Academy of Sciences and Arts, and it is an contribution to the IGCP project 169 "Tectonic evolution and metallogeny of Mediterranean and SW Asia".

References

ADAMIA, SH. A., LORDKIPANIDZE, M. B., ZAKARIADZE, G. S. (1977): Evolution of an active continental margin as exemplified by the Alpine history of the Caucasus. – Tectonophysics, *40,* 183–199.

AKIN, H. (1978): Geologie, Magmatismus und Lagerstättenbildung im Ostpontischen Gebirge/Türkei aus der Sicht der Plattentektonik. – Habilitationsschrift, Technische Universität Berlin, Berlin.

AKINCI, O. T. (1980): Major copper metallogenic units and genetic igneous complexes of Turkey. – European copper deposits, Edit.: JANKOVIĆ, S., 199–208, Belgrade.

AZIZBEKOV, S. A., DZODZENIDZE, G. S. (1970): Magmatizm Kavkaza, Irana i Turcii. – Izvestija Akad. Nauk SSSR, Ser. geol., *12;,* 15–24, Moskva.

BAZIN, D., HÜBNER, H. (1969): Copper Deposits in Iran. – Geol. Survey of Iran, Rep. Nr. 13, 232 p., Teheran.

BOGDANOV, B., Ed. (1983): Magmatizm i metallogenija Karpato-Balkanskoj oblasti. – B. A. N., 300 p., Sofia.

BROWN, G. C. (1982): Calc-alkaline intrusive rocks: their diversity, evolution, and relation to volcanic arcs. – Andesites, edit.: THORPE, R. S., 437–461, John Willey & Sons.

CHAPELL, B. W., WHITE, A. J. R. (1974): Two contrasting granitic types. – Pacific Geol., *8,* 173–174.

CIOFLICA, G., VLAD, S. (1984): Alpine metallogeny in Romania. – An. Inst. Géol. Géoph., *64,* 175–184, Bucuresti.

FÖRSTER, H. (1974): Magmentypen und Erzlagerstätten im Iran. – Geol. Rundschau, *63, 1,* 276–292.

FÖRSTER, H., FESEFELDT, K., KÜRSTEN, M.(1972): Magmatic and Orogenic Evolution of the Central Iranian Volcanic Belt. – 24th IGC, Sect. 2, 198–210.

IANOVICI, V., VLAD, S., BORKOS, M., BOSTINESCU, S. (1977): Alpine porphyry copper mineralization of West Romania. – Mineral. Deposita, *12,* 307–317.

JANKOVIĆ, S. (1977): The Copper Deposits and Geotectonic Setting of the Thethyan Eurasian Metallogenic Belt. – Mineral. Deposita, *12,* 37–47.

Janković, S. (1980): Porphyry-copper and massive-sulfide ore deposits in the northeastern Mediterranean. – Proc. Fifth IAGOD Symposium, 431–444, Stuttgart.

Janković, S., Jovanović, M., Karamata, S., Lovrić, A. (1981): The isotopic age of some rocks from the Timok eruptive area. – Glas 329 Acad. serbe sci. et arts, Class. sci. nat. et math., *48*, 87–94, Belgrade.

Karamata, S. (1974): Beziehungen zwischen den metallogenetischen, petrographischen und geochemischen Provinzen der Balkanhalbinsel und Kleinasiens. – Metallogenetische und Geochemische Provinzen, Edit.: Petrascheck, W. E., Schriftenreihe Erdwiss. Komm. Österr. Akad. Wiss., Band 1, 106–119, Wien.

Karamata, S.(1974a): Geochemical, Petrologic and Metallogenic provinces on the Balkan Peninsula and in Asia Minor. – Serbian Academy of Sci. and Arts, Vol. CDLXXV, monographs, Section for nat. and math. sci., 55 p., Belgrade.

Karamata, S., Djordjević, P. (1980): Origin of the Upper Cretaceous and Tertiary magmas in the Eastern parts of Yugoslavia. – Bull. de l'Acad. Serbe des Sci. et des Arts, Classe des Sci. Nat. et Math., No. 20, 99–108, Belgrade.

Karamata, S. et al. (1967): Les roches magmatiques crétacées-tertiaires des Carpato-Balkanides Yougoslaves. – Acta Geologica Acad Sci. Hungaricae, *11* (1–3), 115–138, Budapest.

Karamata, S., Milojković, R., Čuturić, N., Milanović, B. (1979): Lead and copper contents of igneous rocks from the Pontides and the Anatolides and their significance. – Proceedings, GEOCOME – *1*, 347–366, MTA Enstitüsü, Ankara.

Kazmi, Ali H., Rana Riaz A. (1982): Tectonic map of Pakistan. – Geological Survey of Pakistan, Quetta.

Moore, W. J., McKee, E. H., Akinci, O. (1980): Chemistry and chronology of plutonic rocks in the Pontid Mountains, Northern Turkey. – European copper deposits, Edit.: Janković, S., 209–216, Belgrade.

Stanisheva-Vassileva, G. (1980): The Upper Cretaceous Magmatism in Srednogorie Zone, Bulgaria: A Classification Attempt and Some Implications. – Geologica Balcanica, 10/2, 15–36, Sofia.

Taylor, S. R. (1968): Geochemistry of Andesites. – Origin and Distribution of the Elements, Edit.: Ahrens, J. H., 559–571, Pergamon Press, Oxford.

Taylor, S. R. (1969): Trace element chemistry of andesites and associated calc-alkaline rocks. – Proceedings of the Andesite Conference, Edit.: McBirney, A. R., Bull. State of Oregon, Dept. Geol. Miner. Ind., *65*, 43–63, Portland, Oregon.

Vlad, S. (1984): Alpine porphyry copper occurences in Romania. – Bull. de l'Acad. Serbe des Sci. et des Arts, Classe des Sci. Nat. et math., No. 25, in press, Belgrade.

Geotectonic evolution and metallogeny of the Hercynides in southern marginal folded belt of the North Afghan Platform –

A brief review

By I. N. Tomson*) and R. B. Baratov**)

With 2 Figures

Zusammenfassung

Die variszische Faltungszone des südlichen Tien-Shan und des Nord-Pamir liegt am Südrand der Eurasischen Platte. Die Zone entwickelte sich von einer Karelischen Protogeosynklinale über ein Plattformstadium zur Variscischen Geosynklinale mit einer carbonischen Diabas-Porphyritformation, die auch Gabbro-Granitintrusionen enthält. Die Molasse führt jungpermische Evaporite und Kupfersandsteine. Das Neogen war mit magmatischer Aktivierung verbunden, die Pb-Zn-, F- und Hg-Vererzungen bewirkte. Die axiale Karakul-Zone führt Pegmatite mit seltenen Metallen.

The folded zone of Hercynides in the Southern Tien-Shan and the North Pamir is situated on the Eurasian plate, along its southern border with the Alpine Mediterranean belt. The Hercynides form a comparatively narrow zone in Western Hindukush, North Pamir and Kunlun mountains (Fig. 1); in the west, they are overlain by the cover of the North Afghan platform.

The zone of Hercynides in the North Pamir has been the best studied. Fig. 2 shows the geological map of the North Pamir area.

---

*) Prof. I. N. Tomson, Staromonetny per. 35, IGEM AN SSSR, 109017 Moscow.

**) Acad. R. B. Baratov, Vice President Tajikistan Academy of Sciences, Prospect Lenina 33, 734025 Dushanbe 25, Tajik SSR, USSR.

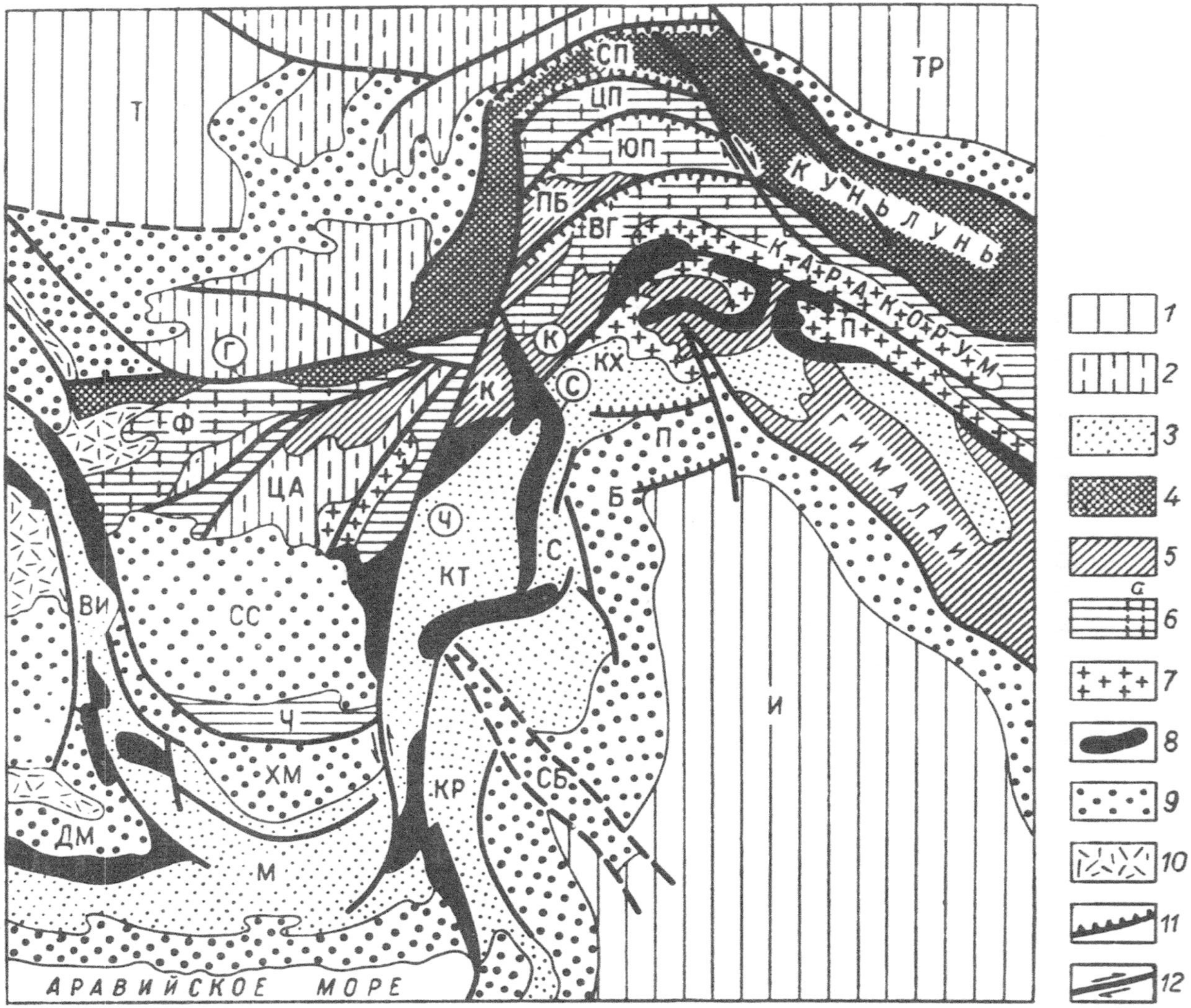

Fig. 1. Tectonic sketch of the Afghan-Pamir folding zone

1. Platforms;
2. the Turan platform;
3. Carbonate and flysh; outer troughs of the Mediterranean belt;
4. Hercynides;
5. and 6. The Alpine belt: Inner zones with dominance of rocks (5), and Mesozoic and Palaeogene sediments and volcanic rocks (6);
7. Cretaceous and Paleogene volcanics;
8. Ophiolites;
9. Molassa;
10. Volcanics; Neogene;
11. Thrust fault;
12. Displacements

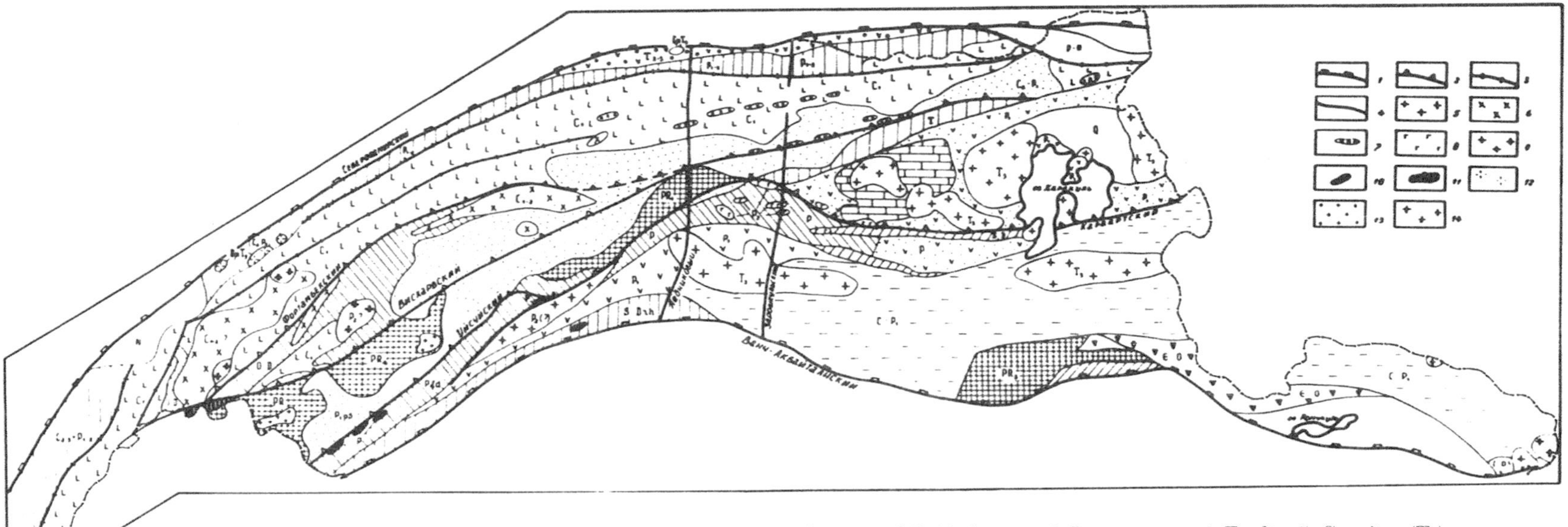

Fig. 2. Geological map of the North Pamir. – 1. Marginal sutures; 2. Sutures of divided zones; 3. Inner suture; 4. Faults; 5. Granites (T_3); 6. Gabbro, diorite (C_{1-3}); 7. Porphyric granite (C_1); 8. Gabbro (P_z); 9. Granite (P_z); 10. Ultramafics (P); 11. Ultramafics (C); 12. Quartz porphyries (T_3); 13. Gneiss; 14. Granodiorite (P – T_3)

Main features of geotectonic evolution

The zone of the Hercynides is bounded in the North Pamir area from the north by the North Pamir fault, and from the south by the Central Pamir fault. BARATOV, FINOGENOVA and VOLYNOV (1981) recognised five geotectonic cycles of the geotectonic evolution of this zone:

1. Karelian proto-geosyncline consists of a thick metamorphic rock sequence intruded by layered gabbro-anorthosite and peridotite complex.

2. The epi-Karelian platform cycle is characterized by the Vend terrigenous and the Ordovician-Devonian carbonate sequence.

3. The Hercynian geosyncline was superimposed on the platform basement in Early Carboniferous time; it is restricted to a narrow strip in the northern part of Pamir (the Kalaihumb-Sauksay – the Karakul zone). Scattered Carboniferous terrigenous troughs and carbonate-terrigenous sequences are developed. In the southern part of the North Pamir (the Darvaz-Sarycol zone) carbonate-volcanogenic sequences and terrigenous sediments are found. The Carboniferous-Permian spilite diabase and porphyritic formations, accompanied by gabbro-granitoid intrusives, are known (the Kalaihumb-Sansay zone which extends into Afghanistan – the Darvas-Zaalay zone).

In the final stage of development, the molasse sediments several kilometers thick were deposited; their upper part consists of Late Permian salt-bearing sediments. Late Permian sandstones contain stratiform copper mineralization.

4. The Mesozoic-Cenozoic carbonate – terrigenous platform sediments were deposited.

5. The neogen is characterized by Late Alpine activization. Some ore deposits are associated with magmatic complexes resulted in the process of activation (hydrothermal lead-zinc, fluorite and celestine – ROMANENKO and MOGAROVSKY, 1984). Locally, mercury mineralization was found in diagonal faults.

Similar geo-tectonic development was observed in the southwestern sector of the Hercynides – in the western Hindukush and NW Badakhshan (DRONOV, 1983).

Principal ore deposits and major metallogenic zones

The common metallogenic feature of the Afghan-North Pamir Hercynian belt is the development of Cu–Zn–Pb ± Au mineralization, characterized by multiple introduction of copper mineralization, starting from the Proterozoic until Late Permian. The copper deposits occur as massive nickel-copper sulphides and stratiform copper-bearing sandstones.

Several major metallogenic zones are distinguished in the Northern Pamir – West Hindukush – NW Badakhshan (CHMYRIOV and STASHILO-ALEKSEEV, 1980):

1. The Kalaihumb-Sayksay/Darvas-Zaalay zone contains numerous occurrences and small deposits of Ni-Cu sulphides, massive sulphides and skarn (VALKOV et al., 1979).

Locally, lead-zinc, copper, tin and bismuth as well as porphyry copper mineralization (the Obinhingou river) are found.

2. The axial Kurgovoz-Karakul zone of the North Pamir contains rare-metal, pegmatite and skarn mineralization associated with orogenic granites.

3. The southern metallogenic zone of the North Pamir is characterized by the predominance of hydrothermal veins of polymetallic composition.

References

BARATOV, R. B., AFINOGENOVA, L. N., and SALNIKOV, R. N. (1981): Osobenosti metallogenii skladchatykh oblastey Tsentralnogo Tadjikistana i Pamira. – Zakon. razm. pol. iskop., v. V, 13, Nauka, Moskow (in Russian).

CHMYRIOV, V. M., and STASHILO-ALEKSEEV, K. F. (1980): Geologia i poleznie iskopaemie Afganistana. – Vol. 1 and Vol. 2, Nedra, Moscow (in Russian).

ROMANKO, E. F., and MOGAROVSKIY, V. V. (1984): Pozdealpiskaya metallogenia Juzhnogo Tadjikistana i Tsentralnogo Irana. – Sov. geol., 5 (in Russian).

VOLNOV, B. A., STEBLOVA, V. M., NARIYNEV, V. V. (1979): Osnovnie cherti metallogenii Darvaza. – DAN Tadj. SSR, T. 22, Vol. 7 (in Russian).

Rare metallic Pegmatite belts of the Hindukush, Eastern Afghanistan

By L. N. Rossovsky, Dushanbe USSR*)

Zusammenfassung

Die Pegmatitreviere mit seltenen Metallen liegen im Zentralmassiv des Pamir, das von archäischen kristallinen Schiefern aufgebaut ist; das Alter der Pegmatite ist End-Kreide bis Eozän. Sie sind an Zweiglimmergranite gebunden. Ihre tektonische Stellung ist durch die Ränder von Mulden und Gräben bestimmt, die mit Sedimenten von Perm-Jura gefüllt sind. Die Mächtigkeit der mineralisierten Pegmatite liegt zwischen 5 und 60 m, ihre Länge zwischen einigen hundert bis 2000 m, die vertikale Tiefe ist bis 1000 m aufgeschlossen. Der Gehalt an Spodumen beträgt zwischen 15 und 25 %, der Li-Gehalt 1–2 %. Die Vorräte werden auf 3 Millionen Tonnen Li_2O geschätzt. Diese Li-Konzentration wird auf eine Spätkristallisation in der anatektischen Granitschmelze zurückgeführt. – Die Edelstein-Pegmatite führen Kunzit, Beryll und edlen Turmalin; sie sind durch große und reichliche miarolithische Hohlräume ausgezeichnet.

Introduction

The numerous ore fields of rare metallic granite pegmatites were discovered in Afghanistan by Soviet geologists during 1971–1976. These deposits are located in the alpine region of the Hindukush in the valleys of Kabul, Alingar, Kunar and Daray-Pich. The ore fields in this part of Afghanistan are occuring in the absolute altitudes ranging from between 800 m to 6000 m above sea level. The revealed pegmatites are characterized by huge reserves, diverse associations of elements and minerals, including gem stones. From the economic point of view these deposits, or at least some of them, are evaluated as being economically valuable.

---

*) Geological Institute, Tajik Acad. of Sciences, Dushanbe – 43 USSR.

Geological Structure of Eastern Afghanistan

Rare metal pegmatite fields are distributed in the Nuristan-Pamir middle massif, composed of the three main formations. The lower Archaean formation is composed of various crystalline schists, gneisses, migmatites, marbles metamorphised in conditions of granulate and amphibolite metamorphic facies (Saunglich series). The middle Proterozoic series consists of strongly displaced crystalline schists of amphibolite metamorphic facies (Nuristan series). The Upper Permian-Triassic sequence is represented by quarzzmica and granatstaurolite schists along with small quantities of limestone and quarzites; the state of .metamorphism of the above corresponds to epidote-amphibolite and green schists facies (Kashmund series). This middle massif is broken into a number of large blocks. The granitoids, both ancient early Proterozoic and Cretaceous-Palaeogene, are widespread. The largest granite massifs (Lagman complex) have been formed by the process of Alpine activization (1).

Rare-metall pegmatites on the territory of Nuristan and Badakhshan are spatially and genetically associated with granites of the Lagman complex. Granite bodies of the Lagman complex are emplaced between sediments of the lower and upper structural sequences. At the foot of the granite massifs lies Precambrian gneiss, at the roof quarz – mica-granat – stawrolite schists of Permian-Triassic age.

The largest and most complicated among Lagman complex massifs is Alingar granite pluton. This intrusive is stretching from the area of the town Jalalabad northeastwards at a distance of about 400 km in the river valley of the Pyanj up to the confluence of the Pamir-Darya with the Vakhan-Darya. The width of a compact granite field of the Alingar intrusive at the level of present erosion varies from 10 to 50 km.

The Lagman complex massifs have been formed during three main phases: 1. granodiorites, plagiogranites, quarz diorites; 2. strongly porphyrie biotite granites; 3. biotite and bimicaceus granite and sillimanite granite, with which rare-metall pegmatites are spatially and genetically associated. The age of Lagman complex granites is considered to be of the Upper Cretaceous-Paleogene. The Lagman complex granites and their rare-metall pegmatites are believed to be associated with Alpine activization (8).

The main features of pegmatites

The pegmatites genetically associated with the third phase of the Lagman complex (bimicaceous granites) occur as 1. syngenetic (schlieren) deposits in granites; 2. epigenetic deposits at the roof of the granite massifs. Gradual transitions have been established: bimicaceous granite – syngenetic (schlieren) pegmatite – epigenetic pegmatite. Rare-metal mineralization, as well as precious stones only occur in the epigenetic pegmatites.

The majority of rare-metal pegmatite veins are located near the exocontact of intrusive bodies, where only small (from 1 × 2 to 3 × 10 km) granite massifs of the third phase are exposed; they are regarded as satellites or large apophysis of the main intrusive.

Wallrocks of rare-metal pegmatites are mostly quarz-micaschists containing andalusite, cordierite, granate, staurolite. A certain correlation of the degree of metamorphism of wall rocks and the localization of rare-metal pegmatite bodies is observed: the rare-metal pegmatite ore bodies are associated with the rocks mainly metamorphised in the condition of epidote-amphibolite facies.

The morphology of rare-metal bodies of Eastern Afghanistan is diverse, but as a whole veins and lenses are prevailing. The dimensions of pegmatite bodies range from 1 to 60 m wide and from 10 meters up to 2–5 km long. Certain rare-metal pegmatite bodies are of very large dimensions, so for example vein Nr. 8 from the Nilan-Kulam field is 4000 m long with an average thickness of 7 m; the spodumene-albite vein of the Drumgal is over 1500 m long and 30 m wide; the "Kulam" vein, composed of albitized microline, with lepidolite and spodumene is 5 to 40 m thick and over 3000 m long.

The inner structure of the pegmatite bodies is massive, stripy or weakly zoned.

Rare-metal pegmatites of Afghanistan have been subdivided with respect to their composition and tipomorphic minerals, as well as to their practical importance into the following types: 1. oligoclase–microline biotite–muscovite with rare beryl; 2. albitized microline with scherle, muscovite and beryl; ores of crystalline beryl; 3. albitized microline ores with nests of blue clevelendite, lepidolite, spodumene, polychromium tourmaline-deposits of precious stones like kunzite, vorobyevite, tourmaline; 4. albite ores with nests of lepidolite, tantalite, spodumene, pollucite, tantal ores; 5. spodumene–microline–albite and spodumene–albite (spodumene pegmatites): lithium ores; 6. lepidolite–spodumene–albite with polychromium tourmaline, tantaline, pollucite: caesium and tantal-caesium ores.

Pegmatite fields and belts

Rare-metal pegmatite veins in the territory of Nuristan and Badakhshan are forming ore fields over an area of 10 to 800 square km. All together 21 fields have been discovered. The number of individual rare-metal bodies within the fields usually are tens and rarely hundreds. Three structural types of ore fields are being distinguished: 1. the fields of veins, dipping at low angles, are the most typical ones for the gabbro diorite massifs. They are mainly confined to contraction joints in the endocontact area. 2. ore fields of subvertical veins are confined to schists in the contact zones. 3. The ore fields composed of lenses are occuring in the cuppolas of granite massifs.

The pegmatite fields in the territory of Eastern Afghanistan are grouped into three clearly defined zones: Badakhshan, Hindukush and Nuristan. These pegmatite belts are related to linear graben–syncline structures, composed of Permian Triassic and locally Jurassic sediments. The presence of pegmatite-granite is often being observed.

The formation of pegmatite-granites and pegmatits is linked with the period of a powerful and dynamic tension, which took place in the conditions of stress, thus causing cataclasm and mylonitization of both granites and pegmatites (Rushan belt of the Southern Pamirs). In other folding zones pegmatite-bearing granites are formed by the activization process, when folding zones have already been constituted by relatively hard tectonic blocks (Yasgulem belt of the Central Pamir).

The rare-metal pegmatite belts of Eastern Afghanistan and the Pamirs are regarded to be the largest in the world (3). Based on our studies, this province is characterized by its clearly defined Li, Be, Ta and Sn specialization which leaves an imprint on geochemical features of certain belts and ore fields of the region.

The study of ore-bearing belts and ore fields of rare-metal pegmatites of the Hindukush has shown that they are characterized by many specific metallogenic features, due to the influence of the following factors: 1. geochemical composition of the province as a whole; 2. geochemical peculiarities of pegmatite bearing granites in various tectonic zones; 3. tectonic regime of the formation of pegmatites; 4. structural and morphological type of pegmatite veins.

Depending upon the intensity of the ore forming factors, some pegmatite belts turn out to be specialized for lithium, others for tantalium, beryl and precious stones.

Mineral Resources of Pegmatite belts

Of the numerous metals and minerals so far discovered in the pegmatites of Eastern Afghanistan, lithium and precious stones are the most interesting ones. Lithium deposits are represented by large and numerous veins of spodumene pegmatites in the roof of granite massifs. Mainly quarz-muscovite-biotite schists with garnet and staurolite are the wall-rocks of the deposits. The spodumene pegmatites are located in various structural environments of pegmatite-bearing granites, but the largest deposits are related to Permian-Triassic graben-synclines in the area of Precambrian bed exposures, as mentioned above.

As far as the structure and morphology are concerned, the deposits under consideration constitute a specific type of vein series, which is characterized by a linear-extended form and steep-dipping. In this series spodumene pegmatites are deposited echelon-like, in the form of extended plates or lenses. The thickness of large bodies varies from 5 to 60 m, and the length from several hundred to 2000 m. The thickness of the veins varies from 20 to 500 m. The vertical extension of spodumene pegmatites varies from 100 to 1000 m. Certain large deposits, containing commercial lithium ore (Jamanak, Drumgal, Tsamgal), are exposed by the relief along their dip for 400–800 m. Spodumene pegmatites which are forming the lithium deposits in the Hindukush, belong to the lithium subtype of sodium-lithium pegmatites according to A. E. Fersman's classification (9), in which mean spodumen content varies between 15–25 %. The lithium deposits of Afghanistan are similar to spodumene deposits of other regions, but they have very close similarity with the deposits of tin-spodumene belts of North Carolina in the U.S.A.

Spodumene deposits of Afghanistan are characterized by their high content of lithium oxide (1–2 %). Six of the deposits only (Drumgal, Pasgushta, Tsamgal,

Jamanak, Paski and Jarigul) are estimated to contain reserves of 3 million tons of lithium oxide to a depth of 100 m. They belong to the largest deposits of the world. The formation of spodumene pegmatite deposits of the Hindukush should be explained by some specific features of the parent magmas and concentration of lithium.

The high-alumine granites are characterized by 2—3 times increased lithium content against clark (z); the sharpest incerease (30—50 times) of lithium content in comparison with granites and plagioclase-microline (ore-barren) pegmatites is observed in spodumene pegmatites. The spodumene pegmatites and their parent granites are regarded as differentiators of a granite magma, highly enriched with lithium.

Special studies of lithium geochemistry have led to the conclusion that granite magmas which are forming pegmatite-bearing massifs, are enriched with lithium by the anatexis process of sedimentary-metamorphic series. Further concentration of lithium in residual granite melts, is conditioned by differentation of magma and re-distribution of lithium by the magmatic alteration, particularly by replacement of biotite by muscovite.

Deposits of Precious Stones

Deposits of kunzite, jewellery tourmaline, beryl, are associated with the rare-metal pegmatite ore fields, where they occupy a specific geological position: 1. they are located in the zones of tension only; they are however absent in the zones of compression; 2. the most productive pegmatites occur either in gneiss series around graben-synclinal structure, or in gabbroid massifs within these structures; 3. the largest deposits are always occuring in subhorizontal or gentle-sloping bodies, but the small ones are connected with bodies of any morphology and any steepness of dipping.

Most productive for gemstones are albitized microline pegmatites with lepidolite, spodumene, polychrome tourmaline. These pegmatites form the unique kunzite deposit of Kulam. Pegmatite bodies containing gemstones are characterized by inhomogeneous inner structure, by slightly expressed differentation, by large minerals and mineral aggregates, by abundance of minerals containing volatile components, and by extensive development of miarols, etc.

The gemstones are occuring in the miarols and near them among mineralized zones, consisting of blue clevelandite, lepidolite, white spodumene, pollucite. The size of a miarol varies from 10 cm to 3 m in diameter. The largest of them are located under the hanging contact of gentle-sloping bodies. Miarols with gemstones only occur in pegmatites formed under the condition of crystallization. The formation of the minerals took place at the end of differentation. Along with free crystallization there are widespread products of regeneration of spodumene and pollucite, as well as tourmaline, beryl, topaz, quarz around the miarols and partially within the miarols itself. Regeneration has an important practical significance: the majority of jewellery kunzite was formed just by the process of regeneration of primary spodumene.

Temperature of mineral formation in miarols, determined by homogenization of gas–liquid inclusions, range mainly within 120–280° C. Regeneration of kunzite took place within the temperature of 160–270° C.

In general, rare-metal pegmatites of the Hindukush containing gemstones, exhibit a wonderful similarity with the pegmatites of Pala and San-Diego in California as to their character of mineralization and geological position.

Conclusion

1. Eastern Afghanistan is a part of the global Himalayan pegmatite belt, stretching 3000 km from Iran in the West to Birma in the Southeast.

2. Rare-metal pegmatites of Eastern Afghanistan are spatially and genetically associated with bimica highly-alumina granites of Upper Cretaceous-Paleogene age. The formation of granites and pegmatites is conditioned by the Alpine activization.

3. The rare-metal pegmatite belts extending from 200 to 400 km in the territory of Nuristan–Pamir, are related to young Permian-Jura graben-synclines.

4. The pegmatite deposits of Eastern Afghanistan are characterized by a remarkably expressed Li, Be, Ta and Sn anomaly. The huge spodumene deposits, the deposits of tantalium and beryl, as well as unique deposits of gemstones-kunzite and jewellery tourmaline, have recently been discovered within this regional metallogenic unit.

References

Denikayev, S. Sh., Feoktistov, V. P., Rossovsky, L. N.: „Skhema Tectonicheskogo Raionirovania Vostochnogo Afghanistana". Dokl. AN Taj. SSR, 1975, T. XVIII No. 9, S. 42–47.

Magarovsky, V. V.: Geochemicheskaya i Metallogenicheskaya assymetria pegmatitonosnogo Pamirsko-Shugnanskogo Kompleksa. Dokl. AN SSSR, 1973, T. 209, No. 1, S. 181–184.

Rossovsky, L. N.: Pamirsko-Hindu-Kushskaya Redkometalnaya Pegmatitovaya Provincia. V kn. Metallogenia Tyan-Shana. Frunze, izdat. ILIM, 1968, S. 396–398.

Rossovsky, L. N.: Redkometalniye Pegmatiti s dragotsennimi Kamnjami i uslovie ikh obrazovaniya (Na primere Hindu-Kusha). Zapiski Vses. Miner. o-va, 1980, ch. 109, No. 3, S. 301–311.

Rossovsky, L. N., Mogarovsky, V. V.: K Probleme Istochnika Litiya pri obrazovanii Spodumenovikh Pegmatitov. Geol. rudn. Mestorozhd., 1984, No. 6 (v pechati).

Rossovsky, L. N., Chmirev, V. M., Salakh, A. S.: Geologia i Usloviya Formirovaniya Spodumenovikh Mestorozhdeniy Hindu-Kusha. Geol. rudn. mestorozhd., 1976, No. 6, S. 19–33.

Rossovsky, L. N., Chmirev, V. M., Eromenko, G. K.: Litiyevie Mestorozhdeniya Afghanistana. Izv. vissh. uchebn. zaved. Geologia i Razvedka, 1977, No. 7, S. 85–94.

Stazhilo-Alexeyev, K. F., Dovgal, You. M., Chalyan, M. A.: O Svyazi Magmatisma s Tektonikoy v Afghanistane. Sov. Geol., 1973, No. 6, S. 102–116.

Fersman, A. E.: Pegmatites. T. 1, Izd-vo AN SSSR, 1940, S. 712.

Metallogenetic features of the lead-zinc and copper deposits in Iran

By J. G. HADITSCH*)

With 9 Figures

Zusammenfassung

Obwohl die Blei-Zink- und Kupfer-Lagerstätten des Iran in den vergangenen Jahren verstärkt untersucht wurden, blieben bis heute noch gewisse Fragen nach dem Zusammenhang dieser Mineralisationen untereinander und mit magmatischen Ereignissen unbeantwortet. Eine kritische Gegenüberstellung der betreffenden Vererzungen ergab

1. keine auffallende zeitliche Übereinstimmung der Pb-Zn- mit den Cu-Mineralisationen, besonders im Paläozoikum und in der unteren und mittleren Trias,
2. daß magmatische Gesteine oft nur die Erzträger, aber nicht die Erzspender darstellen,
3. daß, abgesehen von Metallisationen, die von magmatischen Ereignissen abgeleitet werden können, es verschiedentlich auch solche gibt, die keine aszendente Herkunft erkennen lassen, und
4. daß vielfach noch die Rolle der Paläokarstentwicklung für die Lagerstättenbildung unberücksichtigt blieb.

Bewußt wird in dieser Arbeit nicht auf den Zusammenhang der Mineralisationen mit der plattentektonischen Entwicklung eingegangen.

M. TARKIAN, M. LOTFI & A. BAUMANN (1984) were able to demonstrate that in the Central Lut region all magmatic events were connected with copper and, sometimes, also with lead-zinc mineralizations: The Jurassic magmatism with its intrusions of granites was followed by a disseminated Cu mineralization; the Laramidian uplift and magmatism (at the Upper Cretaceous/Paleocene boundary), by

*) Author's address: Univ.-Prof. Dr. Johann Georg HADITSCH, Mariatroster Straße 193, A-8043 Graz.

porphyry copper ores, and, consequently, the Pyreneean volcanism (at the Eocene/Oligocene boundary) by vein-type mineralizations, copper in andesites, and lead and zinc in dacitic rocks.

The same multi-phase development, like in the Central Lut, can be demonstrated for the other metallogenic units of Iran. A comparison of the type of magmatism, the petrology of the country rocks and the metallisations show at least 8 different metallogenic phases, beginning with the Late Precambrian, and ending in the Early Miocene (Tab. 1). This sequence may give reasons for the assumption of a close connection between copper and lead-zinc mineralizations. It was the aim of this work to examine this presumption.

Table 1. Magmatic activity and related metallic occurrences in Iran (after D. Bazin & H. Hübner, 1969a, modified)

Probable age	Region	Type of Magmatism	Petrology	Metallisation
Early Miocene – Late Paleogene	Azerbayjan, Kashan, Kerman	Subvolcanic	Acid to intermediate	Cu (Pb, Zn, Mo, Au)
	Azerbayjan, Tarom, Kashan, Balutchestan	Plutonic	Acid to intermediate	Cu, Mo, W (Pb, Zn)
Paleogene, Upper Cretaceous	Wide spread	Volcano-sedimentary	Dacitic to basaltic, mainly andesitic	Cu, Pb, Zn, Mn, Ba (Ni, Co, Bi, U)
Upper Cretaceous	Zagros, Azerbayjan, Khorassan, Nain, Balutchestan	Submarine-volcanic to plutonic	Ophiolite, gabbro	Cr, Mn (Ni, Cu)
Late Early Cretaceous	Alborz, Golpaygan, Balutchestan	Volcanic	Basic extrusives	Pb, Zn (Cu)
Early Cretaceous – Late Jurassic	Azerbayjan, Hamadan, Central and SE-Iran	Plutonic, subvolcanic	Granodiorite, quartz porphyry, andesite porphyry	Cu, Pb, Zn, W, Mo
Infracambrian? (Taknar Form.)	Khorassan	Volcano-sedimentary	Mainly acid	Pb, Zn, Cu
Infracambrian (Rizu Series)	Central Iran	Volcano-sedimentary	Quartz porphyry, acid tuff	Pb, Zn (Cu)
Late Precambrian	Soltanieh, Golpaygan, Central Iran	Plutonic	Granite	Au

In Iran now more than 200 copper occurrences are known, most of them only being indications, which only partly have been investigated in a profound way. Nevertheless, the results encourage to give a brief synopsis.

The country rocks of the copper mineralizations show different ages (Fig. 1). While till up to the Triassic, and even in the Jurassic and Cretaceous sequences only a few metallisations are established, the most frequent and the most important mineralizations are developed in Tertiary rocks (Tab. 2). The occurrences in Triassic and older rocks are mainly bound to metamorphic formations, those in the Jurassic and Cretaceous to sedimentary and volcanic rocks, and the youngest deposits mainly to a volcanic environment. Drawing conclusions from this geological situation, it

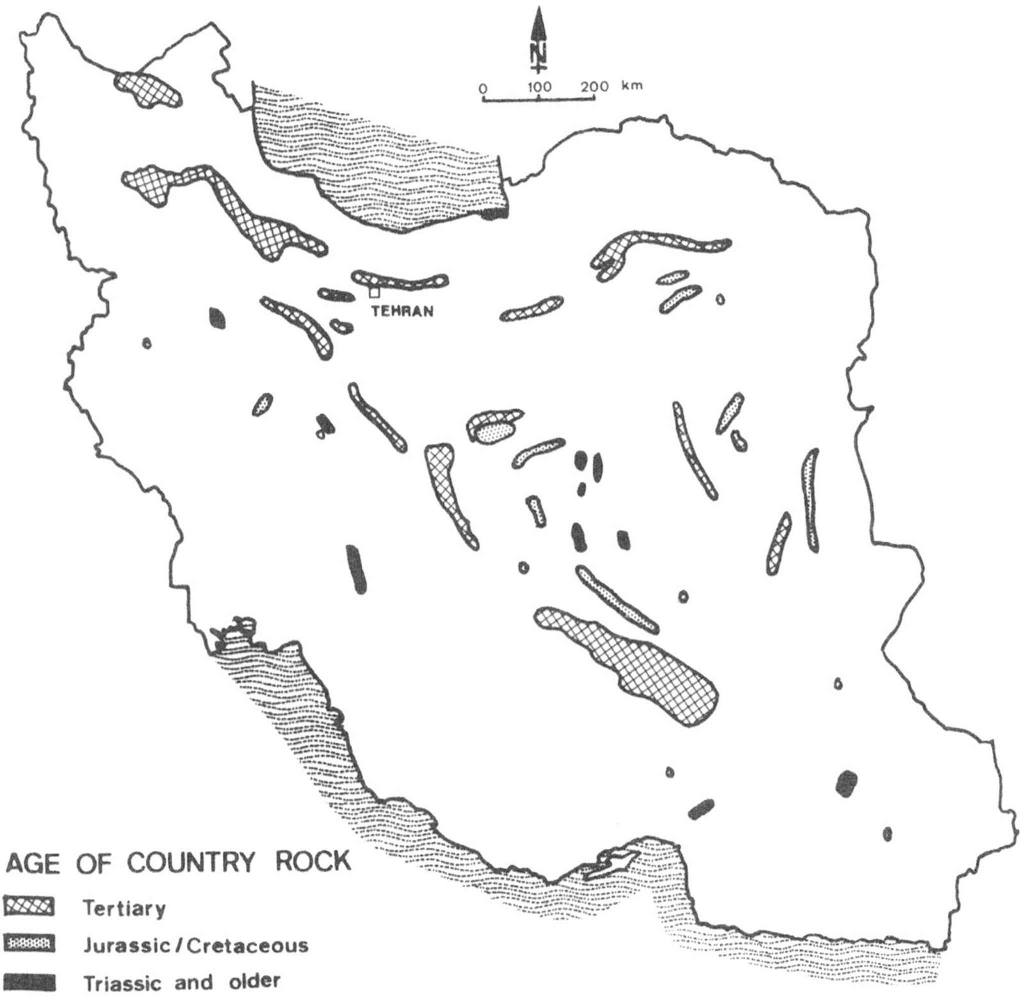

Fig. 1. Copper mineralizations in Iran: Age of the country-rocks

can be stated that all copper districts are showing a strong Tertiary volcanism, the different magmatic products being the main host-rocks for the mineralization. Only a few ore occurrences are connected with older country rocks (p. e., in the Anarak district Talkhe, Sebarz, Kan Mess, Chah Mille, Baqeroq).

Table 2. Copper deposits in Iran and the age of the host-rocks

Age	Size				Sum
	Large		Small		
Triassic and older		1	1		2
Jurassic			1	10	11
Tertiary	1	6	8	34	49

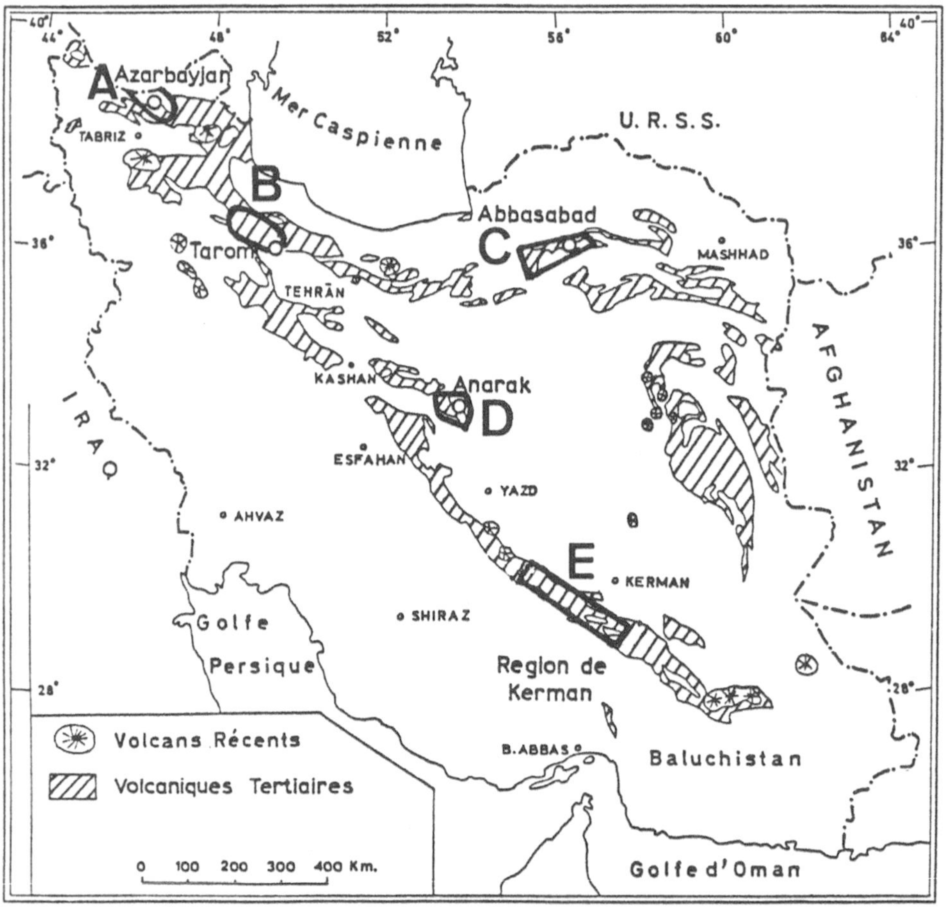

Fig. 2. Main copper districts in Iran (A–E) and their relation to Tertiary volcanic rocks (after D. Bazin & H. Hübner, 1969, modified).

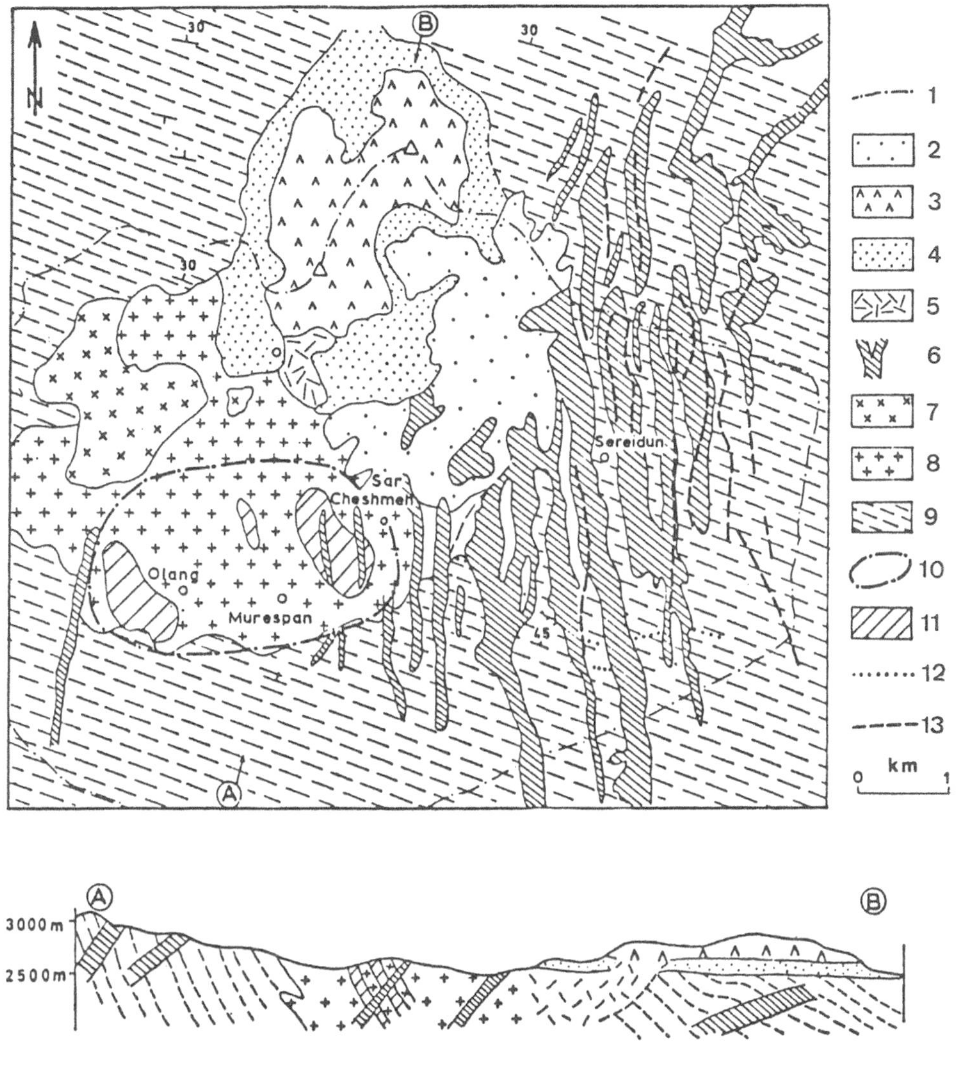

Fig. 3. Map and geological profile of the deposit Sar Cheshmeh. – 1 = Limit of the basin, 2 = Recent terrace with volcanic material, 3 = Dacitic lavas and ignimbrites, 4 = Volcanic conglomerates, cemented by ash tuff, 5 = Crystal tuffs, 6 = Dykes, 7 = Adamellite, 8 = Granodiorite, 9 = Paleogene lavas and pyroclastics, 10 = Boundary of primitive altered and mineralized intrusive, 11 = Enrichment area, 12 = Pyroclastic beds, 13 = Siliceous dykes (after D. BAZIN & H. HÜBNER, 1969)

Five different main copper districts can be distinguished easily (Fig. 2): Azerbayjan (A), Tarom (B), Abbasabad (C), Anarak (D), Kerman (E). The shape of these relatively small metallogenic areas is caused by very strong tectonic structures, which bound the Arabian and Euroasian plates and the Central Iranian fragments (A. A. NOWROOZI, 1976).

Among these five metallogenic units, the Kerman district is the most important one, with the greatest copper deposit of Iran, Sar Cheshmeh (D. BAZIN & H. HÜBNER, 1969 a, b). Here, Paleogene lavas, pyroclastics and sediments are cut by

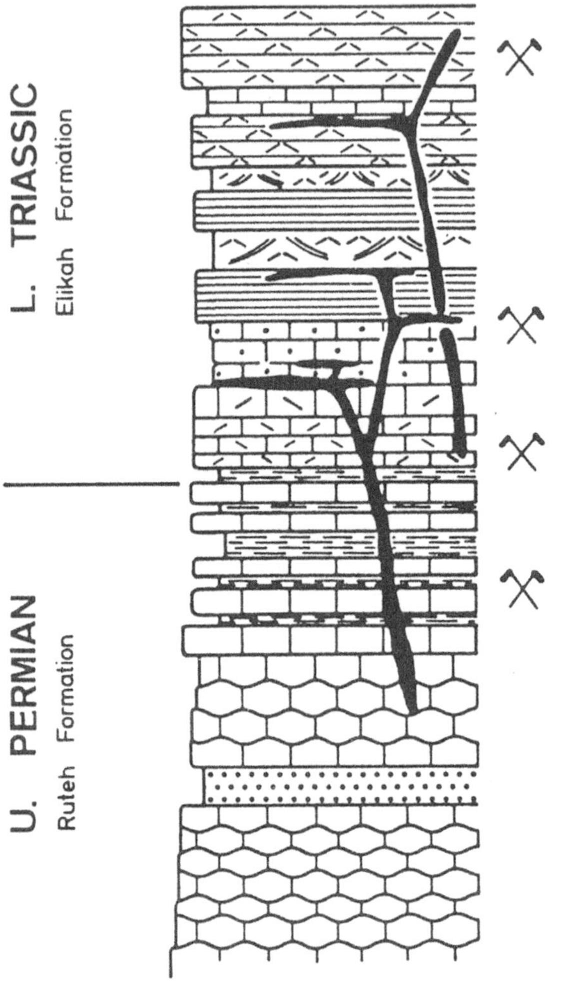

Fig. 4. Profile of Rochak near Rudbarak (Alborz) (after R. BRANDNER et al., 1981). – This figure demonstrates the stratigraphic position of the mineralization, which affected Upper Permian and Skythian strata. Some parts show stratiform mineralizations, but the discordant ones predominate. The synchronous dolomitization verifies the epigenetic character of the metallisation.

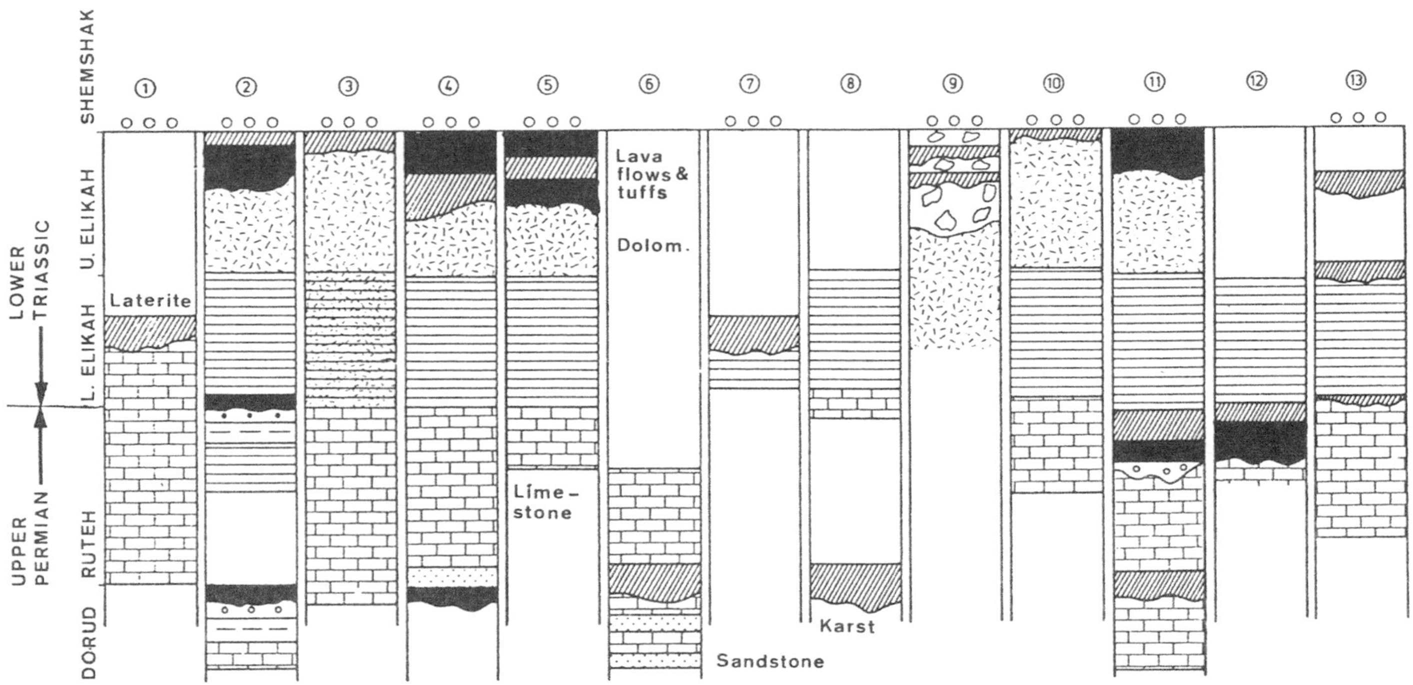

Fig. 5. Stratigraphic position of basaltic volcanic rocks and the karst development in the Alborz region (after R. Brandner et al., 1981, modified). – 1 = Rudbar, 2 = Gashnarud, 3 = Kelesther, 4 = Elikah, 5 = Kamerbun, 6 = Dizin, 7 = Gachare, 8 = Ruteh, 9 = Noor Valley, 10 = Mangol, 11 = Arruh, 12 = Gaduk, 13 = Goshnavi. – Ba-F-Cu deposits are bound to the karst and the subaeric basalt volcanism. At the different localities volcanism and karstification and, consequently, the mineralizations, show different stratigraphic positions.

Neogene subvolcanic intrusions (dykes, sills and stocks of a porphyritic granodiorite). The copper mineralization occurs mainly in the form of disseminations around altered stocks and within these bodies, and as veins, stockworks and impregnations in the surrounding lavas and pyroclastic rocks (Fig. 3).

Also the less important deposits are bound to Tertiary magmatic rocks: Sungun and Mazraeh (Karadagh) being skarn mineralizations, and Zahbad, Meskani, Talmessi and Chahar Gonbad such of vein-type. Only the stratiform deposits of Taknar are situated in Precambrian green schists (B. Razzaghmanesh, 1968).

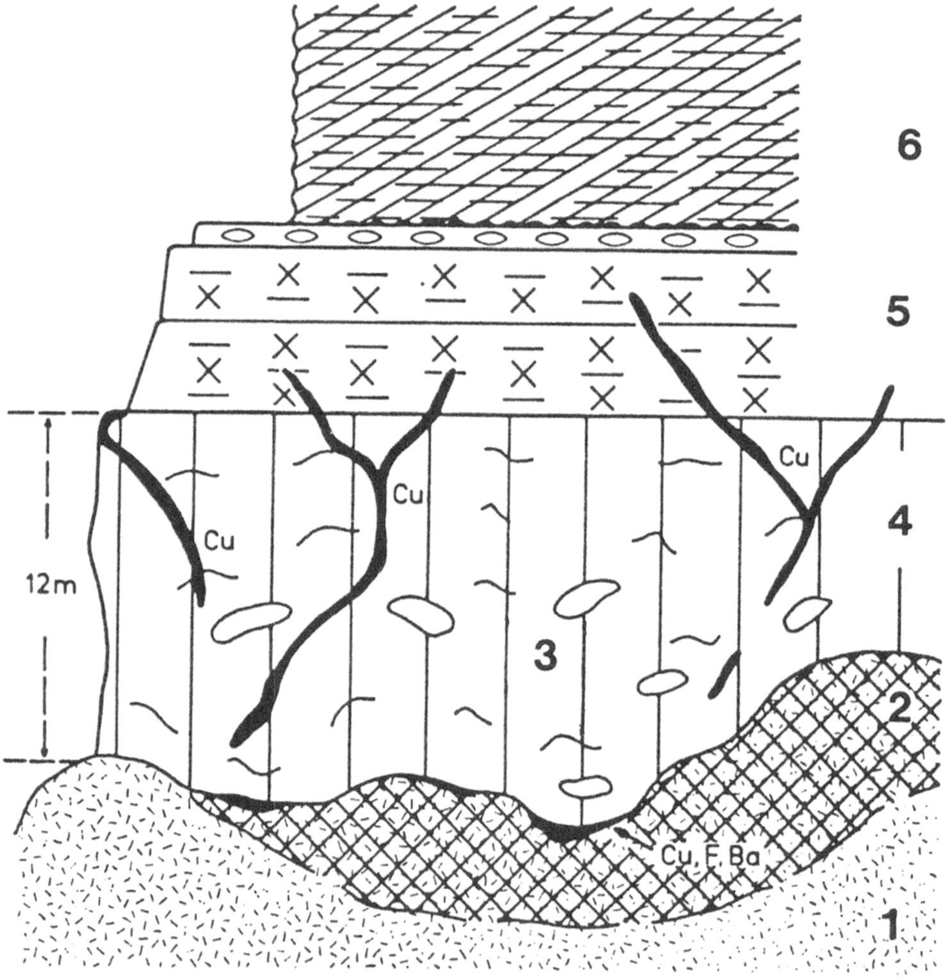

Fig. 6. Profile of Kamerbun (Alborz) (after R. Brandner et al., 1981). – 1 = Triassic dolomitized limestone, 2 = silicified beds, 3 = Laterite, 4 = Laterite with mobilized chalcopyrite, 5 = Crystal tuff, 6 = Slates (Shemshak Formation). The chalkopyrite veinlets originate from the small copper content of the laterite.

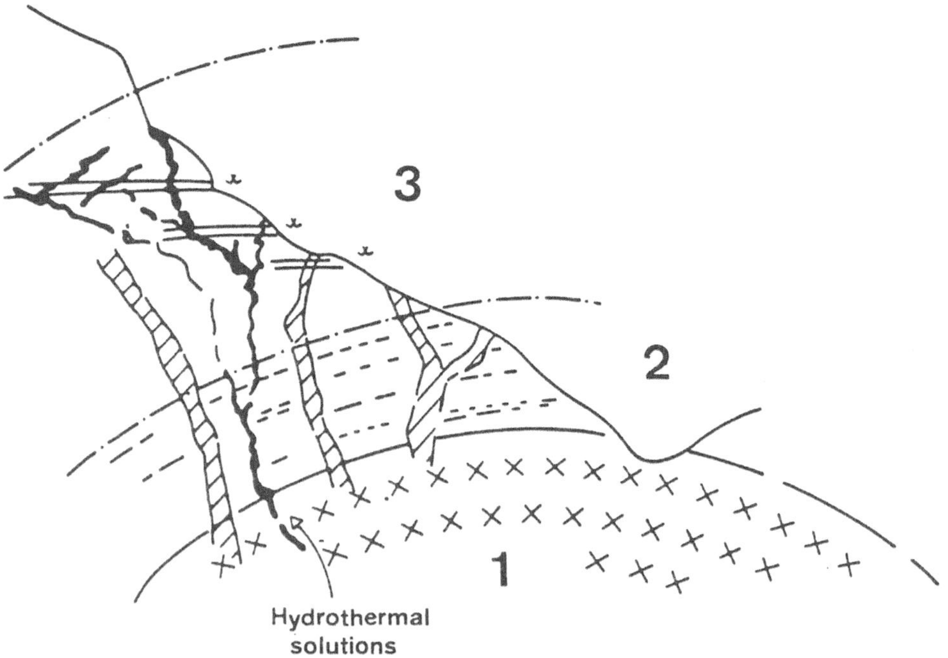

Fig. 7. Guret Mine (after R. BRANDNER et al., 1981). – 1 = Granite, 2 = Inner zone of contact metamorphosis (with hornblende, garnet, epidote, magnetite), 3 = Outer zone of contact metamorphosis. – Limestones and marls were metamorphosed by a Mesozoic granite. Aplitic, granodioritic and dioritic dykes cut the metamorphosed rocks. The chalcopyrite-galena-sphalerite mineralization is younger than the dykes, and was formed by hydrothermal solutions.

Among the smaller deposits most of those, which occur in Jurassic and Cretaceous rocks, have a sedimentary origin–with one exception: Dahaneh Siah–, while those of Tertiary age have magmatic, mostly volcanic country rocks. Also the Upper Cambrian Khanegah (Kuh-e-Dena) deposit in the Zagros range has limestone beds as host rocks.

Summarizing, the fact can be stressed, that there exists a remarkable difference between the metallisations in Paleozoic and Mesozoic rocks and those which affected Tertiary rocks: Nearly all of the mineralizations in the older rocks are connected with sedimentary rocks, and all in the Tertiary with magmatic ones.

The connection with sedimentary rocks does not necessarily imply that in any case the related ore deposits are of syngenetic origin, even if these mineralizations show stratiform features. This can be demonstrated easily in the section of Rochak in the Alborz chain (Fig. 4).

The extrusives may also have covered a karstic landscape; the karst phenomena too, can be of different ages (Upper Permian, and Lower Triassic, resp.; Fig. 5). Sometimes the uppermost parts of the Upper Permian paleo-karst are characterized by iron-rich carbonates, like ankerite, sideroplesite and siderite, and a weak copper mineralization, which later on could have been mobilized on some places; this, for instance, can be observed, at Kamerbun, where the karst is developed in Triassic limestones, associated with an intensive silicification, a lateritization and a Cu-F-Ba mineralization, followed by a postgenetic copper mobilization, evolved in the laterites (Fig. 6). The basic volcanic rocks, which at Kamerbun were affected by the lateritization, most probably have a lower Jurassic age, like in Elikah, where the Liassic age is proven.

It is not very often possible to prove the magmatic origin of the ore-bearing solutions in such a clear way like at the Guret Mine, near Kelardasht, in a series of limestones, marls and cherty shales, the age of these sediments most probably being Carboniferous. In Mesozoic times this series was affected by a granite, thus forming two zones of contact-metamorphism. Later, aplites, and granodiorite and diorite dykes crossed the metamorphic zones, and hydrothermal solutions had as a result Pb-Zn-Cu ores (Fig. 7).

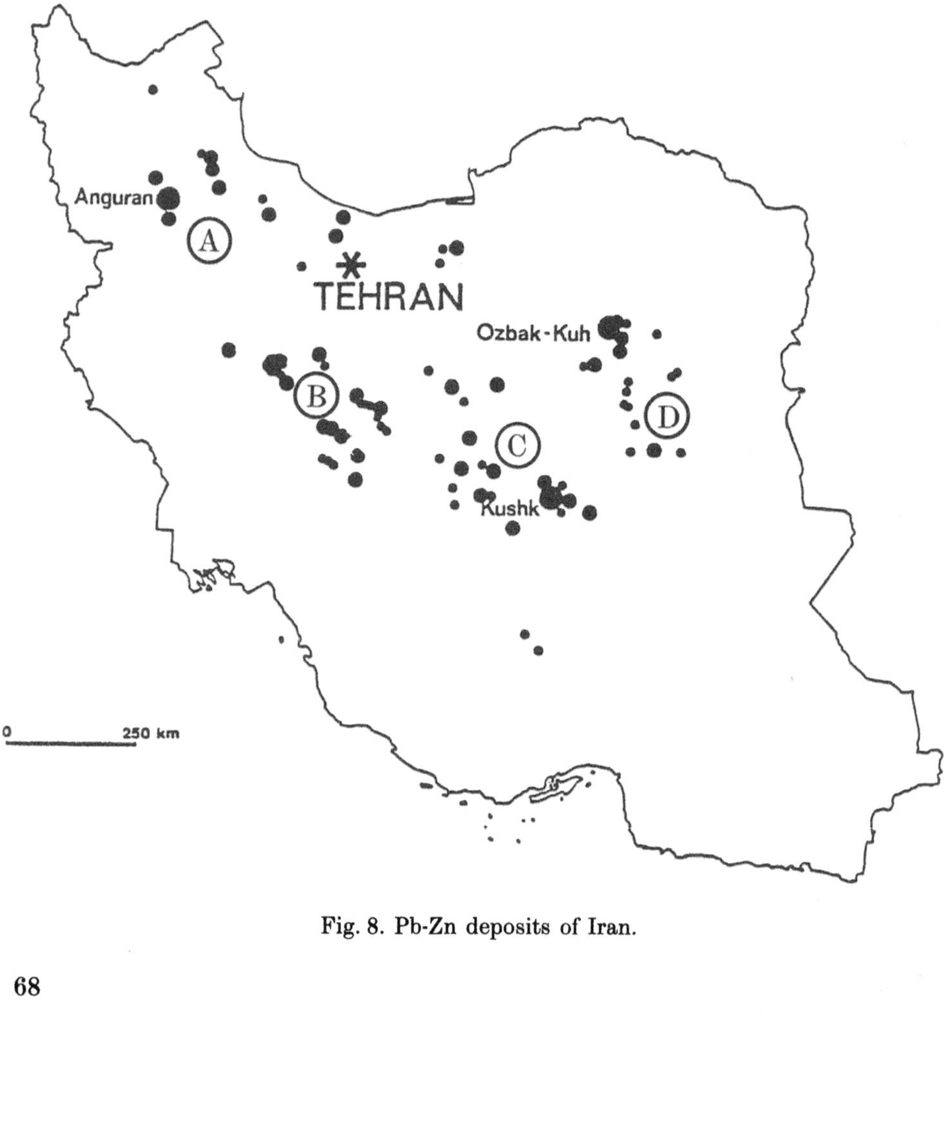

Fig. 8. Pb-Zn deposits of Iran.

The Iranian Pb-Zn deposits are concentrated mainly in the ranges, which are bound with the Great Salt Desert (Kavir). Four different regions can be distinguished (P. BARIAND et al., 1965, L. BURNOL, 1968):

1. Azerbayjan and the Alborz chain (Fig. 8, A),
2. the region of Esfahan and Arak (B),
3. the area of Anarak and Yazd (C),
4. the eastern ranges of the Kavir between Ferdows and Tabas (D).

In the first area (A) the most important Pb-Zn deposit of Iran is situated (Anguran), partly showing stratiform and partly karstiform features. Between Anguran and the deposits in the Alborz chain volcanic rocks of Tertiary age (Eocene), near Rasht, show vein mineralizations with a remarkable proportion of copper and gold (p. e., Zahbad, Ghare Djangal).

In the central part of Iran (B) some well-known mineralizations are situated around Esfahan and some more to the NW, i. e. between Golpaygan and Arak (p. e., Lakan, Hosseinabad).

In the region of Anarak and Yazd (C) most varying deposits are developed, namely Kushk, Nakhlak, Merjard, and Zirakan.

Of all the metallisations in the chain east of the Kavir (D), the deposit of Ozbak-Kuh is the most interesting one.

For a lot of these deposits epigenetic (i. e. hydrothermal-metasomatic) events were postulated. Meanwhile for some of the most important mineralizations, indications of an originally syngenetic sedimentary protore formation, could be found. These originally stratiform mineralizations were overwhelmed epigenetically by multistage hydrothermal mobilizations (p. e., J. G. HADITSCH, 1982, 1984). By this, e. g., the Pb-Zn deposits in Upper Paleozoic and Mesozoic strata in the east of Iran could be defined as inherited or traced ore deposits ("Vererbungslagerstätten" and "durchgepauste Lagerstätten", resp.).

L. BURNOL (1968) distinguished different types of Pb-Zn deposits, according to their ore minerals, the temperature of their formation and the type and the amount of the gangue:

Type I: Deposits with sphalerite, galena, pyrite, chalcopyrite and minerals of elevated formation temperature ("mesothermal" after W. LINDGREN). These deposits show veins with quartz ± baryte ± dolomite as gangue minerals, some arsenopyrite and pyrrhotine; the latter mineral and chalcopyrite may also occur in the form of segregated grains in the blende. The deposits were formed during different stages, which demonstrate a normal mineral succession, according to the successive decrease of temperature, and an alteration of the country rock.

Type II: Deposits with sphalerite, galena, pyrite, chalcopyrite, without minerals of elevated formation temperature. – Sub-type II a: Mesothermal deposits, with a normal mineral succession, deposited in one phase, and with a remarkable amount of various gangue minerals. Quartz is the main gangue, some carbonate minerals (siderite, dolomite, calcite) may also occur. The host-rocks were altered with different intensities. – Sub-type II b: Volcano-sedimentary stratiform deposits, with sphalerite (schalenblende), pyrite (melnikovite-pyrite), galena ± chalcopyrite, and with only gangue accessories.

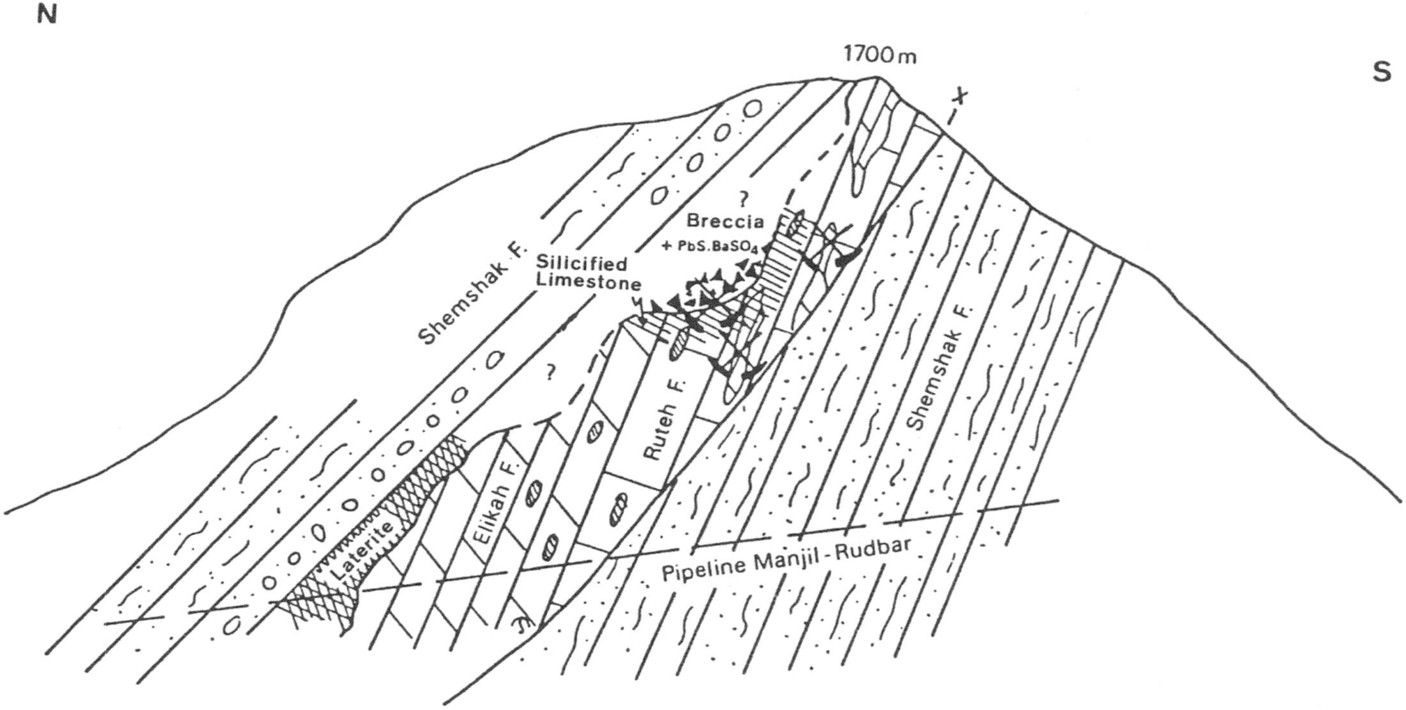

Fig. 9. Himmelmine (Manjil) (after R. BRANDNER et al., 1981). – Laterite, in sedimentary contact with Triassic dolomites. Limestones of the Ruteh Formation (Permian) and the hanging breccia were partly, resp. totally silicified, and mineralized by galena and baryte. The silicification and ore mineralization originate from solutions, which were transported by a river system, and precipitated at the water table.

Type III: Pb-Zn deposits, with either dominating galena or sphalerite, and few or even missing gangue minerals.

Type IV: Epigenetic galena deposits, partly stratiform, and sometimes also bearing sphalerite, and with quartz, fluorine, baryte as gangue minerals. – Sub-type IV a: Quartz is the main gangue mineral, accompanied by more or less baryte and calcite. – Sub-type IV b: With baryte or fluorine, one of these gangue minerals dominating, often together with some quartz and calcite. – Sub-type IV c: The gangue is mainly baryte, sometimes associated with some quartz, calcite, and fluorine.

According to this classification, the Type III (with the mineralizations of Anguran, Kelardasht, Ozbak-Kuh, Zirakan, Nakhlak, Lakan I, Shah-Kuh and Anjireh-Tiran) is by far the dominating among the Iranian deposits, the relating mineralizations only rarely being stratiform, but in any case of epigenetic origin.

The Sub-types II a and II b are represented by the interesting mineralizations of Zahbad and Seh Changi, and Kushk, respectively.

The hypothermal deposits (Type I) with, p. e., the mineralization of Merjard, are of less importance, and the metallisations of the remaining Sub-types IVa, b, c, represented by p. e., Lakan II, Sar-e-siah, Ormodeh, Mehdiabad, are unimportant.

There is no doubt about the existence of sedimentary Pb-Zn mineralizations, as, e. g., the deposits of Kushk and Zirakan in the Infracambrian sequence of the Saghand-Bafq area. Only a few of these deposits are of economic interest. Many other deposits are strata-bound, or even stratiform, like, e. g., Ozbak-Kuh, Lakan II, Anjireh, but in these cases the original sedimentary metallisation is of no economic value in contrast to the related epigenetic vein mineralizations. In these cases, mobilizations generated economic Pb-Zn concentrations, this mainly by hydrothermal alteration (e. g., Ozbak-Kuh) or by processes in a karstic environment (Anguran, Manjil, etc.).

The most important Iranian Pb-Zn deposit, Anguran, shows at its base a stratiform sulphide mineralization covering mica schists. It is not yet proven, that this contact is caused by an overthrust fault. The dolomitic limestones in the hanging wall of the above-mentioned and less important sulphide mineralization were affected by a strong karstification. The karst cavities were filled epigenetically by Pb and Zn oxide minerals, as, e. g., cerussite, anglesite, smithsonite, hydrozincite, willemite, hemimorphite. Only this high-graded mineralization actually is worked in an open pit.

At Manjil, Lower Triassic dolomites of the Elikah Formation and limestones of the Permian Ruteh Formation show an Upper Triassic paleo-karst development, with a silicification and a mineralization, mainly consisting of galena and baryte (Fig. 9).

A comparison with the copper mineralizations only partly shows a coincidence, and not in the case of the most important Pb-Zn deposits, namely Anguran, Ozbak-Kuh and Kushk.

H. FÖRSTER (1974) tried to give a first synopsis of the Iranian deposits, based on plate-tectonics. Since then, it became possible to give another review, related to the metallogenic evolution.

Table 3. Metallogenic evolution of the Pb-Zn and Cu deposits in Iran

METALLOGENIC EPOCHS	SUB-EPOCHS	EXAMPLES	FORM	Cu	Pb-Zn
Upper Cretaceous – Pliocene	Eocene – Pliocene	Sar Cheshmeh	O /	■	
		Kuh-e-Panj	O /	■	
		Merjard	/	■	
		Meskani	/ .	■	
		Talmessi	/ .	■	
		Qaleh Zari	/	■	
		Dahaneh Siah	O / =	■	
		Abbasabad	/	■	
		Neshapour	/	■	
		Seh Changi	/	■	
		Kuh-e-Garmab	/		■
		Nakhlak	/ =		■
		Zahbad	/		■
	U. Cret. – Paleocene	Nohdeh	O /	■	
		Garmab	=		■
Jurassic – Lower Cretaceous	Lower Cretaceous	Natanz	/		■
		Akhlamad	/ =		■
		Boghestan	/ =		■
		Darreh zanjir	=		■
		Chek Chekou	=		■
		Mehdiabad	=		■
	Jurassic	Sorkhkuh	O	■	
Permian – Upper Triassic	Late Triassic	Manjil (Himmelmine)	=		■
		Alborz	/ =	■	
	L/M Triassic	Ahmadabad	/ =		■
		Kuh-e-Sormeh	=		■
	Permian	Anjireh	/ =		■
		Harzabil (N)	=		■
Upper Devonian		Ozbak-Kuh	/ =		■
		Galeh	/		■
Infracambrian		Koshami	=	■	
		Narigan	=	■	
		Kushk	=		■
		Zirakan	=		■

/ Vein mineralization.

O Porphyry copper.

. Impregnation.

= Stratiform mineralization (Karst mineralizations included).

Concerning the Pb-Zn and Cu mineralizations, now in Iran five metallogenic epochs are distinguishable; three of them also can be subdivided into sub-epochs (Tab. 3. To be on the safe side, in this figure only verified examples are given).

The metallogenic evolution from the Infracambrian to the Tertiary shows a distinct development from stratiform mineralizations and impregnations to such of vein-type.

A comparison of Tab. 1 with Tab. 3 demonstrates that

1. there is no strong temporal coherence between the Pb-Zn and Cu mineralizations, especially in the Paleozoic, and in the Lower and Middle Triassic;

2. magmatites often are only the country-rocks of the mineralizations, and were not the donors of the ore-forming solutions;

3. there exist not only metallisations, which have their origin in magmatic events, but also mineralizations without any visible connection with a magmatic source. For these mineralizations a formation by descending solutions is to assume with good reasons;

4. in many cases the important role of paleo-karst for the genesis of ore concentrations has been undervalued or even neglected.

References

BARIAND, P., ISSAKHANIAN, V., & SADRZADEH, M. (1965): Preliminary metallogenetic map of Iran. – Geol. Survey of Iran, Rep. 7.

BAZIN, D., & HÜBNER, H. (1969): Copper Deposits in Iran. – Geol. Survey of Iran, Rep. 13, 6+ 232+4 p. (1969a).

BAZIN, D., & HÜBNER, H. (1969): La région cuprifère á gisements porphyritiques de Kerman (Iran). – Mineral. Deposita, *4,* 200–212 (1969b).

BRANDNER, R., HADITSCH, J. G., & MOSTLER, H. (1981): Beiträge zur vortertiären Pb-Zn-Cu-Metallogenese im Raum zwischen Rasht und Chalus (Alburs, Iran). – Geol. Paläont. Mitt. Innsbruck, *10,* 7, 257–285.

BURNOL, L. (1968): Contribution a l'Etude des gisements de plomb et zinc de l'Iran. Essais de classification paragenetique. – Geol. Survey of Iran, Rep. 11, 113 p.

FÖRSTER, H. (1974): Magmentypen und Erzlagerstätten im Iran. – Geol. Rundschau, *63,* 1, 276–292.

HADITSCH, J. G. (1982): Geologie und Erzmineralisation der Lagerstätte Ozbak-Kuh (Chorassan, Iran). – Geol. Paläont. Mitt. Innsbruck, *11,* 10, 309–353.

HADITSCH, J. G. (1984): Geology and Mineralization of the Ozbak-Kuh Mine and the Genesis of the East Iran Pb-Zn Deposits. – Syngenesis and Epigenesis in the Formation of Mineral Deposits, 391–400.

NOWROOZI, A. A. (1976): Seismotectonic Provinces in Iran. – Bull. Seismol. Soc. Am., *66,* 4, 1249–1276.

RAZZAGHMANESH, B. (1968): Die Kupfer-Blei-Zink-Erzlagerstätten von Taknar und ihr geologischer Rahmen (Nordost-Iran). – Diss. RWTH. Aachen, 131 p.

TARKIAN, M., LOTFI, M., & BAUMANN, A. (1984): Magmatic Copper and Lead-Zinc Ore Deposits in the Central Lut, East Iran. – N. Jb. Geol. Paläont. Abh., *168,* 2/3, 497–523.

WALTHER, H. W., & MOMENZADEH, M. (1983): Mineral deposits and metallogenic epochs in the area of geotraverse in Iran. – Geol. Survey of Iran, Rep. 51, 489–499.

The Pb-Zn-Deposits of Keban (SE-Taurus Mountains) and its Position in the Aegean-Tauriden Metallogenetic Province

By ERICH SCHROLL*) and HIDIR CAGLAYAN**)

With 9 Figures

Zusammenfassung

Der Vergleich der geochemischen Parameter, also vorwiegend der Verhältnisse der Spurenelemente und der Isotopen, zeugt für die Verwandtschaft der Pb-Zn-Lagerstätten von Laurion und von Keban und bestätigt damit die von DE LAUNAY (1911) und W. E. PETRASCHECK (1955) auf lagerstättengeologischer Basis getroffene Zuweisung zu einer tertiären, magmatogenen ägäisch-tauridischen Erzprovinz.

1. Introduction

It is one of the objectives of this paper to test the application of geochemical characterization (geochemometry***)) of lead, zinc ore mineralizations hosted in carbonatic rocks. The principles of geochemical characterization are presented by SCHROLL (1979, in press). The geological observations, the petrological and mineralogical investigations, including the geochemical data of the ore mineralization of Keban are published in more detail by CAGLAYAN (1984). Another point is the contribution to a better knowledge of the lead-zinc bearing ore mineralization of the Aegean-Tauriden metallogenetic province.

2. Geographical and geological situation

The ore occurrences of Keban are situated in the South-Eastern Taurus Mountains in the vicinity of the town Elazig in Turkey. The ore mineralizations are bound

*) Geotechnisches Institut der BVFA-Arsenal, A-1031 Wien, POB 8.
**) University of Ankara, Department of Geology, Tandógan, Ankara, Turkey.
***) This technical term is used in this paper for the first time.

mainly on a Permian carbonatic rock series like other ore deposits of the Taurus Mountains, such as Bolkardag or Karalar, which are known as strata-bound ore mineralizations (VACHE, 1966, LAZNICKA, 1981).

The carbonatic rock series consists of Permocarboniferous limestones and marbles, Permian limestones and calcareous schists, Triassic calcareous phyllites and meta-conglomerates, followed by Jurassic sedimentary rocks and Quaternary sediments. Permean and Triassic rocks are weakly metamorphosed by tectonic events in Jurassic time. The Paleozoic and Triassic strata are penetrated by subvulcanic syenitic intrusions and trachytic effusives, partly forming skarn mineralizations within the carbonatic rocks.

Fig. 1 shows the localization in the metallogenetic province and the geological sketch of the area of Keban.

3. Ore mineralizations

The geological sketch Fig. 1 shows three localities of mining activities:

1. Nalliziyaret, named shortly Keban in this paper. The skarn mineralization caused by the intrusion of syenit-porphyr is situated in the valley of the river Firat.

The mineralizations bear scheelite, fluorite, pyrite, galena, sphalerite and chalcopyrite; diopside, actinolite, vesuvian, epidote, mica, titanite etc.

2. Bamas, in the North of Keban, is a skarn type too, characterized by a dominant mineralization of molybdenite, fluorite and chalkopyrite.

3. Firat-Derebaca, named shortly Firat in the South of Keban. This is an intensive ore mineralization of galena, sphalerite, chalcopyrite and pyrite. These ores are the object of the present mining activity. However scheelite and molybdenite is not found.

4. Magmatism

The syenitic and trachytic rocks are contemporary with granodioritic intrusions in this area. The alcali rocks are rich in K 5–10% (K/Rb ~ 500), Ba 0,3%, Pb 30 ppm, and Th 67 ppm (Th/U ~ 10), but relatively poor in Rare Earth elements and zirconium.

5. Preceding genetic conceptions

Most geologists who visited the ore deposits of Keban, like KOVENKO (1938), MAUCHER (1938), PETRASCHECK (1955, 1960), OELSNER (1958, 1960) emphasized an epigenetic origin.

ZUSSERMANN (1969) only decided in favour of stratiform deposition.

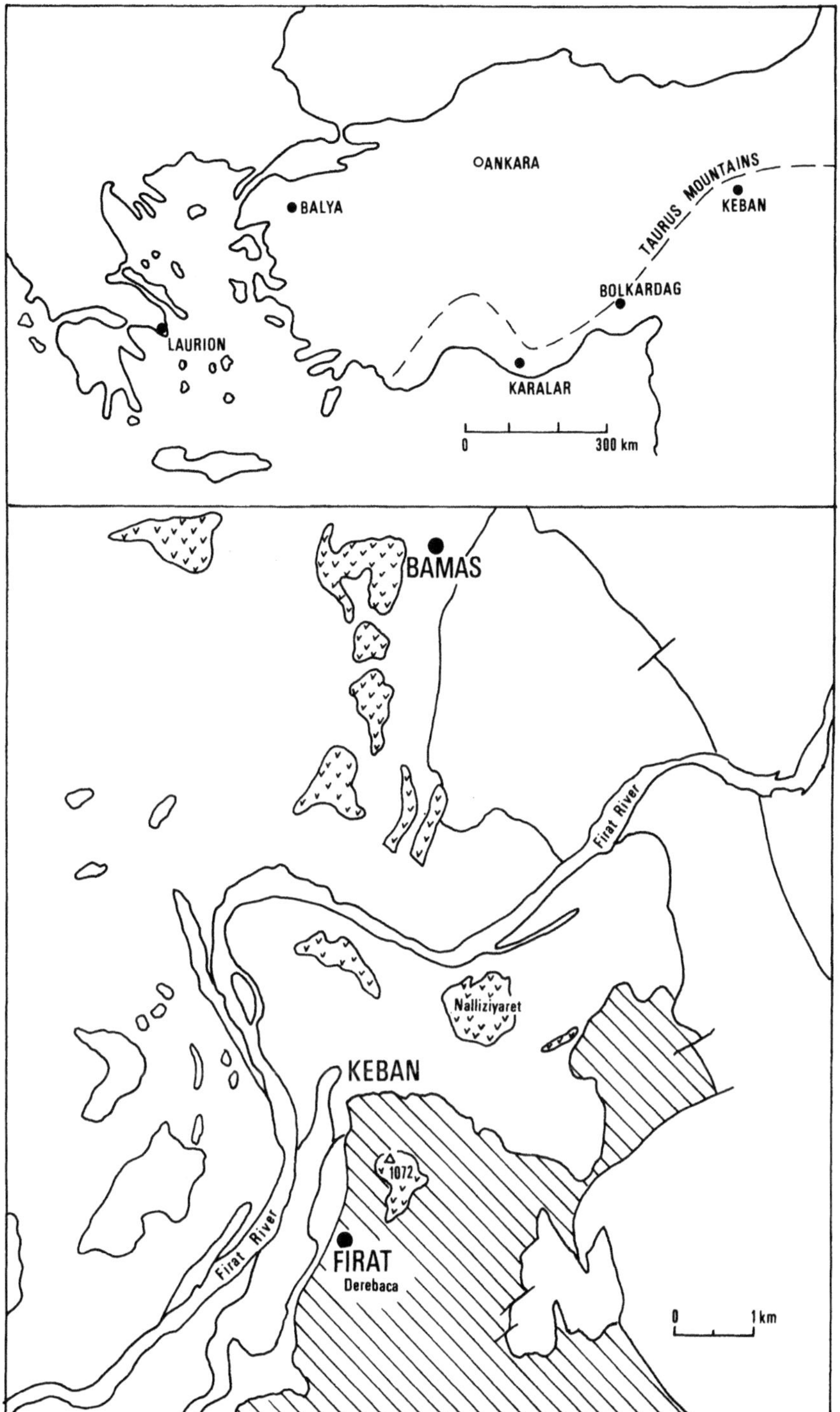

Fig. 1. Sketches of the localization of the Keban ore deposits in the Aegean-Tauridean metallogenetic province and the geological environment of the ore mineralizations. The area of Triassic rocks is hatched and the occurrences of trachytes and syenitic rocks are indicated. The Permian is not marked. The marls are separated by border lines. The younger covering sediments are neglected.

6. Geochemometric investigations

A minimum of numbers of samples, but non less than three for each mineral species of each locality, has been selected. The specific sampling based on field work and mineralogical observations. The objects were mainly the carbonatic wall rocks, calcites, fluorites, pyrites, sphalerites and galenas.

Chemical and isotopic data are necessary for geochemometric work: trace- and minor element distribution in ore minerals and gangue minerals, and the isotopic composition of carbon, oxygen, sulphur and lead.

6.1. Chemical data

Locality	Pb	Zn	Cu	Ag	
Keban	3,5%	4,0%	1,4%	20 g/t	Fe > Pb + Zn
Firat	6,0%	4,2%	1,0%	20 g/t	Fe < Pb + Zn

Minor and trace element distribution. Some selected examples of the distribution of Sb/Bi in galena, Fe/Cu in sphalerite of Ni/Co in pyrite (Ni/Co) and of Tb/La in fluorite (Tb/La) are represented in Figs. 2–5.

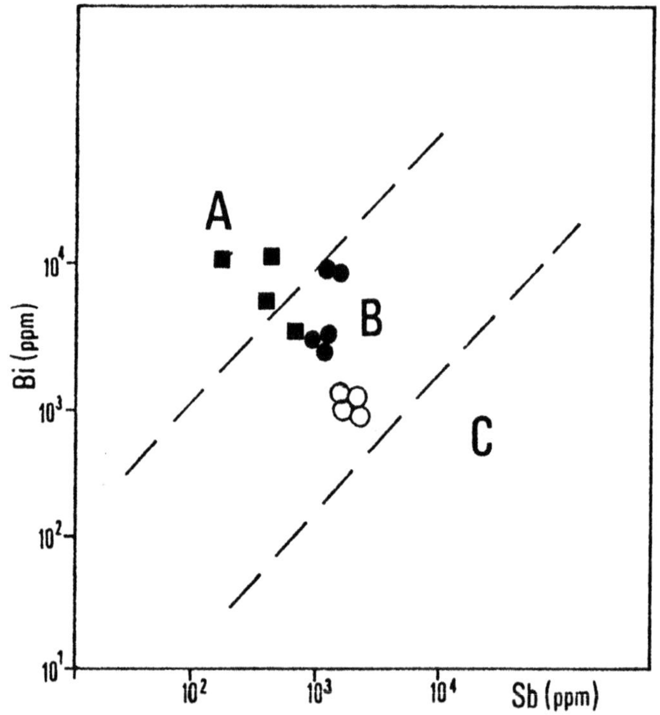

Fig. 2. Sb/Bi-rations of galenas. Ranges of deposition temperatures after MALAKHOV (1969): A Bamas (200–300° C), B Keban-Nalliziyaret 140–220° C, C Firat (100–150° C). Black squares: Bamas, black points: Nalliziyaret and open circless: Derebca-Firat.

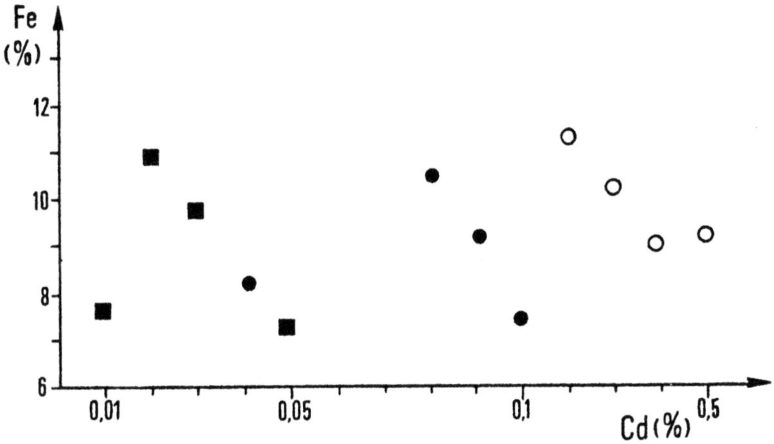

Fig. 3. Fe/Cd-Variation diagramm of sphalerites. The cadmium concentrations decrease from Bamas to Firat following the formation temperature.

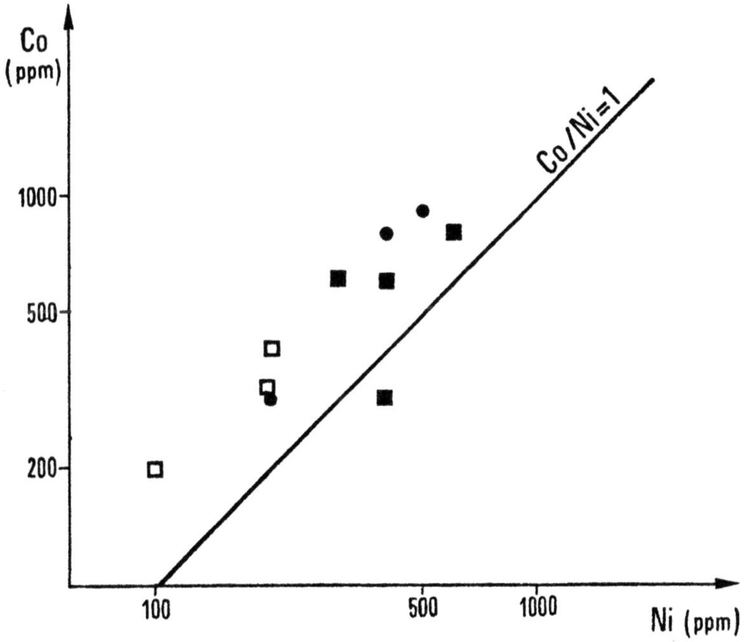

Fig. 4. Ni/Co-rations of phyrites. Co dominates in all samples except one. The absolute concentrations of Ni and Co are lowest in Firat.

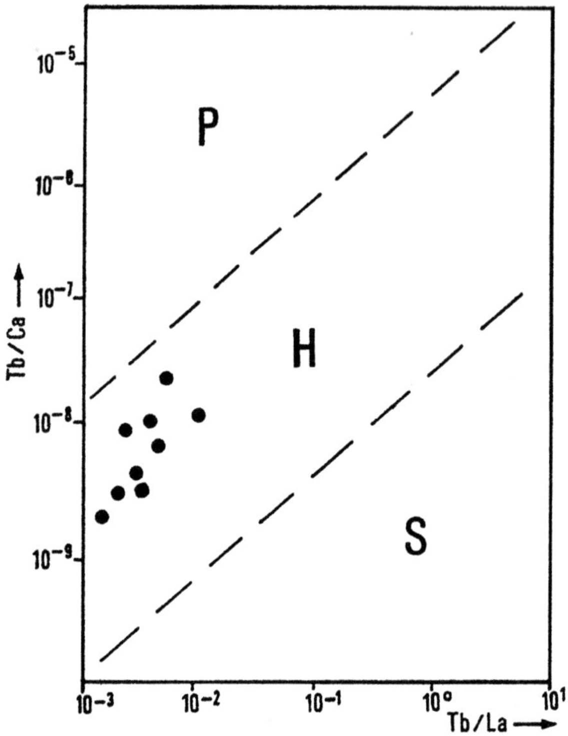

Fig. 5. The Tb/La–Tb/Ca-diagram after MÖLLER et al. (1977) suggests the hydrothermal formation. Remobilization processes can't be excluded. – P = pegmatitic or pneumatolytic; H = hydrothermal; S = sedimentary.

The sphalerites contain relatively high concentrations of Fe (up to 10%), Mn (up to 1%), low contents of Cd (0,02–0,2%) and other rare metals.

6.2. Isotopical data

6.2.1. Isotopes of carbon and oxygen of carbonates

The hydrothermal-metasomatic calcites of the ore mineralization are significantly different from the limestones. The isotopic composition shows large differences on the sharp bounderies between the ore calcite and the wall rocks (Δ $\delta^{13}C$ 3‰, Δ $\delta^{18}O$ 7‰).

The calcites of the ore mineralization fall within the hydrothermal field of the $\delta^{13}C/\delta^{18}$ O-diagram (Fig. 6).

6.2.2. Isotopes of sulphur

Galenas, sphalerites and pyrites show $\delta^{34}S$ values near zero, but with a slight tendency to light sulphur (–0,8 up to –4,4‰). The lightest sulphur is found in the

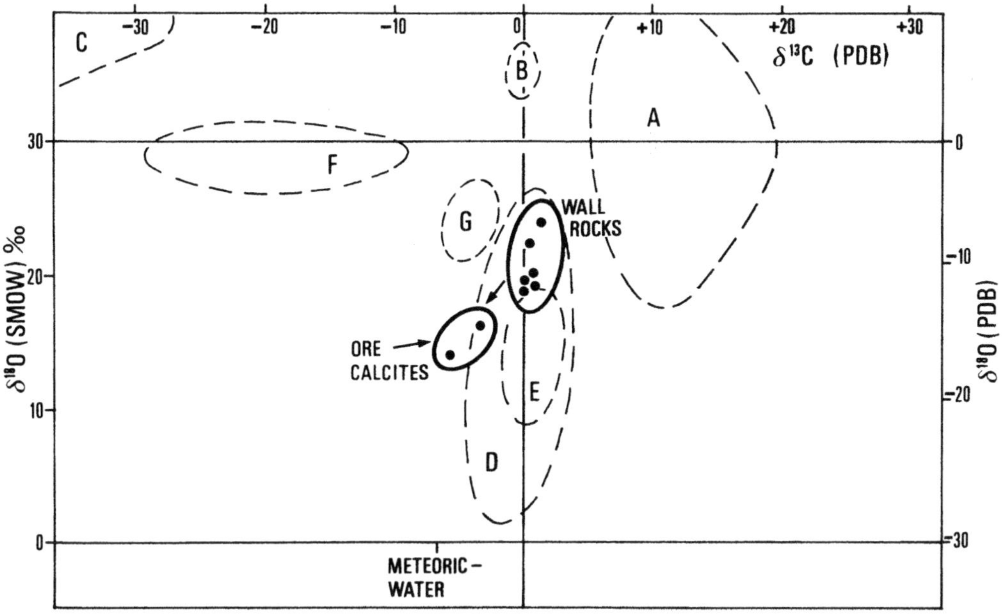

Fig. 6. Carbon and oxygen isotope composition of carbonatic wall rocks and calcites of the mineralization. Fields of carbonatic sediments after Hudson (1977). Dotted field after Hudson (1977): A = heavy carbon sediment; B = deep water limestones; C = methan derived cements; D = late cements; E = marine limestones; F = early diagenetic concretions; G = fresh water limestones.

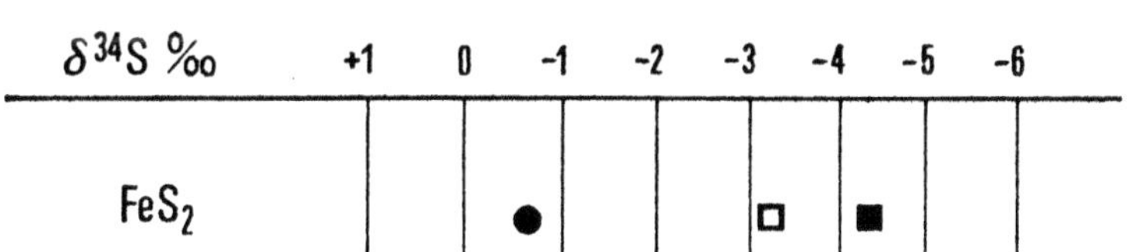

Fig. 7. Sulphur isotope composition of sphalerites, galenas and pyrites.

occurrences of Bamas and Firat. If coexistent mineral pairs being in equilibrium (after OHMOTO, 1972), the differences of $\delta^{34}S$ values should correspond to high temperature ranges.

The sphalerites are showing the heaviest sulphur, the pyrites, except samples of Keban the lightest (Fig. 7).

6.2.3. Lead isotope

The mass spectrometric investigation of two samples of galena and feldspar from one of the trachyte rock samples gives the following results:

	$^{206}Pb/^{204}Pb$	$^{207}Pb/^{204}Pb$	$^{208}Pb/^{204}Pb$
Galena, Firat	19.185 ± 015	15.684 ± 015	39.180 ± 0.030
Galena, Keban	19.146 ± 003	15.695 ± 002	40.438 ± 0.010
Feldspar, Firat	21.868 ± 005	15.807 ± 005	40.438 ± 0.010

The leads are average lead of the Earth crust mixed with a component of younger lead of the upper crust. The leads of ore and of rock are radiogenic (Fig. 8). If the line drawn through the points of the data for ore and rock would be identical with an isochron, a model age could be estimated approximately with 0 ± 100 Mio years corresponding to the young age of the magmatic rocks. The ore lead is slightly thorogenic, so that the acceptance of consanguinity with the thorium-rich alcali rocks seems to be probabel.

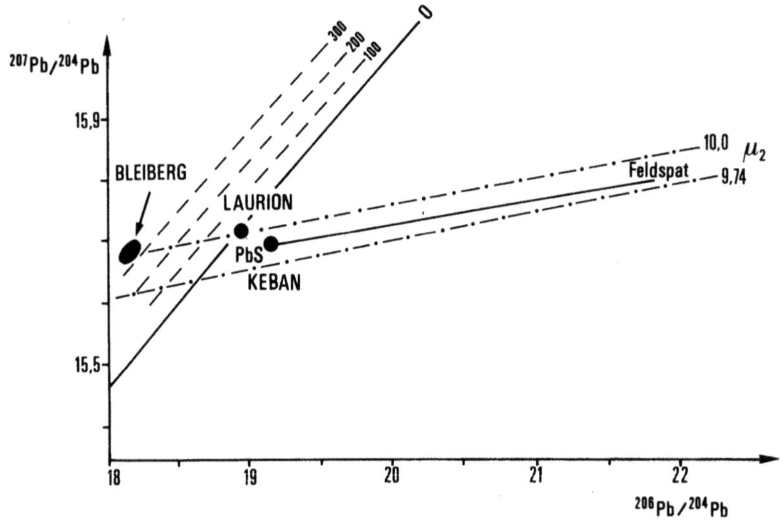

Fig. 8. Lead isotope diagram (lead-evolution model after STACEY/KRAMERS, 1975) showing the radiogenic character of ore lead and rock lead.

6.3. Geochemical parameters

The geochemical data supply strong arguments for epigenetic genesis, caused by high-temperature leaching and hydrothermal deposition. The chemistry of the ore minerals and lead isotopes of the largest metal concentration Firat is identical with the other ore occurrences. The temperature of the deposition of the ore solutions in Firat should be lower than in Keban and Bamas, but around 200° C or more. The negativ $\delta^{13}C$-value of the calcites speaks for endogenic origin.

However it remains still an open question, whether ore mineralizations are partly the product of remobilization processes or not.

7. Position in the Aegean-Tauridean metallogenetic province

The visual comparison of ore samples of Keban (Turkey) and Laurion (Greece) suggests similarities. There is a correspondance of the geochemistry of the sphalerites. SCHROLL (1954) reported analyses of sphalerites of Laurion: Fe 3–10%, Mn up to 1%, Co 0,05–0,3%, Ga up to 10 ppm.

The leads of Laurion showing young model ages are perhaps weakly radiogenic, but are not thorogenic after the data of GALE (1978). The μ_2-values around 10 are comparable with leads derived from the African platform.

A further geochemical study could be of interest. However the Pb-Zn-Ba-mineralizations of the type Bolkardag are the key to the problem. A regional study of lead isotopes is emphazised. Some published values of lead isotopes are plotted in the last Fig. 9.

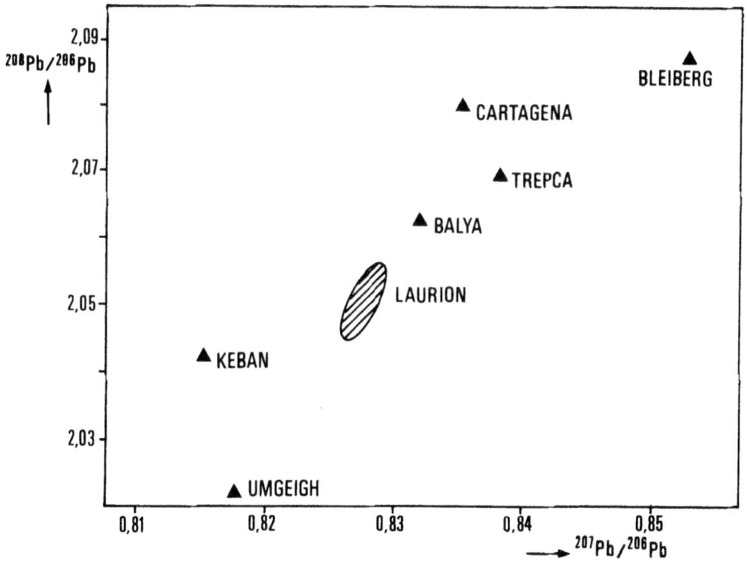

Fig. 9. $^{207}Pb/^{206}Pb$ – $^{208}Pb/^{206}Pb$ diagram using data GALE, 1978.

8. Conclusion

The ore mineralizations of Keban in the SE-Taurus Mountains are products of endogen epigenetic processes. The geochemometric data show some analogy to the lead-zinc-ores of Laurion.

Aknowledgement

We are very grateful to Prof. W. KIESL and Dr. F. KLUGER (Vienna) for the analysis of Rare Earthes in the fluorites, to Dr. H. DIETRICH (Vienna) for micro-probe analysis of ore minerals, to Dr. PAPESCH (Vienna) for the isotope analysis of carbon and oxygen, Dr. E. PAK (Vienna) for the isotope analysis of sulphur, and Prof. Dr. V. KÖPPEL (Zürich) for lead isotope data.

Literature

CAGLAYAN, H. (1984): Die Vererzung der Fluorit-Molybdänglanzführenden Blei-Zink-Lagerstätten von Keban-Elazig im Südost-Taurus (Türkei). Ph. D. Thesis Univ. Vienna.

GALE, H. N. (1978): Lead isotopes and Aegean Metallurgy. *In:* Vol. I. Thera and the Aegean World (W. W. DOUMAS ed., London).

HUDSON, I. D. (1977): Stable isotope and limestone lithification. Q. J. Geol. Soc. (London), *133,* 637–660.

LAZNICKA, P. (1981): Data on the worldwide distribution of stratiform and strata-bound ore deports. *In:* WOLF, K. H. (ed.), Handbook of Strata-bound and Stratiform Ore Deposits. Vol. 10, Elsevier-Amsterdam, 79–390.

LEULEU, M., MORIKIS, A., PICOT, P. (1973): Sur des mineralistions de type skarn au Laurium (Greece). Mineral. Deposita, *8,* 259–263, Berlin.

KOVENKO, V. (1938): Beitrag über Keban-Lagerstätte. Unveröff. Ber. M. T. A., 1095, Ankara (in Turkish).

MALAKHOV, A. A. (1969): Bismuth and antimony in galenas as indicators of some conditions of ore formation. Geochimiya, *11,* 1283–1296 (russisch).

MAUCHER, A. (1938): Über die Erzvorkommen von Keban-Maden. – Z. prakt. Geol., *46,* 79–84.

MÜLLER, P., SCHNEIDER, H. J., PAREKH, P. P., & ZIMMER, E. (1977): Fluorine contents in carbonate sequences and Rare Earths distribution in fluorites of Pb-Zn deposits in East-Alpine Mid-Triassic. Mineral. Deposita, *12,* 23–36, Berlin.

OELSNER, O. (1960): Möglichkeiten zur Mobilisation von Blei-Zinnkerzen aus bestehenden Lagerstätten und Kriterien zu ihrer genetischen Einstufung. N. Jb. Miner. Abh., *94,* 98–120, Stuttgart.

OELSNER, O.: Kriterien zur genetischen Einstufung von Blei-Zink-Lagerstätten. Z. dt. geol. Ges., *110,* 441–448.

OHMOTO, H. (1972): Systematics of sulfur and carbon isotopes in hydrothermal ore deposits. Economic Geology, *67,* 551–578.

PAREKH, P. P., SCHNEIDER, J. H. (1976): The application of Tb/Ca–Tb/La abundance ratios to problems of fluorspar genesis. – Mineral. Deposita, *11,* 111–116, Berlin.

PETRASCHECK, W. E. (1955): Beziehung zwischen der anatolischen und südosteuropäischen Metallprovinz. Bull. M. T. A. *46/47,* 64–74, Ankara.
(1960): Die alpin-mediterrane Blei-Zink-Provinz. Erzmetall, *13,* 189–204, Stuttgart.
SCHROLL, E. (1954): Ein Beitrag zur geochemischen Analyse ostalpiner Blei-Zink-Lagerstätten. Mitt. Österr. Miner. Ges. Sdh. 3, 1–85 (Wien).
SCHROLL, E. (1979): Progress in the knowledge of indicator elements. *In:* AHRENS, L. H.: Origin and Distribution of the Elements. Pergamon (1984) Press, Oxford-New York, 213–216.
SCHROLL, E.: Geochemical indicator parameters of lead-zine ore deposits in carbonate writes. *In:* WAUSCHKUHN et al. eds., Syngenesis and Epigenesis in the Formation of Mineral Deposits. Springer-Verlag Bereich Heidelberg, 294–305.
STACEY, J. S., KRAMERS, J. K. (1975): Approximation of terrestrial lead isotope by a two-stage model. Earth Planet. Lett., *26,* 207–221, Amsterdam.
VACHE, R. (1966): Zur Geologie der Variszíden und ihrer Lagerstätten im Südanatolischen Taurus. Mineralium Deposita, *1,* 30–42.
ZUSSERMANN, A. (1969): Geological and mining study of Keban Maden. – Etibank-Arsivi, Ankara.

Sulphide and Arsenide Mineralization at Limassol Forest, Cyprus*)

By O. THALHAMMER**), E. F. STUMPFL**), and A. PANAYIOTOU***)

Zusammenfassung

Die Fe-Ni-Co-Cu-Erze des Limassol Forest auf Zypern treten in Linsen, Gängen und Imprägnationen in serpentinisierten Duniten und Harzburgiten auf. Die wichtigsten Erzminerale sind Pyrrhotin, Pentlandit, Chalkopyrit, Maucherit, Cubanit, Valleriit, Westerveldit, Magnetit und Chromit. Auflichtmikroskopische Untersuchungen und Mikrosondenanalysen zeigen, daß die Mineralisation durch Remobilisation und hydrothermale Aktivität während früh einsetzender Serpentinisierung gebildet worden ist. Die geotektonische Position des Limassol Forest nahe der Arakapas Faltenzone, einer Transform Störung, unterstützt diese Interpretation.

The Fe-Ni-Co-Cu ores of Limassol Forest occur in form of veins, disseminations and partly in lenticular or irregular bodies in serpentinites which are interpreted as dunites and harzburgites forming the basal part of the ultramafic sequence of the Troodos ophiolite complex (PANAYIOTOU, 1978, 1980).

The mineralization is dominantly concentrated in two localities known as Lakxia tou Mavrou and Pevkos and is associated with small podiform chromite deposits.

The mineral association of Pevkos comprises pyrrhotite with near-stoichiometric composition, pentlandite, maucherite, chalcopyrite, cubanite, valleriite, chromite, magnetite and traces of ilmenite, westerveldite, cobaltite, bornite, molybdenite and gold (Tab. 1). With the exception of the arsenides (maucherite, westerveldite) all the minerals show various alteration textures and transformation into secondary ore minerals such as bravoite.

---

*) This is an extended abstract of a paper given at the final meeting of IGCP Project No. 169, at Leoben, Austria, in October 1984. Analytical data and a comprehensive discussion will be published elsewhere.

**) Institut für Mineralogie der Montanuniversität, A-8700 Leoben.

***) Geological Survey Department, Nicosie, Cyprus.

The mineralization at Lakxia tou Mavrou consists mainly of pyrrhotite, pentlandite, chalcopyrite, cubanite, löllingite, chromite, magnetite and minor amounts of pyrite, maucherite, valleriite and gold (Tab. 1)

At both localities chromite grains are characterized by a "bleached rim" containing many secondary silicate inclusions. The chemical composition of these rims reveal typically high Fe and low Cr_2O_3, MgO and Al_2O_3 contents.

Detailed reflected light and microprobe investigations show that mineralization at both localities is affected by distinct remobilization and hydrothermal activity. Typical features such as the occurrence of ore minerals on cracks and as fissure

Table 1. Mineral assemblages in Limassol Forest

Mineral Type	Mineral Name	Chemical Composition	Occurrence	
			Pevkos	Lakxia tou Mavrou
Sulphides	Pyrrhotite	FeS	××××	×
	Pyrrhotite	$Fe_{1-x}S$	×	××××
	Mackinawite	$Fe_{1+x}S$	×	–
	Pyrite	FeS_2	–	××
	Pentlandite	$(Fe, Ni, Co)_9S_8$	××	×
	Co-rich Pentlandite	$(Fe, Co)_{6.4}Ni_{2.6}S_8$	–	××
	Bravoite	$(Fe, Ni)S_2$	×	×
	Chalcopyrite	$CuFeS_2$	×××	×××
	Cubanite	$Cu\ Fe_2S_3$	××	××
	Bornite	Cu_5FeS_4	×	–
	Chalcocite	Cu_2S	×	×
	Neodigenite	Cu_9S_5	×	×
	Covellite	CuS	×	–
	Cu-Valleriite	$(Fe_{0.65}Cu_{0.50}S_{1.40})\cdot Mg_{0.14}Ca_{0.25}Fe_{0.97}(OH)_2$	×××	×
	Fe-Valleriite	$Fe_2S_2\ .\ Mg(OH)_2$	×××	×
	Molybdenite	MoS_2	×	–
Arsenides	Maucherite	Ni_3As_2	×	–
	Löllingite	$(Fe, Co, Ni)_{2+x}As_{4-x}$	–	××
	Westerveldite	$(Fe_{0.77}Ni_{0.22}Co_{0.04})As_{0.97}$	×	–
	Cobaltite	$(Co_{0.58}Ni, Fe_{0.42})AsS_{0.9}$	×	–
Oxides	Chromite	$(Fe, Mg)_{1.1}(Cr, Al, Fe)_{1.9}O_4$	×××	×××
	Magnetite	Fe_3O_4	××	×
	Rutile	TiO_2	×	–
	Ilmenite	$FeTiO_3$	×	–
Elements	Gold	(Au, Ag)	×	×

× rare
×× common
××× very common
×××× main constituent

fillings, large amounts of arsenides and stoichiometric pyrrhotite, magnetite which is pure Fe_3O_4, etc., reveal distinct differences to magmatic sulphide deposits. They indicate a complex polyphase origin for ore mineralization at Limassol Forest which was linked to serpentinization and hydrothermal processes.

The initial formation of magnetite as a byproduct of serpentinization is followed by liberation of metallic elements like Fe, Ni, Co and Cu, from primary magmatic olivine. These elements tend to form sulphides or arsenides during various stages of the serpentinization process (RAMDOHR, 1967, ECKSTRAND, 1975, and others). Finally, the formation of löllingite and maucherite by reaction of hydrothermal fluids with preexisting niccolite (OEN et al., 1971, 1977, IXER et al., 1979) has taken place in a temperature range of about 500–300° C.

During the late Upper Cretaceous the Troodos ophiolite was obducted onto the Afro-Arabian continental margin. This tectonic event resulted in extensive deformation and fracturing of parts of the ophiolite facilitating percolation of sea water which is considered responsible for the continuation of serpentinization processes. Serpentinizing solutions under slightly oxidizing conditions lead to the following features: peripheral resorption of stoichiometric pyrrhotite and its partial substitution by pyrrhotite (Fe_9S_{10}), formation of pyrite, decomposition of Cu-minerals resulting in the formation of valleriite and magnetite, distinct deformation of pyrrhotite at Lakxia tou Mavrou. In addition, they signify tensional stress at temperatures of about 200° C (ANTUN et al., 1966, RAMDOHR, 1967, ECKSTRAND, 1975, ATKINSON et al., 1975, TEWARI & CAMPDELL, 1976).

Apart from the cores of chromites no member of the complex mineral association at Limassol Forest can be linked to "magmatic" processes. Quite on the contrary, mineralization is attributed to hydrothermal activity in an area which, due to its vicinity to a fossil transform fault, the Arakapas fault zone, was exposed to particularly intense tectonism and serpentinization. These processes probably commenced in an early stage of Troodos ophiolite formation and continued long after the obduction of the ophiolite.

References

ANTUN, P., EL GORESY, A., & RAMDOHR, P. (1966): Ein neuartiger Typ „hydrothermaler" Cu-Ni-Lagerstätten mit Bemerkungen über die Mineralien Valleriit, Mackinawit, Oregonit. – Mineralium Deposita 2, 113–132.

ATKINSON, B. K. (1975): A preliminary study of the influence of temperature and strain rate on the rheology of a polycrystalline pyrrhotite ore. – N. Jb. Miner. Mh., 483–499.

ECKSTRAND, O. R. (1975): The Dumont Serpentinite: A model for control of nickeliferous opaque mineral assemblages by alteration reactions in ultramafic rocks. – Econ. Geol. 70, 183–201.

IXER, R. A., STANLEY, C. J., & VAUGHAN, D. J. (1979): Cobalt-, nickel- and iron-bearing sulpharsenides from the North of England. – Mineral. Mag. 43, 389–395.

OEN, J. S., BURKE, E. A. J., KIEFT, C., & WESTERHOF, A. B. (1971): Ni-arsenides, Ni-rich loellingite and (Fe, Co)-rich gersdorffite in Cr-Ni-ores from Màlaga Province, Spain. – N. Jb. Miner. Abh. 115, 123–139.

OEN, J. S., BURKE, E. A. J., & KIEFT, C. (1977): Westerveldite from Igdlanguaq, Ilimaussaq alkaline massif, South Greenland. – Mineral. Mag. 41, 77–83.

PANAYIOTOU, A. (1978): The mineralogy and chemistry of the podiform chromite deposits in the serpentinites of the Limassol Forest, Cyprus. – Mineralium Deposita 13, 259–274.

PANAYIOTOU, A. (1979): Cu-Ni-Co-Fe sulphide mineralisation, Limassol Forest, Cyprus. – Ophiolite Proc. Int. Ophiolite Symp. Cyprus 1980, 102–117.

RAMDOHR, P. (1967): A widespread mineral association, connected with serpentinisation. – N. Jb. Miner. Abh. 107, 241–265.

TEWARI, P. H., & CAMPDELL, A. B. (1976): Dissolution of iron sulphide (troilite) in aqueous sulphuric acid. – J. Phys. Chem. 80, 1844–1848.

Bemerkungen zu mineralogisch-geochemischen und lagerstättenkundlichen Studien an Vererzungen der Gebiete Brskovo/Jugoslawien und Lakonia/Griechenland

Von Andreas G. Vgenopoulos*)

Mit 6 Abbildungen, 2 Tabellen

Abstract

The base metal deposits of Lakonia (Peloponnes) and Brskovo (Bosnia) are associated with Triassic porphyrites, both related to the crustal tension phase in early Mesozoic time. Striking chemical and structural similarities in the volcanic rocks can be observed, particularly where the carbonate country rock was assimilated.

Die Geologie der beiden untersuchten Gebiete

Brskovo/Jugoslawien und Lakonia/Griechenland sind zwei Gebiete, die große Ähnlichkeit in bezug auf den eruptiven Gesteinstypus und die umliegenden Gesteine aufweisen und wahrscheinlich eine analoge Erzgenese haben.

Brskovo liegt innerhalb der Karbonatgesteinserien Montenegros in den Dinariden (vgl. Pamic, J. J., 1982 und Strucl, I., 1981). Es enthält den mitteltriassischen Geosynklinal-Porphyrit-Vulkanismus, und die dort vorkommenden Pb-Zn-Lagerstätten befinden sich in einer Serie von Porphyriten und Keratophyren, zwischen die schiefrige triassische Sedimente eingeschaltet sind. In den sericitisierten Eruptivgesteinen treten Erzimprägnationen von unterschiedlicher Art und Intensität auf, die stellenweise auch in die Sedimentgesteine übergreifen. Neben den angeführten Eruptivgesteinen besteht die vulkanogen-sedimentäre Serie noch aus Lavabreccien, Tuffen und Tuffiten, Dolomiten und dolomitischen Kalken. Hauptgemeng-

---

*) Mineralogisch-Petrographisches Institut der Technischen Hochschule, Patission 42, Athen/Griechenland.

Tabelle I. Chemische und mineralogische Zusammensetzung von Proben aus Brskovo/Jugoslawien und Lakonia/Griechenland.

a/a	SiO_2	Al_2O_3	Fe_2O_3	MgO	CaO	K_2O	Na_2O	TiO_2	Zn	Pb	Cu	S	H_2O	Mineralogische Zusammensetzung
1. Brskovo No 1 Pb-Zn-Erz	47.0	5.6	9.4	0.9	0.17	3.8	0.3	0.35	7.6	4.3	SP	9.4	11.3	Quarz, Sphalerit, Galenit, Pyrit, Muskovit, Albit
2. Brskovo No 2 Pb-Zn-Erz	60.8	9.6	5.9	0.2	0.76	2.5	0.6	0.2	7.4	2.3	SP	4.3	4.9	Quarz, Sphalerit, Galenit, Pyrit, Muskovit, Albit
3. Brskovo No 3 Spilit-Keratophyr	58.0	17.6	7.0	2.9	4.0	1.6	4.6	1.02	SP	SP	SP	–	3.3	Albit, Quarz, Chlorit
4. Brskovo No 4 Spilit-Keratophyr	56.5	16.7	6.4	3.6	5.3	1.7	5.9	0.88	SP	SP	SP	–	2.9	Albit, Quarz, Chlorit
5. Lakonia No 3 Weich, Apidia bei Agios Ioannis	58.2	18.1	8.9	0.2	10.1	0.24	0.8	0.4	SP	SP	0.4	–	3.0	Epidot, Quarz, Azourit, Kaolinite, Aragonit, Chalcedon
6. Lakonia No 3 Hart, Apidia bei Agios Ioannis	49.3	18.7	10.3	0.5	1.7	1.8	5.3	0.55	SP	SP	4.6	–	7.3	Albit, Chlorit, Malachit, Kaolinite, Muskovit
7. Lakonia No 9 Apidia bei Agios Ioannis Neben No 3	58.4	21.1	4.9	0.7	1.5	3.2	4.0	0.78	0.085	0.0047	0.061	–	4.4	Albit, Quarz, Chlorit, Muskovit
8. Lakonia No 4 Gagania b. Molai	51.4	19.6	6.2	0.6	5.6	6.8	1.2	0.83	–	–	0.090	–	7.7	Muskovit, Albit, Quarz

9. Lakonia No 4 Gagania b. Molai	49.4	14.2	10.1	4.0	10.5	1.9	1.4	0.31	–	–	0.080	–	6.4	Quarz, Muskovit, Gips, Chlorit
10. Lakonia No 5 Hart, Gargania b. Molai	42.5	–	10.6	0.4	18.7	0.2	–	–	–	–	0.024	–	27.3	Quarz, Calcit
11. Lakonia No 6 Gagania Schwarz bei Molai	53.7	17.9	7.7	0.5	6.1	4.1	2.3	1.06	–	–	0.084	–	6.4	Quarz, Albit, Chlorit, Muskovit, Montmorilonit
12. Lakonia No 7 (1) Kritsova bei Apidia Forschungsstollen	47.1	15.6	10.4	3.1	7.0	2.7	4.3	0.58	0.004	0.0024	0.080	–	7.8	Montmorilonit, Quarz, Albit, Chlorit, Muskovit
13. Lakonia No 7 (2) bei Apidia Forschungsstollen	48.2	19.8	11.2	0.3	8.9	0.56	3.7	0.62	0.007	0.003	0.0655	–	7.6	Albit, Quarz, Montmorilonit
14. Lakonia No 7 (3) bei Alpidia Forschungsstollen	48.6	16.9	9.7	2.0	5.0	2.8	5.9	0.57	0.0045	0.0032	0.212	–	–	Montmorilonit, Albit, Quarz, Muskovit, Kaolinit
15. Lakonia No 7 (4) bei Apidia Forschungsstollen	56.8	15.8	8.2	1.5	2.7	0.8	5.2	0.22	0.009	0.003	0.43	–	7.6	Chlorit, Albit (Malachit)
16. Lakonia No 10 Kritsova bei Apidia Neben No 7 (1)	50.6	17.4	11.8	1.0	5.9	3.5	2.5	0.61	SP	SP	0.037	–	6.2	Montmorilonit, Quarz, Albit, Chlorit, Muskovit

Tabelle II. Semiquantitative Spurenelementanalyse einiger Proben der Gebiete Brskovo/Jugoslawien und Lakonia/Griechenland

	Cu	Sb	Sn	Ag	Cd	Zr	As	Au	Re	Ga	Os	Yb	Hf	Lu	Dy	Ba	Rb	Tc	Se	Ge	Zn	Pb	Ni	Sr	In	Pd	U	Y	Th	Br	Mn	Co
1. Brskovo No 1 Pb-Zn-Erz	500	++ +	+	+	+	+	+	+	+	+	+	+	+	+	+	+																
2. Brskovo No 2 Pb-Zn-Erz	120	+	+	+	+	+		+	+	+			+			+	+															
3. Brskovo No 3 Spilit-Keratophyr	++ ++		+	+	+	+	+										+	+	+	+	++ +	++ +	+	120								
4. Brskovo No 4 Spilit-Keratophyr	60		+			+											+					++ ++ +	++ +	180	+							
5. Lakonia No 3 Weich, Apidia b. Ag. Ioannis	4.200	+	+	+	+	+	+	(+)		+			+			+	+		+		+	+	+	1000	+	+	+	+	+	+	+	+
6. Lakonia No 3 Hart, Apidia b. Ag. Ioannis	50.000	+	+	+		+				+			+				+				+		+	+++	+	+					+	+
7. Lakonia No 5 verwittert, schwarz, Gagania b. Molai	+	+			+	70											++ +				++			50		+						

8. Lakonia No 5 Hart, Gagania b. Molai	+	+		+	+	60	+	+	+		+		+	+		90	+			+	
9. Lakonia No 7 (1) Kritsova b. Apidia, Forschungs-stollen	220	+				++ +	+		+		+		+		+	160			+		
10. Lakonia No 7 (2) Kritsova b. Apidia, Forschungs-stollen	700	+			+		+		+	+	+				+	80	+	++ +	+	+	
11. Lakonia No 7 (3) Kritsova b. Apidia, Forschungs-stollen	2.300	+				++	+				+	+	+		+	++ +	+				
12. Lakonia No. 7 (4) Kritsova b. Apidia, Forschungs-stollen	4.300	+	+			++							++ +		+		+			+	+

+ = je Kreuz 1 bis 10 ppm

teile der Erzkörper sind Pyrit und Zinkblende; Bleiglanz tritt nur in untergeordneter Menge auf.

Lakonia befindet sich im südlichen Teil des Peloponnes, und zwar im Geosynklinalbereich der karbonatischen Gesteinsserien der Gebirge Parnon und Taigetos. Die Vulkanite tauchen bei Gythion in der mittleren Halbinsel des Peloponnes aus dem Meer und erstrecken sich mit Unterbrechungen nach Nordosten. Die vererzten Zonen umfassen hauptsächlich die Gebiete Molai und Apidia, wo die Pb-Zn-Cu-Aufschlüsse vorkommen. (Vgl. geol. Karte, Blatt Molai 1 : 50.000 IGME 1984.)

Der vulkanitische Gesteinskomplex grenzt im NO an mittleren bis oberen triassischen Kalk, dolomitischen Kalk und Dolomit, im SW an carbonische bis permische Schiefer und Marmor sowie Sedimente des Pliozäns. Er überlagert abwechselnd carbon-permische Schiefer oder Marmor. Sehr oft besteht er aus Lavablöcken, die hauptsächlich basischen Charakter haben, oder aus gemischten Serien, die aus Lava-Breccien, Tuffen und Tuffiten in Wechsellagerung mit dolomitischen Kalken und Dolomiten bestehen. Die Vererzung wird entweder diffus verteilt im Vulkanit selbst oder zwischen Vulkanit und carbonischen bis permischen Marmoren gefunden.

Im vulkanischen Komplex und in den umliegenden Gesteinen liegen mehrere Brüche, deren Richtung parallel zu der Richtung der umliegenden Gebirge ist. Außerdem verlaufen sie fast parallel zum großen Bruch des Hellenischen Grabens, welcher südlich von Kreta verläuft und sich bis zum südwestlichen Rand des Peloponnes erstreckt (vgl. International Symposium on the Hellenic Arc and Trench April 1981, Athens); der Hellenische Graben soll nach neueren geologischen Angaben die Grenze zwischen der Afroarabischen und der Euroasiatischen Platte in der Fortsetzung der großen tektonischen Linie sein, die sich zum Fernen Osten erstreckt und die Indische Platte umrahmt.

Brüche, die parallel zum Gebirgsstreichen verlaufen, sind gegen Ende der Orogenese entstanden, und zwar hauptsächlich durch Kräfte, die eine Hebung zur Folge hatten. Diese Deformationen mit entsprechender Hochhebung verursachten höchstwahrscheinlich den Ausbruch aller im Peloponnes existierenden Vulkane. Die vulkanischen Gesteinslagen haben fast alle das gleiche Alter wie diejenigen von Lakonia, d. h. Mittlere Trias, und dieses Alter ist identisch mit dem Alter der vulkanischen Gesteine Bosniens bei Brskovo.

Methodisches

Um die Mineralogie, Geochemie, Vererzung und Lagerstättenbildung der beiden Gebiete näher zu studieren, sammelten wir eine Reihe von Proben aus beiden Gebieten, welche im Labor für Mineralogie und Petrographie der Technischen Hochschule Athen analysiert wurden.

Die erhaltenen Daten für den Chemismus und die mineralogische Zusammensetzung sind in Tabelle I angegeben, die semiquantative Analyse der Spurenelemente in Tabelle II. Die Auswertung beider Tabellen und Bemerkungen zu den Dünn- u. Anschliffen werden im nächsten Kapitel diskutiert.

Diskussion der Resultate

Nach der Zusammenstellung der Resultate (chem. Analyse und Mineralanalyse) der Proben beider Gebiete (vgl. Tab. I und II) und dem Vergleich des Chemismus, der mineralogischen Zusammensetzung und des Gefüges der Proben aus dem gleichen Gebiet, besonders von Lakonia, können wir sagen, daß im gleichen vulkanischen Komplex verschiedene Gesteinstypen existieren. Grund für diese Annahme ist der verschieden chemische und mineralogische Charakter der Proben sowie in vielen Fällen das unterschiedliche Gefüge. Eine Ähnlichkeit des Chemismus, der Mineralogie sowie des Gefüges treffen wir hauptsächlich an Proben, die den ursprünglichen vulkanischen Charakter bewahrt haben, d. h. an Gesteinen, die von Nebenwirkungen oder postgenetischen Wirkungen nicht betroffen wurden. Beispiele dafür zeigen Abb. 1 und 2. In beiden Gebieten zeigen die Dünnschliffe den Charakter von typisch vulkanischen Gesteinen. Es handelt sich hauptsächlich um albitische Einsprenglinge, die in einer Matrix von Mikrokristallen und Glasphasen schwimmen. Solche Gesteine sind arm an Spurenelementen, zeigen keine Störung oder Umwandlung der Mineralkörner und haben fast immer die gleiche mineralogische und chemische Zusammensetzung.

An Stellen mit teilweiser Assimilation der Kalke in der Lava stellten wir neben der Änderung des Gefüges eine Anhäufung von Erzen, besonders des Kupfers, fest (Abb. 3). (Vgl. Vgenopoulos, 1980.) (Vgl. Proben a/a 5, 6, 7 und 12, 13, 14 und 15 der Tabelle I.) Es handelt sich dabei um Proben von Lakonia. An den Proben der ersten Gruppe, und da besonders an den Proben 5 und 6, zeigt sich eindeutig, daß eine Veränderung der chemischen und mineralogischen Zusammensetzung stattge-

Abb. 1. Andesitischer Vulkanit/Lakonia/Griechenland. Dünnschliffphoto, Vergr. 8×10+N.

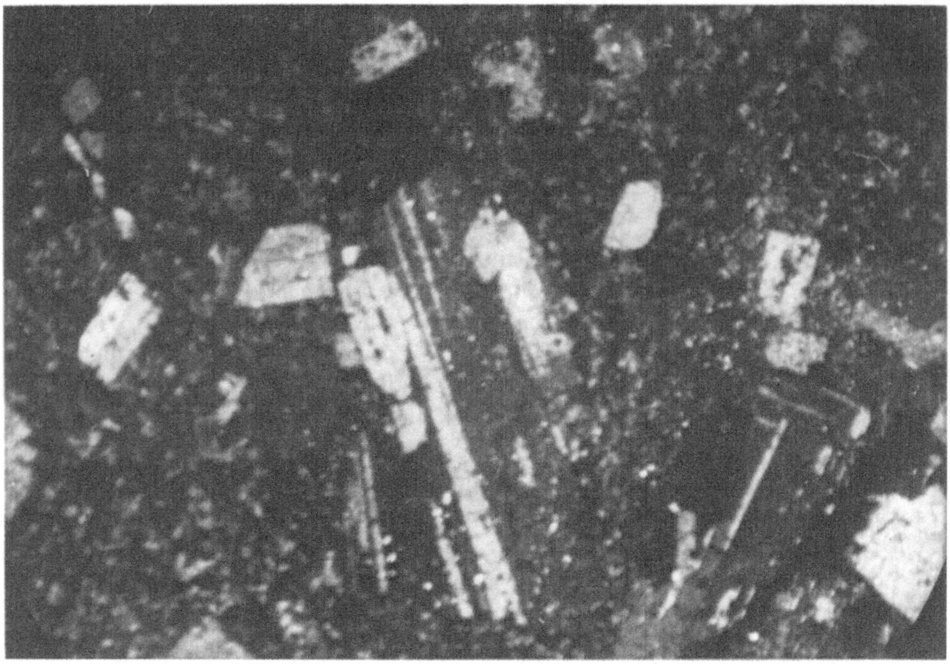

Abb. 2. Spilit-Keratophyr von Brskovo/Jugoslawien. Dünnschliffphoto, Vergr. 8×10+N.

Abb. 3. Calcitische Gesteinsreste, die von vulkanischer Lava nicht assimiliert wurden. Dünnschliffphoto von Lakonia, Vergr. 8×10+N.

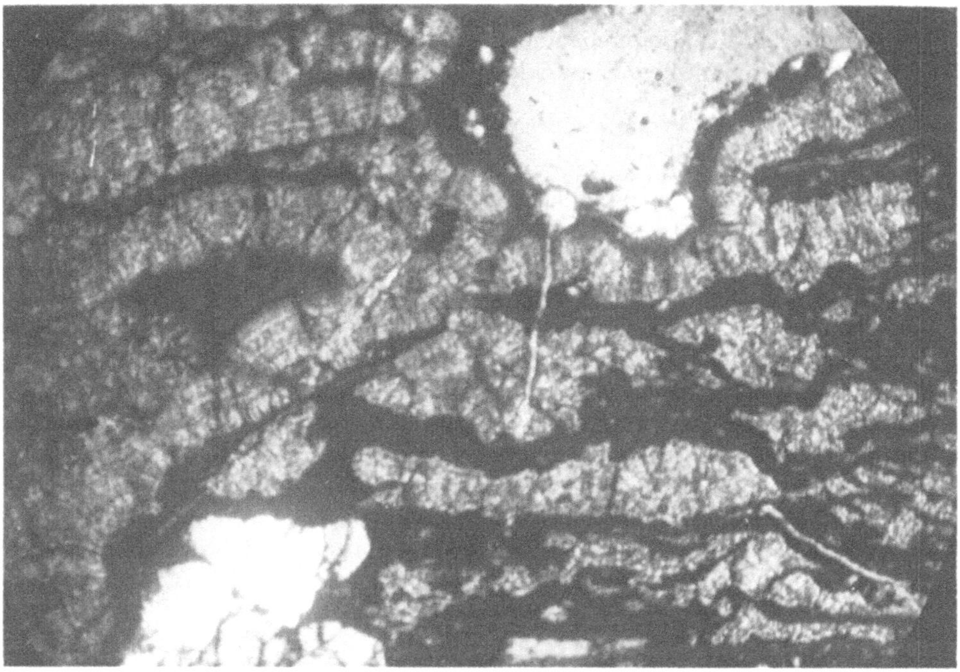

Abb. 4. Gelform in wurmartiger Entwicklung des Epidots mit etwas Chalcedon, durchzogen von Erzadern (Lakonia).

funden hat. Probe 5 z. B. ist reicher an Silicium, Calzium und Spurenelementen, besonders an Strontium, Probe 6 hingegen ärmer an Silicium und Calcium, dafür reicher an Natrium, Kupfer und flüchtigen Bestandteilen, welche in der Tabelle I als Totalwasser angegeben werden. Probe 7 nähert sich im Chemismus, der mineralogischen Zusammensetzung und im Gefüge dem typischen Vulkanit, wie er in Abb. 1 und 2 gezeigt wird.

Der Abstand von einer Probe zur anderen betrug nicht mehr als 40 cm, und trotzdem konnten wir die erwähnten deutlichen Unterschiede feststellen, besonders im Gefüge der Probe 5. Abb. 4 zeigt eine gelförmige bis wurmartige Entwicklung des Epidots und eine Einlagerung von Eisenerz sowie die Bildung von neuen Quarzkristallen. Abb. 5 zeigt die gelförmige Entwicklung der Kieselsäure in Form von Chalcedon, welcher die Epidotkristalle umrahmt; ferner sieht man daneben auch alte Quarzkristalle.

Die Proben 12, 13, 14 und 15 der zweiten Gruppe variieren im Chemismus und der mineralogischen Zusammensetzung des entsprechenden Gefüges nicht so stark, trotzdem nimmt auch hier mit der Zunahme des Kupfergehaltes der Gehalt an Calcium deutlich ab.

Wenn wir nun vergleichend die stark vererzten Lagen in Brskovo betrachten, dann stellen wir ein reichlicheres Vorkommen der Spurenelemente sowie eine deutliche Zunahme der Kieselsäure mit entsprechender Abnahme des Aluminiums und einem fast völligen Verschwinden des Calciums fest. Dort, wo größere Erzanhäufungen auftreten (Abb. 6), kann man deutlich eine verworrene, parallel-strahlige oder manchmal rosettenförmige Entwicklung der Quarzkristalle aus der Mikrokristallmatrix sehen.

Schlußfolgerungen

Aus den Beobachtungen läßt sich eindeutig folgern, daß die Erzbildung erstens bei viel tieferen Temperaturen als denen einer vulkanischen Schmelze stattfand und zweitens, daß sie ein eigenes chemisches Milieu erforderte. Die postvulkanische Mobilisation und Mineralumwandlungen im Vulkanit und die Vererzung wurden dort festgestellt, wo die Bedingungen für die Erzbildung erfüllt waren, d. h. wo hauptsächlich eine Änderung des pH rund um die sauren, erzhaltigen Lösungen stattfand. Die pH Änderung war successiv, denn in den meisten Fällen folgte zuerst eine Calcitverdrängung und anschließend ein Absatz des Erzes.

Daß die Temperaturen während der Erzbildung abnahmen, verrät Abb. 2, welche eine nicht vollständige Assimilation des Calcits zeigt und dazu eine topochemische Änderung des pH wegen der partiellen Löslichkeit des Calcits. Der Ausbruch

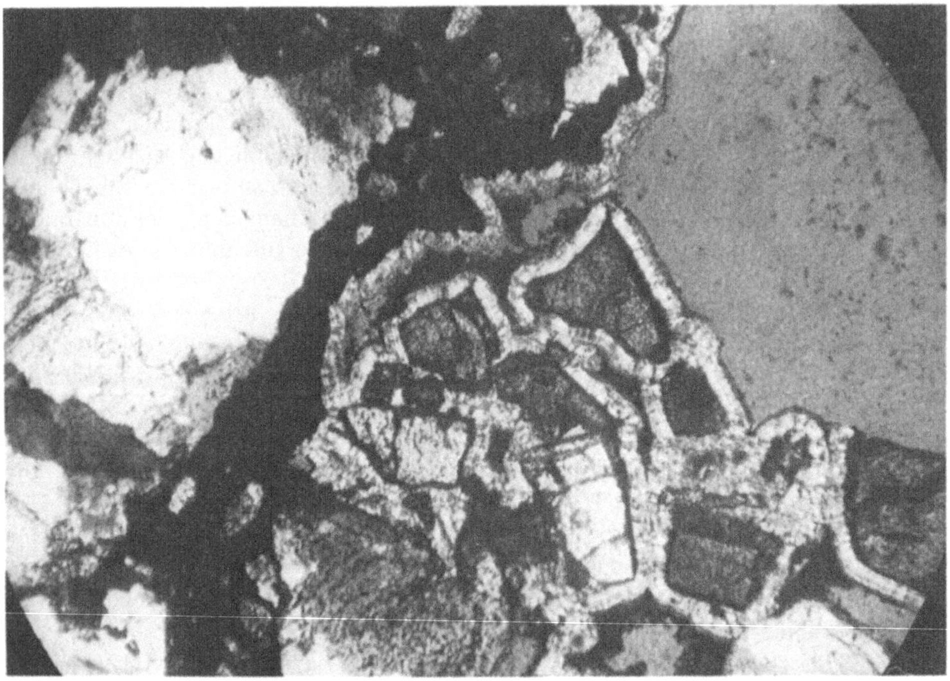

Abb. 5. Chalcedon umrahmt Epidotkristalle, links frische, rechts alte Quarzkristalle, Mitte Erzgang. Dünnschliffphoto, Vergr. 8×8+N (Lakonia).

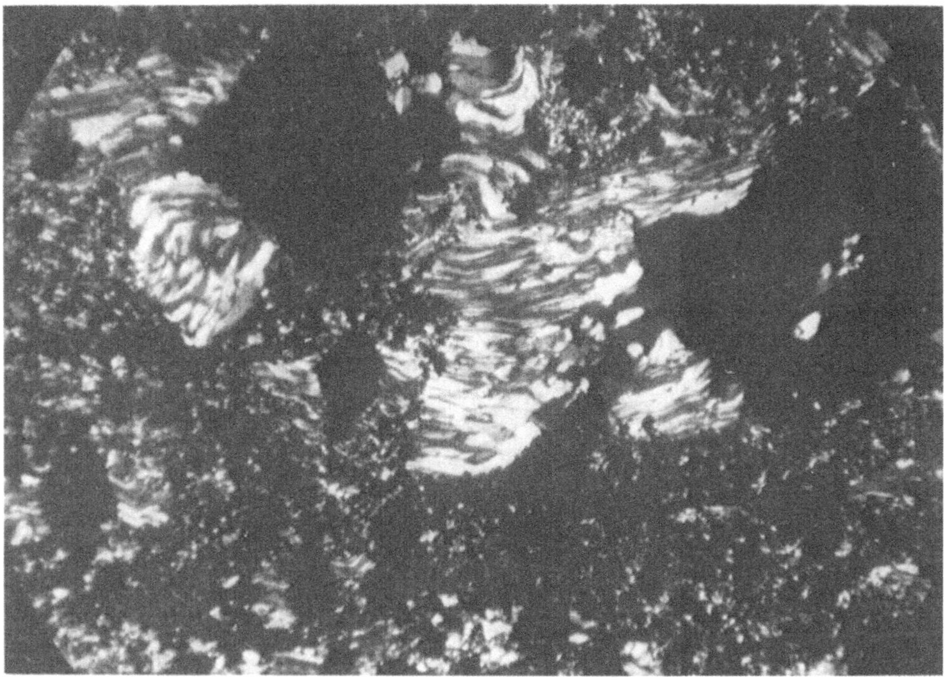

Abb. 6. Verworrenes, parallelstrahlige bis rosettenförmige Entwicklung der Quarzkristalle um große Anhäufungen von Erz. Dünnschliffphoto, Vergr. 8×8+N (Brskovo).

einer vulkanischen Lava dauert eine gewisse geologische Zeit. Während dieser Zeit gibt es öfters Bewegungen, Umformungen und Mobilisationen. Während dieser Phasen hat die Ausscheidung der Erze aus dem Magma stattgefunden, zumal die Bruchtektonik den Durchfluß der flüchtigen Bestandteile begünstigte.

Literatur

PAMIC, J. J. (1982): Triassic Magmatism of the Dinarides. Zagreb.

PIPER, G. (1983): The Triassic Volcanic Rocks of Tyros, Zarouhla, Kalamai and Epidauros, Peloponnes, Greece. Schweiz. Mineral. Petrogr. Mitt., Bd. 63.

STRUCL, I. (1961): Die schichtgebundenen Blei-Zink-Lagerstätten Jugoslawiens. Mitt. Oesterr. Geol. Ges. Wien.

VGENOPOULOS, A. (1975): Zur Mineralchemie einiger Hellglimmer-Pegmatite aus dem Gebiet Westthrakiens/Griechenland. Praktika Akademias Athinon, Bd. 5.

– (1980): About the Genesis of Copper Mineralisation in Andesitic Basalt of Apidia Lakonias. UNESCO Intern. Symposium on Metallogeny of Mafic and Ultramafic Complexes: The Eastern Mediterranean-Western Asia Area and its Comparison with similar Metallogenic Environment in the World. Vol. 2, Athens 1980.

Copper ore Deposits in Bulgaria

By B. BOGDANOV*)

With 1 Figure

Zusammenfassung

Die Kupfererzlagerstätten Bulgariens liegen in der Srednogorje-Zone, die als Rift-Zone bezeichnet wird, vor allem in ihrem mittleren Abschnitt um Panagjurište (Fig. 1). Der oberkretazische Magmatismus ist teils andesit-basaltisch, teils trachybasaltisch (im Osten). Wenig jünger sind gabbro-syenitische Intrusionen. An diesen Magmatismus sind Porphyrie-Kupferlagerstätten, massive Sulfiderzkörper, polymetallische Gänge (bei Burgas) und Skarnerze gebunden. Alle Lagerstättentypen werden charakterisierend beschrieben.

The copper ore deposits on the territory of Bulgaria have been known for ancient times. According to the archaeological data, the oldest copper mines in Europe were located in the central part of the country, near the town of Stara Zagora (CHERNIH, 1978). The archaeological investigations discovered 11 open pits, situated along the vein type ore body, known under the Turkish name "Aibunar", where 4 thousand years B.C. the oxide copper ores were yielded. The length of the open pits varied from several tens to 100 m, the width–from 3 to 10 m, and the depth was up to 20 m from the surface. Two meltings pots filled with malachite and azurite were discovered in the vicinity of this oldest mine (KONYAROV, 1953), which was exploited till 2000 years B.C.

Besides this famous copper mine, more than 100 points with traces of ancient mining and metallurgical activities have been fixed on the territory of the country. One of the largest coppermelting centers, existing from II to I century B.C., was located near to the Black Sea coast, north from the contemporary active Rossen copper ore deposit. It is considered (KONYAROV, 1953) that the bronze statue of Appolo-

*) Higher Institute of Mining & Geology, 1156 Sofia, Bulgaria.

nius, erected in the ancient Greek colony Appolonia (the present-day town of Sozopol) was made of copper from this metallurgical center.

People's Republic of Bulgaria occupies today one of the first places in copper production in Europe. The discovered new large copper ore deposits during the last 20 years made possible developing the modern copper mining, refining and smelting industry.

Brief information concerning the genetic types of the industrial copper deposits in Bulgaria, their geological setting and mode of formation is given below.

Genetic types of Copper ore Deposits

The majority of the industrial copper ore deposits in Bulgaria occurs in the Srednogorie metallogenic zone (Magm. & Metall. of CBA, 1983). It is a part of the global Thetyan-Eurasian metallogenic belt (JANKOVIĆ, 1976). During the Upper Cretaceous a secondary geosynclinal trough was formed in Srednogorie zone, along the rift structures between Moesian plate to the north and Rhodope massif to the south (BONCHEV, 1974; TVALCHRELIDZE, 1972; BOGDANOV, 1977). In connection with the magmatic activity which took place during the Senonian time, the various types of copper ore deposits were formed. The Srednogorie metallogenic zone is characterized by intensive development of predominantly Ca-alkaline volcanism represented mainly by two volcanic formations—basalt-andesite-rhyolitic in the central part of the zone and trachybasalt-trachyandesitic in the eastern part, where the K-alkaline-basaltoid formation is locally deveoped as well (STANISHEVA, 1971, 1980; POPOV et al., 1979). The intrusive type of magmatism is manifested in various volcano-plutonic centers, in which almost simultaneously with the volcanic activity two intrusive formations were developed—gabbro-diorite-granodiorite and gabbro-monzonite-syenite formations.

The Upper Cretaceous copper mineralizations are the main peculiarity of the Srednogorie metallogenic zone. The copper ore deposits can be subdivided into the following genetic types or ore-formations (Fig. 1, Table 1): 1. porphyry copper, 2. massive sulphide, 3. vein copper and copper-gold-polymetallic and 4. aposkarn copper molybdenum-sheelitic. Besides the four main types of copper deposits there are also several non-commercial mineralizations which have to be mentioned, such as native copper-zeolitic in andesitic rocks, copper-nickel type in ultrabasic rocks, copper-bearing sandstones, stratiform lead-zinc-copper deposits in Triassic limestones and stratabound copper-polymetallic deposits in Paleozoic volcanic-sedimentary rocks. The last two types are subject to exploitation as copper-polymetallic deposits and have very low significance for the copper production. The present paper deals with the Upper Cretaceous copper deposits of the above mentioned four ore formations. Among them the porphyry copper deposits are of greatest economic importance. They supply more than 90% of the copper in the country and are considered as most perspective type for increasing the copper production in the future. The other types are of less industrial significance, but the higher copper content and presence of other valuable metals makes them interesting for the copper industry.

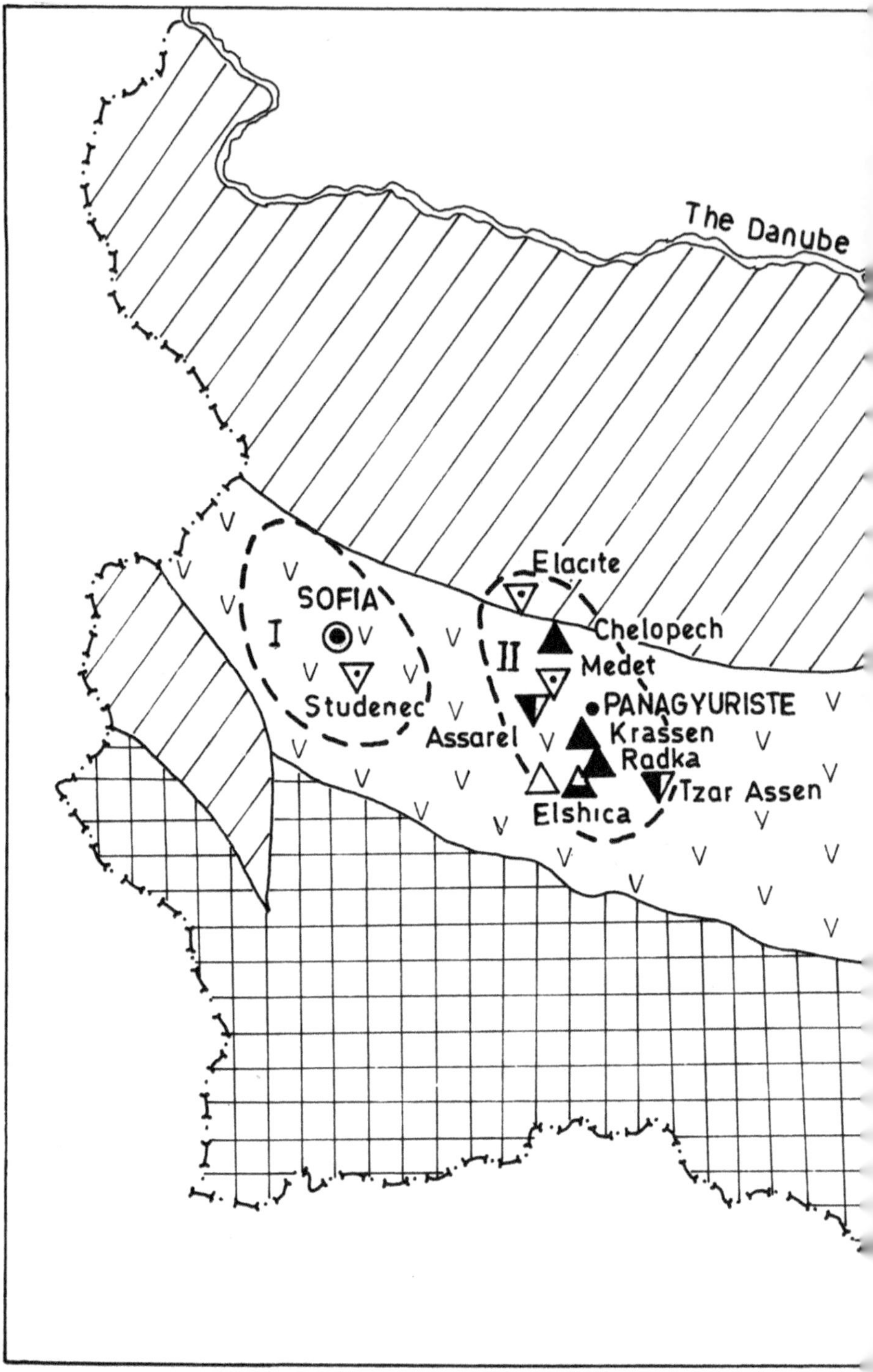

Fig. 1. Upper Cretaceous industrial copper deposits in Srednogorie zone. – 1 = Moesian pl
4 = Copper ore districts: I = Sofia, II = Panagyurishte-Etropole, III = Jambol, IV = Burga
horovo type; 8,9 and 10 = Massive sulphide deposits: 8 = pyrite type, 9 = chalcopyrite-
11 = copper type, 12 = copper molybdenum type, 13 = copper-gold-polymetallic type;

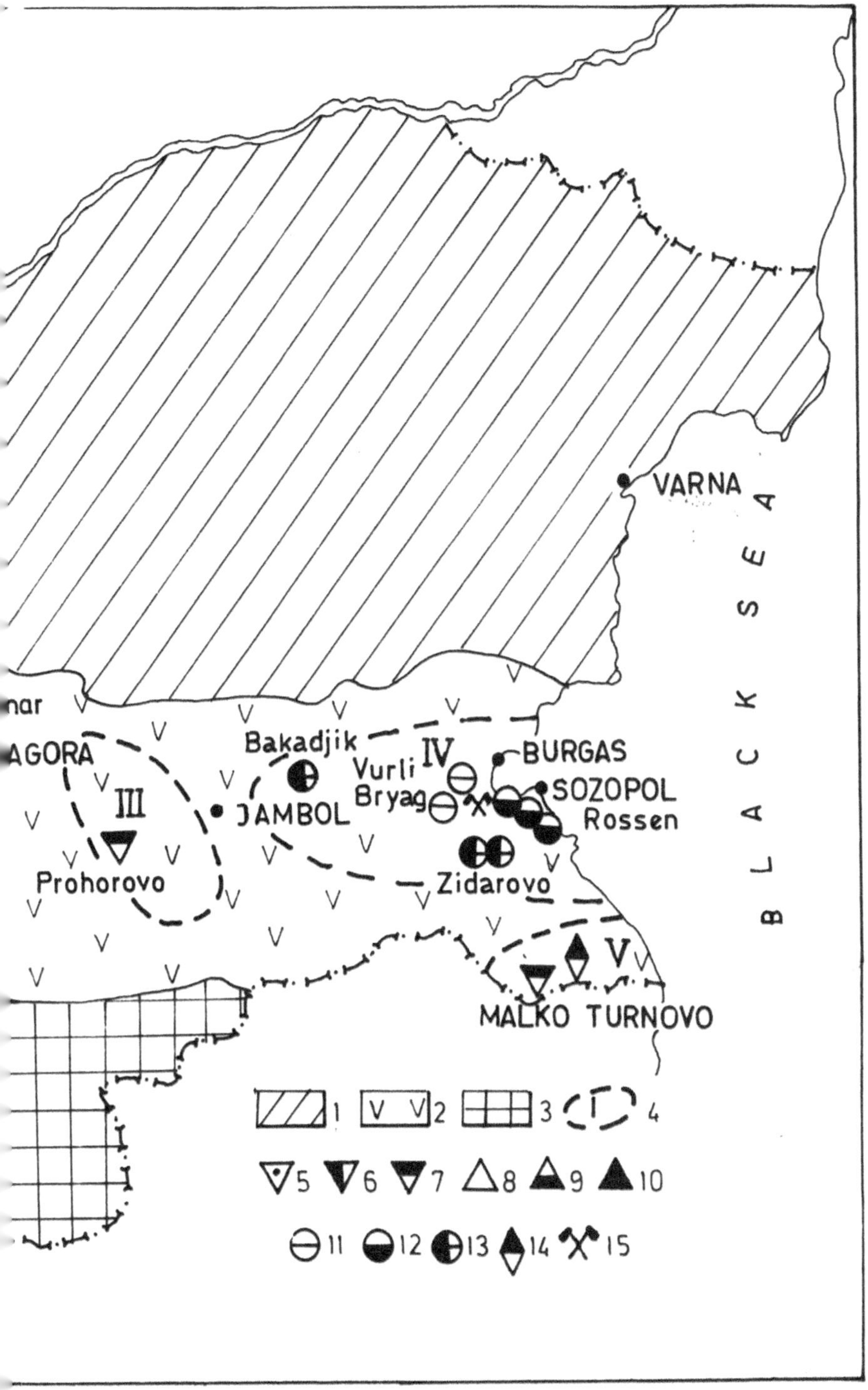

rebalkan, Stara Planina and Kraishtide zones; 2 = Srednogorie zone; 3 = Rhodope zone;
rnovo; 5, 6 and 7 = Porphyry copper deposits: 5 = Medet type, 6 = Assarel type, 7 = Pro-
= chalcocite-bornite-tennantite gold bearing type; 11, 12 and 13 = Vein copper deposits:
opper-molybdenum-sheelitic deposits; 15 = ancient mines and metallurgical centers.

Table 1. The Main Formation Types of Copper Ore Deposits in Srednogorie Zone

Formation types of copper deposits	Subformation and mineralogical types	Magmatic associations	Representative or deposits and orefields	Metals & Elements of industrial significance
1	2	3	4	5
1. Porphyry copper	1. Porphyry-copper-molybdenum	Hypabyssal Q-monzodiorites	Medet, Elacite, Studenec	Cu, Mo, S
	2. Porphyry-copper-alunite	Subvolcanic Q-diorite porphyrites and andesites	Assarel, Tzar Assen	Cu, S
	3. Porphyry-copper-skarn	Q-diorite-granite small intrusions	Prohorovo, Burdceto	Cu, S (Mo)
2. Massive sulphide	1. Massive-pyrite	Andesite-dacite volcanites and their pyroclastics	Elshica	S
	2. Massive-chalcopyrite-pyrite	The same + subvolcanic	Elshica, Radka, Krassen, Chelopech	Cu, S, (Pb, Zn)
	3. Massive-chalcocite-bornite-tenantite (gold bearing)	rhyodacite dykes		
3. Vein copper and copper-gold-polymetallic	1. Vein copper	Dykes of monzodiorite	Vurli Bryag,	Cu, S (Bi)
	2. Vein copper-molybdenum	porphyrites and volcanic rocks of	Rossen	Cu, Mo (Co)
	3. Vein copper-gold-polymetallic	trahybasalt-trahyandesite and andesite formations	Zidarovo, Bakadjik	Cu, Bi, Pb, Zn
4. Aposkarn-copper-molybdenum-sheelitic	?	Granodiorite-granite porphyrites	Malko Turnovo	Cu, Mo, W (Bi)

Some Peculiarities of the Upper Cretaceous Copper ore Deposits

Porphyry copper deposits

The majority of the porphyry copper deposits is situated in Panagyurishte-Etropole ore district in the central part of the country. Medet, Assarel, Elacite and Tzar Assen deposits are under exploitation now by open pits. In the same area there are also several small and low grade deposits or mineralizations of this type which are under prospecting and economic evaluation, such as Popovo dere, Petelovo, Tzar Assen-2, Kominsko Chukarche and others. Moreover some new porphyry copper deposits were discovered in the western part (the Studenec deposit in Sofia ore district) and in the eastern part of the Srednogorie zone (the Prohorovo deposit in Yambol ore district and the Burdceto deposit in Malko Turnovo ore district).

The oxide ores from the upper parts of Tzar Assen and Vlajkov Vruh deposits are subjected to acid-bacterial leaching and some quantity of copper has been produced over the last ten years.

The formation of the porphyry copper deposits is related to the last stage of the Upper Cretaceous magmatic activity, manifested by intrusion of subvolcanic to hypabyssal small bodies and dykes of quartz-monzodiorite, granodiorite to granite porphyric rock types. A typical feature of them is the stock-like shape of the ore bodies, tapering out in depth. The ore mineralizations occupy the apical parts of the hypabyssal intrusions (the Medet, the Studenec, the Prohorovo, the Burdceto ore deposits), ore penetrate into the cover of effusive rocks over subvolcanic bodies (the Asserel, the Tzar Assen, the Petelovo, the Kominsko Chukarche deposits). In some cases the ore mineralization impacts the upper parts of the small intrusions and penetrates into the Paleozoic and Mesozoic metamorphic intrusive and sedimentary rocks from the cover (the Elacite, the Prohorovo, the Burdceto deposits). The ore mineralization is of veinlet-impregnated type.

The main ore minerals are pyrite and chalcopyrite, the subordinate ones—bornite and molybdenite, but there are also small quantities of magnetite, specularite, sphalerite, galena, tenantite, native gold, electrum etc., which can be found in all porphyry copper deposits. The gangue minerals are represented by quartz, K-feldspar, sericite, chlorite, epidote, carbonates, zeolites, anhydrite etc.

The ore forming processes in the porphyry copper deposits can be subdivided into three stages—preore stage, during which the hydrothermal alteration of the country rocks took place, ore stage, characterized by deposition of copper, iron, lead, zinc and other sulphide minerals accompanied mainly by quartz as a gangue mineral and postore stage, represented by low temperature sulphate minerals (anhydrite, gypsum), zeolites (lomontite, stilbite, habasite, natrolite etc.), carbonates (calcite, dolomite, rhodochrosite, Mn-calcite, ankerite), superimposed on the sulphide mineralization of the ore stage.

Three subformation types of porphyry copper deposits can be distinguished among them in Srednogorie zone: 1. porphyry-copper-molybdenum deposits or Medet type; 2. porphyry-copper-alunite deposits or Assarel type and 3. porphyry-copper-skarn deposits or Prohorovo type. The main peculiarities of subformation types of porphyry copper deposits are given in Table 2.

Table 2. Subformation Types of Porphyry Copper Deposits

Type and representative deposits	Structural-geological setting	Magmatic associations	Typical mineral parageneses			Temperature of formation of the main parageneses
			Preore stage	Ore stage	Postore stage	
1. The Medet type (Medet, Elacite, Studenec)	Horst-anticlinal structures of Precambrian and Paleozoic metamorphic and intrusive rocks	Hypabyssal intrusions of quartz-monzodiorites, quartz-diorites and quartz-syenodiorites	Q-felspar-biotite, Q-magnetite	Q-molybdenite, Q-chalcopyrite, Q-pyrite, Q-sphalerite-galena	Zeolite-carbonate	$Q–C_{py}$–320–350° C Q–Mo–300–320° C
2. The Assarel type (Assarel, Tzar Assen, Petelovo, Kominsko Chukarche)	Volcanogenous graben-synclinals of Upper Cretaceous andesites and their pyroclastics	Subvolcanic small bodies and dykes of quartz-dioritic to granodioritic porphyrites	Alunite-secondary quartzites, Q-sericite, Q-kaolinite	Q-chalcopyrite, Q-pyrite, Q-sphalerite-galena	Anhydrite-gypsum, zeolite-carbonate	$Q–C_{py}$–290–310° C
3. The Prohorovo type (Prohorovo, Burdceto)	Anticlinal structures built-up of low-metamorphic and sedimentary rocks of Mezozoic age	Hypabyssal intrusions and dykes of granodiorite-granite composition	Epidote-scapolite-garnet skarn, K-felspar + magnetite, Q-sericite	Q-molybdenite, Q-chalcopyrite, Q-pyrite, Q-sphalerite-galena	Zeolite-carbonate, anhydrite-gypsum	Q-Mo–340–360° C Q–Py–320–340° C

The Medet type of porphyry-copper deposits originated in cases when the small hypabyssal intrusions of Ca-alkaline type were imbedded into a crystalline core of residual horst-anticlinal structures (the Medet deposit) or were intruded into the metamorphic rocks from the framework of the zone (the Elacite deposit). Typical for them are the commercial contents of molybdenum, potash-alkali alteration of the rocks, represented by K-feldspathization and biotitization and zeolite-carbonate type of postore mineralization.

Porphyry-copper deposits of Assarel type were formed in close connection with the central part of a satellite volcanic structure, the central core of which is built-up by subvolcanic small intrusions of quartz-dioritic and granodioritic porphyrites.

The most characteristic peculiarities of these deposits are the following: localization in graben-synclinal volcanogenic structure, absence of commercial content of molybdenum, acidic type of preore hydrothermal alteration of the volcanic and subvolcanic rocks, represented by alunite secondary quartzites, accompanied by diaspore, kaolinite, dickite, pyrophyllite, and surrounded by zones of quartz-sericite rocks and propylites. The postore stage is manifested by deposition of anhydrite-gypsum and zeolite mineralizations.

The Prohorovo type of the deposits is typical for the area built-up by low grade metamorphosed and sedimentary rocks with calcareous layers, intruded by small hypabyssal granodiorite-granite bodies. The copper-porphyry mineralization there is superimposed on the skarns and affects the metamorphic halo around the intrusive bodies.

These three types of porphyry copper deposits show some differences in temperature and depth of formation. The Medet and Prohorovo types of ore deposits were formed at greater depth and higher temperature than the Assarel type of porphyry copper deposits (Table 2).

By the author's opinion the differentiation of three subformation types of porphyry-copper deposits results mainly from their various geological and structural settings and the different character of the intrusive bodies, associated with them.

Massive Sulphide Deposits

They are of hydrothermal-metasomatic origin. Their formation is related to the development of the hydrothermal system at the end of the andesite-dacite stage of volcanic activity in Panagyurishte-Etropole ore district (Bogdanov, 1984). They are represented usually by massive ore bodies surrounded by haloes of veinlet-disseminated ores. The ore bodies are of stock-like and lense-like shape and occur amidst volcanic rocks, andesite agglomerate tuffs (Chelopech), ash tuffs or along the contact and tectonic fractured zones between andesite and dacite volcanic rocks (Radka). Some of them are controlled by the contact of the rhyodacite dykes cut off through the volcanites (Elshica).

Three types of massive sulphide ore bodies can be distinguished in Panagyurishte-Etropole ore district (Table 1): 1. massive-pyrite, 2. massive-chalcopyrite-pyrite and 3. chalcocite-bornite-tenantite (gold-bearing) type. They were formed during two stages of development of hydrothermal system, connected with the same Upper Cretaceous volcanic chamber. The first stage was manifested at the end of the volcanic activity, when the massive pyrite ores formed. The fragments of these ores, the so called "oreclastics" are included in dacite agglomerate tuffs (BOGDANOV et al., 1970). The second stage of massive sulphide ore deposition occured after the end of the volcanic activity and penetration of the rhyodacite dykes into the volcanic rocks.

The massive chalcopyrite-pyrite and the massive chalcocite-bornite-tenantite ore bodies were formed during the second stage of development of the hydrothermal system (BOGDANOV, 1985). The massive chalcopyrite-pyrite ore bodies are typical for the Elshica deposit, while the chalcocite-bornite-tenantite ore bodies represent the main type of massive sulphide mineralization in Radka, Krassen and Chelopech deposits. The last two types for ore mineralizations are superimposed on the massive pyrite ore bodies. The following mineral paragenetic associations were formed during them: chalcopyrite-pyrite, enargite-bornite-pyrite, bornite-chalcocite-tenantite, sphalerite-galenite, quartz-pyrite-chalcopyrite (vein type), marcasite-pyrite and anhydrite-gypsum. They cut off or replace one another and determine the complex type of the ores from these deposits. Gold and silver are disseminated as rare minerals in complex enargite-bornite-chalcocite-tennantite ores (BOGDANOVA, 1973).

Vein Copper and Copper-polymetallic Deposits

They are wide-spread in the eastern part of the Srednogorie zone in the framework of the Burgas ore district and can be grouped in four ore fields, closely connected with the volcano-plutonic centers of the Upper Cretaceous magmatic activity. These are the Rossen, the Vurly Bryag, the Zidarovo and the Bakadjik ore fields. The Rossen ore field is the best known and very well studied among them.

Three subformation (mineralogical) types of vein-copper and copper-polymetallic deposits can be distinguished, belonging to the various ore fields: 1. vein chalcopyrite-pyrite in Vurly Bryag ore field; 2. vein chalcopyrite-molybdenite in Rossen ore field and 3. vein chalcopyrite-gold-polymetallic in Zidarovo and Bakadjik ore fields.

The geological structure of the Rossen copper-molybdenum ore field was a subject of investigation of several authors (BOGDANOV et al., 1969; POPOV, 1980; STANISHEVA-VASSILEVA, VASSILEFF, 1981).

According to the last investigations it can be interpreted as a large volcano-plutonic structure of central type, formed in Upper Cretaceous time. The central caldera part of the volcano-plutonic structure is occupied by the Rossen pluton, built up of gabbro-monzonite-syenite intrusive formation. This intrusive core is encompassed by trachyandesite and andesite volcanic rocks and their volcanoclastic materials, represented by interbedded agglomerate tuffs, clastolavas, ash tuffs and lava flows. The volcanic and intrusive rocks are cut off by dykes of monzodiorites, syenodiorites, leucosyenites and the subvolcanic bodies of trachydiorite porphyrites.

The ore veins are built up of chalcopyrite, cobalt-containing pyrite, molybdenite, magnetite, hematite, scheelite, lineite, bornite, nickeline, galenite, sphalerite etc. Vein minerals are represented mainly by carbonates and quartz. The ore veins were formed in two stages: 1. pneumatolytic, with deposition of scapolite, biotite, phlogopite, apatite etc., and 2. hydrothermal, when all above mentioned minerals were deposited (Bogdanov et al., 1968).

The vein chalcopyrite-pyrite deposits of Vurly Bryag ore field are in many respects similar to those ones of Rossen ore field. The differences are connected with the mineral composition—the main minerals are chalcopyrite and pyrite. In the Zidarovo and Bakadjik ore fields the main peculiarities of the veins consist in the higher content of lead, zinc, gold, silver and bismuth mineralizations (Table 1).

In the frame of the Burgas ore district the regional lateral zonality was established, expressed by the replacement of the copper-molybdenum mineralization from east to west by copper-bismuth and copper-gold-polymetallic-bismuth mineralizations (Vassileff).

Skarn-Copper Deposits

Skarn-copper deposits are situated in the Malko Turnovo ore district in the southeastern part of the country (Vassileff et al., 1964; Magm. & metall. of KBA, 1983). They are related to the imbedding of the Malko Turnovo hypabyssal intrusive body into the Paleozoic and Mezozoic marbles, shists and marls. On the contact zone of the intrusive body with the carbonate rocks several iron-skarn deposits as Burdceto, Mladenovo, Propada, Gradishteto and s. o. were formed. The copper mineralization accompanied by some molybdenite and sheelite is superimposed on the iron-skarn ore bodies. The iron-skarn mineralization is connected with the first gabbro-monzodiorite stage of magmatic activity, while the sulphide copper-molybdenum and sheelite mineralizations were formed later on, during the next granodioritic and granitic intrusive stages. They are of hydrothermal origin.

The skarn bodies are of andradite-grossular-pyroxen composition with wollastonite, skappolite, forsterite and other minerals. They are accompanied by magnetite and hematite. The superimposed hydrothermal copper-molybdenum-scheelite mineralization form the nest-like, column-like and irregular ore bodies. As a result of replacement of the iron-skarn ore bodies by later hydrothermal mineralization, the chalcopyrite-magnetite, chalcopyrite-specularite, chalcopyrite and chalcopyrite-bornite ore bodies were formed. They are accompanied usually by pyrite and disseminated molybdenite, scheelite, bismuthinite, vitehinite and other rare minerals. The copper ore mineralization can be traced up to 100 m from the surface.

Sequences of the Copper ore Deposition in Srednogorie Zone

The copper mineralizations in Srednogorie zone are connected with Paleozoic (?) and Upper Cretaceous magmatic activity. During the Upper Cretaceous the commercial copper ore deposition appeared in several stages, when the massive sulphide, porphyry copper, vein-copper and skarn copper deposits were formed. According to the available data of absolute age determination and development of the magmatism in Srednogorie zone, the following four types of Upper Cretaceous copper ore mineralizations, connected with various stages of magmatic activity, can be distinguished: 1. massive sulphide, related to the andesite-dacite stage of magmatic activity in Panagyurishte ore district, with absolute age of 82–77 my (Chipchakova, Lilov, 1976); 2. porphyry-copper, associated with subvolcanic stage of granodiorite and quartzdiorite porphyrites in Srednogorie zone with absolute age of 75–72 my (Bogdanov, 1982); 3. skarn-copper, connected with the granodiorite-granite porphyrites in Malko Turnovo gabbro-diorite-granite intrusive body (no data about absolute age determination are available), and 4. vein-copper and copper-gold polymetallic, related to the last postdyke stage of development of subalkaline magmatic activity in Burgas ore district, with absolute ages of 50–60 my (Amov et al., 1976).

References

Amov, B., Bogdanov, B., Baljieva, Ts. (1976): Isotopic composition of lead and some problems of the origin and age of South Bulgaria ore formations. – Probl. Rudoobraz., Sofia, *2*, 13–25 (in Russian).

Bogdanov, B. (1977): Metallogeny of the Sredna Gora zone in the context of plate tectonics. – *In:* Metall. and Plate Tecton. in NE Mediterranean. Univers. Belgrade, 493–504.

Bogdanov, B. (1982): Porphyry copper deposits in Bulgaria. – Proceed. VUZ, ser. Geol. and Surveying, Moscow, *6*, 37–52 (in Russian).

Bogdanov, B. (1984): Hydrothermal systems of massive sulphide, porphyry-copper and vein copper deposits of Sredna Gora zone in Bulgaria. – Proceed. of the Sixth IAGOD Symposium, Stuttgart, 63–67.

Bogdanov, B., Rashkov, R., Todorov, T. (1968): Stages of mineralization in copper-molybdenum deposits of Rossen ore field. – *In:* Anniversary book of the Geol. Inst. of Bulg. Acad. of Sciences, Sofia, 291–307 (in Bulgarian).

Bogdanov, B., Rashkov, R., Iarmov, G. (1969): Structural peculiarities of the Rossen ore field. – *In:* Proceed. of Geol. Inst., Sofia, Ser. Ore and Nonmetal. miner. depos., *13*, 7–25 (in Bulgarian).

Bogdanov, B., Bogdanova, R., Chipchakova, St. (1970): Ore clastics from Radka and Elshica deposits in Panagjurishte ore district. – Sp. BGD, Sofia, *31*, *1*, 97–101 (in Bulgarian).

Bogdanova, R., Bogdanov, B. (1973): Gold in the Panagyrishte ore district. – C. R. Acad. Bulg. Sci., *26*, 2.

Bonchev, E. (1974): General features of the geological structures of Bulgaria. – IV IAGOD Symposium. *In:* Twelve Ore Deposits of Bulgaria, Sofia, Bulg. Acad. Sc.

Chernih, E. (1978): Mining and Metallurgy in Ancient Bulgaria. – Bulg. Acad. Sci., Sofia, 388 p. (in Russian).

Chipchakova, S. Lilov, P. (1976): On the absolute age of Upper Cretaceous magmatites in the Western part of Central Srednogorie and of associated mineralizations. C. R. Bulg. Acad. Sci., *29*, 1 (in Bulgarian).

JANKOVIĆ, S. (1977): Major Alpine metallogenic units of the NE Mediterranean and concepts of plate tectonics. – *In:* Metallogeny and Plate tectonics in NE Mediterranean. Belgrade, 9–18.

KONYAROV, G. (1953): Hystory of ore mining and metallurgy in Bulgaria. Bulg. Acad. Sci., 172 p. (in Bulgarian).

Magmatism and Metallogeny of the Carpathian-Balkan Area. (1983). (Chief editors: B. BOGDANOV, E. DIMITROVA.) Bulg. Acad. Sci., Sofia, 300 p. (in Russian).

POPOV, P. (1980): Structural conditions for the Upper Cretaceous ore depositions in the Balkanides. – XI Congress CBGA, Miner. and geochem., Kiev, 75–83 (in Russian).

POPOV, P., BAIRAKTAROV, I., MARINOV, T. (1979): Magmatism and structure of the Eastern part of the Burgas ore region. I. Characterization of the Magmatic formations. – Geol. Balc. 9.

POPOV, P. (1981): Magmatotectonic features of the Banat-Srednogorie belt. – Geol. Balc. *11, 2,* 43–72.

STANISHEVA, G. (1971): Upper Cretaceous magmatic formations of Burgas synclinorium. – C. R. Bulg. Acad. Sci., *24, 11,* 1509–1512 (in Russian).

STANISHEVA-VASSILEVA, G. (1980): The Upper Cretaceous magmatism in Srednogorie zone, Bulgaria: A classification attempt and some implication. – Geol. Balc., *10, 2,* 15–36.

STANISHEVA-VASSILEVA, G., VASSILEFF, L. (1981): Rossen ore field: possible model of central volcanic structure and its metallogenic significance. – Rudodobiv, Sofia, *2,* 1–6 (in Bulgarian).

TVALCHRELIDZE, G. (1972): Ore provinces in the World. Mediterranean belt. Moscow, Nedra, 344 (in Russian).

VASSILEFF, L., STAYKOV, M., IVANOVA-PANAIOTOVA, V., NECHEV, H. (1964): Skarns and ores in the mantle of the Malkoturnovo pluton, Strandja Mountain. *In:* Jubilee book dedicated to acad. Y. Yovchev. GUGOZN, Sofia, 277–348 (in Bulgarian).

The Romanian Carpathians: Alpine Geotectonic Evolution and Metallogeny

By Şerban Vlad*)

With 10 Figures

Zusammenfassung

Die sehr komplexe Struktur der rumänischen Karpaten wird auf der Basis der plattentektonischen Interpretation dargestellt. Der tektonischen Entwicklung werden die magmatischen und metallogenetischen Phasen und Bezirke zugeordnet. Das triassische Rifting bewirkt die Bildung von Ba-Fe-Lagerstätten in den Ostkarpaten; durch das seafloorspreading zur Unterkreidezeit wurde im Bereich der Südkarpaten ein schmaler Streifen von Ozeanboden frei gelegt, an den kleine Erzvorkommen gebunden sind. Die laramische Subduktion begann mit der Produktion erzfreier Vulkanite und anschließend erzbringender Intrusiva (Banatite), denen wichtige Kupferlagerstätten der Südkarpaten und des Apuseni-Gebirges zu verdanken sind. Eine neogene Subduktionsphase bewirkte die Entstehung von weiteren Cu-Porphyrie-Lagerstätten und von Buntmetallgängen mit Gold. Auch die junge Kontinent-Kontinent-Kollision erzeugte eine schwächere Gangvererzung.

Abstract

The structure of the Romanian Carpathians was accomplished during the Alpine orogenic cycle. Their metallogenetic evolution during Alpine times developed from unevolved intracontinental rifting (Triassic Fe ores, Jurassic-Lower Cretaceous Ba and Mo ores in the East Carpathians) to spreading areas (Lower Cretaceous Cu-pyrite ores in the South Carpathians) and subduction-related settings (Jurassic-Cretaceous Fe-Ti-V, Ni, Cu-pyrite, Mn ores in the Southern Apuseni Mountains, Laramian Mo, Bi, W, Cu, Co, Ni, Pb, Zn, B, Fe ores in the Apuseni Mountains and the western part of the South Carpathians, neogene Au-Ag, Cu and base-metal ores in the East Carpathians and the Southern Apuseni Mountains).

---

*) Ministerul Geologiei/DCRG, str. Mendeleev 34–36, Bucuresti, Romania.

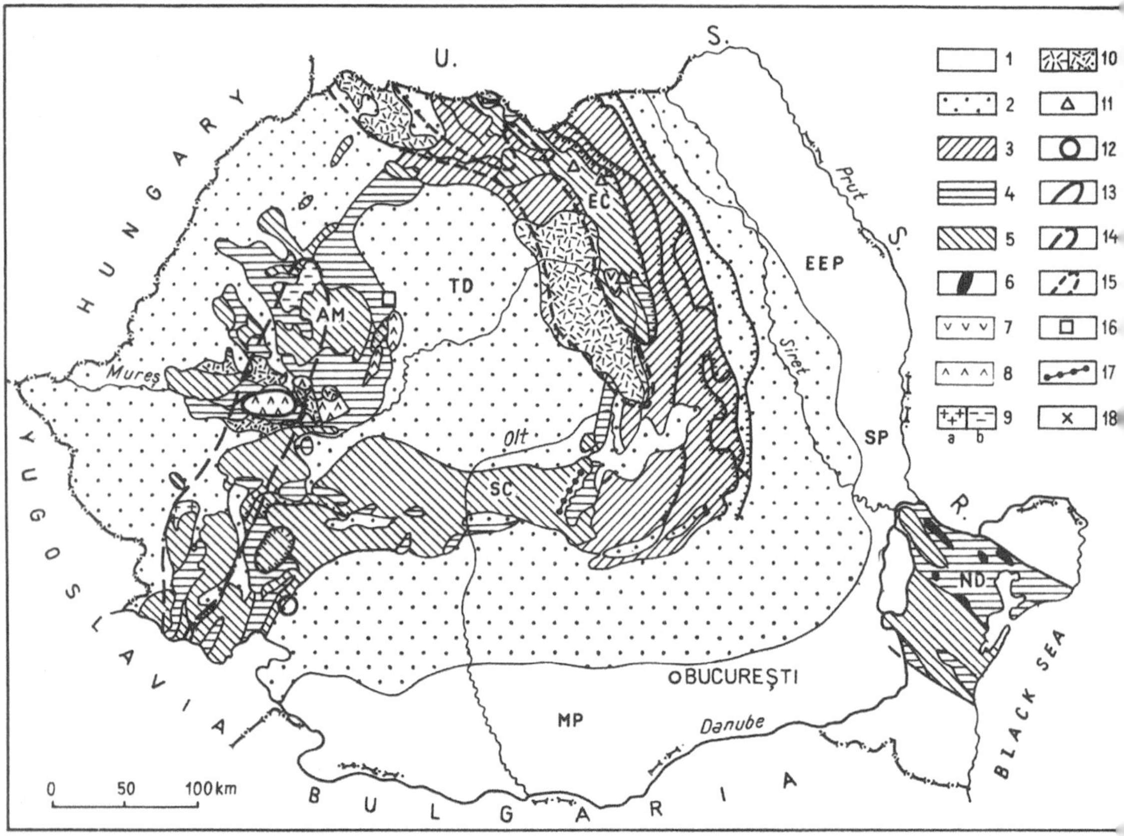

Fig. 1. Alpine structure and metallogeny in the Romanian Carpathians. 1 = foreland, 2 = molasse, 3 = flysch nappes, 4 = Mesozoic-Cainozoic cover, 5 = pre-Mesozoic basement including North Apusenides Nappes, Central East Carpathian Nappes, Supragetic and Getic Nappes, 6 = Triassic basalts of the short-lived intracontinental rift of North Dobrogea, 7 = island arc volcanics and active marginal basin spilites of the South Apuseni Mountains (Jurassic-Cretaceous), 8 = Ditrău alkaline massif related to intracontinental rifting (Jurassic-Cretaceous), 9 = subduction-related Laramian magmatism (Banatites) (a = plutons, b = volcanics), 10 = subduction-related Neogene volcanism (a = volcanics of the external arc; b = volcanics of the back-arc reactivization rifting areas), 11 = metallogenesis related to intracontinental rifting, 12 = metallogenesis related to ocean-floor spreading areas, 13 = metallogenesis related to Jurassic-Cretaceous island arc volcanics, 14 = Laramian (Banatitic) metallogenesis, 15 = Neogene metallogenesis, 16 = metallogenesis in passive margin-related settings, 17 = metallogenesis in continental collision-related settings, 18 = metallogenesis in post-collisional-related settings. EC = East Carpathians, SC = South Carpathians, AM = Apuseni Mountains, TD = Transylvanian Depression, EEP = East European Platform, SP = Scythian Platform, MP = Moesian Platform.

Minor Fe occurrences of Paleogene age are related to a passive margin of the Northern Apuseni Mountains and restricted Helvetian base-metal ores to post-collisional setting in the East Carpathians. A continental collision-related belt with Co, Ni, Bi, Cu mineralization connected with the Cretaceous deformation was inferred from geological-geophysical data in the eastern part of the South Carpathians.

The Structure of the Romanian Carpathians

The Romanian Carpathians, located of the southern boundary of the Eurasian Plate, originated during pre-alpine and especially alpine deformations. This significant orogenic structure is built up of inner units (Apuseni Mountains), moulded by the external Carpathian Arc (the East Carpathians, proceeding beyond the Romanian territory to the West Carpathians, and the South Carpathians, proceeding to Eastern Serbia, Yugoslavia). The Carpathian Foreland is represented by the Moesian Platform in the south, the Scythian Platform in the east and the East European Platform in the north and northeast. The North Dobrogea region lies within this cratonic environment, between the Moesian and Scythian Platforms. It has no connection with the Carpathians, but proceeds to Crimes and the Greater Caucasus, USSR (Fig. 1).

The Internides contain the Southern and the Northern Apusenides. The Southern Apusenides comprise thrust units with ophiolites and Mesozoic sediments covered by Neogene molasse in the west, south of the Mureş Valley, and in the east, within the basement of the Transylvanian Depression. The continuation of this structure towards the Vardar Zone was discussed by various authors and interpreted in different ways (e. g. SĂNDULESCU, 1983). The Northern Apusenides are built up of basement nappes (crystalline schists, pre-Mesozoic granitoids) and in fact represent the prolongation of the Austroalpine structure of the Eastern Alps (SĂNDULESCU, 1983).

The Carpathian Arc with external position is built up of three successive groups of nappes of different ages and origin; besides them, the East Carpathians contain the Transylvanian Nappe Group. The innermost region of the East Carpathians, namely the Crystalline-Mesozoic Zone, contains various nappes formed during Hercynian and Mesocretaceous deformations (BALINTONI et al., 1983). These Central East Carpathian Nappes correlate with Supragetic and Getic Nappes of the South Carpathians, which were involved in Laramian deformations too. The Transylvanian Nappes with Mesozoic ophiolites and sediments resulting from the basement of the Transylvanian Depression (SĂNDULESCU and VISARION, 1978) were obducted upon the above mentioned nappes of the Crystalline-Mesozoic Zone. Beyond the inner nappe assemblage, the Flysch Nappes built up of Upper Jurassic-Lower Cretaceous flysch sediments and ophiolites arose during Mesocretaceous and Laramian deformations. They comprise the Black Flysch Nappe, the Ceahlău Nappe and the Baraolt Nappe within the East Carpathians as well as the Severin Nappe within the South Carpathians (e. g. SĂNDULESCU, 1983). The Moldavides have the outermost position, confined to the East Carpathians; they represent cover nappes built up of Paleogene flysch sediments and Miocene molasse.

It is worth mentioning that the South Carpathians comprise a significant region of continental type (mainly crystalline schists and pre-Mesozoic granitoids), the Danubian Realm, early considered as Autochthon (e. g. CODARCEA, 1940). According to recent interpretations, it can be subdivided into the Lower Danubian Unit and the Upper Danubian Unit, both involved in pre-Alpine and Alpine (Mesocretaceous and Laramian) thrustings (BERZA et al., 1983). On the other hand, new data provide evidence of Danubian thrusting over the Moesian Platform during intra-Sarmatian deformations (MOTAŞ, 1983).

Besides the above mentioned ophiolites, two major petrogenetic associations of calc-alkaline character are found in the Carpathians: the Laramian (Banatitic) igneous rocks in the west (Apuseni Mountains and South Carpathians) and the Neogene volcanics in the South Apuseni Mountains and the East Carpathians.

Alpine Geodynamic Evolution of the Romanian Carpathians

The post-Paleozoic geodynamic evolution which promoted the present-day Carpathian architecture is associated with the development of the Tethyan trench system. Both short-lived and more evolved rifts are characteristic of this setting and yielded various types of magmatic and metallogenetic products. Thus, Triassic intracontinental rifting in the North Dobrogea region gave rise to an aulacogen-like failed arm linked with the Tethys through Crimea and the Greater Caucasus (VLAD, 1978). In places riftings along reactivated proto-Carpathian alignments were rapidly aborted, but commonly promoted ocean-floor spreading. Accordingly, narrow elongated troughs with oceanic crust developed around sialic blocks. It is likely that the distensional tectonics of the Lower Mesozoic period and the subsequent motion of resulting microplates were promoted by triple junctions; the Carpathian curvatures are interpreted as evidence of this kind (VLAD, 1980). According to geological and geochemical data the Severin Nappe ophiolites represent ocean-floor tholeiites (CIOFLICA et al., 1981). Similar recent studies on South Apuseni Mountains ophiolites strongly suggest that the oceanic crust of the paleo-basin was consumed; inferred oceanic crust remnants are to be found in the basement of the Transylvanian Depression (e. g. CIOFLICA et al., 1980, CIOFLICA and NICOLAE, 1981).

Following this, significant westward pulsatory compressions gave rise to Andean or island arc subduction-related magmatic arcs. Neokimmerian deformations promoted tholeiitic and calc-alkaline island arc magmatism in the South Apuseni Mountains (CIOFLICA et al., 1980, CIOFLICA and NICOLAE, 1981).

Mesocretaceous deformations gave rise to numerous thrust nappes accomplished in places during Laramian times when continental collision was reached. These deformations yielded eastward obduction of ophiolites in the East and South Carpathians. On the other hand, a Laramian westward subduction-related belt of calc-alkaline character (partly of Andean type) runs from the Apuseni Mountains to

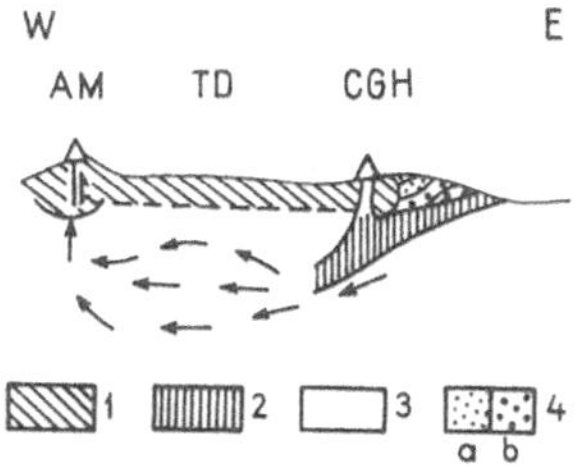

Fig. 2. Plate tectonic model of the Neogene reactivation in the Apuseni Mountains (according to VLAD, 1980). 1 = continental crust, 2 = westward subducted oceanic crust, 3 = upper mantle, 4 = nappes (a = flysch nappes, b = Moldavides), 5 = energetical eddies that promoted heat of the base of the lithosphere and behind-arc diapirism. CGH = Căliman-Gurghiu-Harghita Mountains with island arc volcanics, AM-TD = behind arc environment (TD = Transylvanian Depression with turbidite apron-like sedimentation, AM = back-arc region with extensional tectonics characterized by mainly andesitic volcanism and related metallogenesis caused by heat transfer from external Carpathian subduction and reactivization of the Apuseni Mountains hot spot).

the South West Carpathians ("Banatitic Province") and further south to the Eastern Serbia (Yugoslavia).

Following the suture of the South Apusenides to the North Apusenides sialic block, the north-western part of that internal paleobasin acted as a passive continental margin.

The Neogene calc-alkaline island arc volcanism of the East Carpathians (BOCCALETTI et al., 1973, BLEAHU et al., 1973) is related to deformations that yielded the Moldavides Nappe System. The similar volcanism of the South Apuseni Mountains cannot be satisfactorily explained by subduction. The South Apuseni Mountains contain three superposed magmatic products with distinctive lineation, that is Jurassic-Lower Cretaceous island arc volcanics, Laramian (Banatitic) igneous rocks and Neogene volcanics. The pulsatory magmatism that acted within a restricted area was presumably controlled by a deepseated hot spot (VLAD, 1980). It seems likely that hot spot reactivization yielded rifting along Laramian transverse faults and promoted the Neogene volcanicity. The reactivization resulted during the above mentioned westward subduction in the East Carpathians; energetical eddies promoted heat of the base of the lithosphere and back-arc diapirism with volcanism (Fig. 2).

As an significant post-collision event, intra-continental rifting yielded basaltic volcanicity during late Neogene times (RĂDULESCU et al., 1981).

Alpine Metallogenesis and Plate Tectonics in the Romanian Carpathians

The metallogenetic characteristics of the Romanian Carpathians have recently been taken into account by various authors; among them IANOVICI and BORCOŞ (1982) studied in detail the time-space distribution of numerous ore deposits and

Cioflica and Vlad (1984) discussed the Alpine metallogeny of Romania in terms of plate tectonics.

The structure of the Romanian Carpathians, accomplished during the whole Alpine cycle, provides opportunities to outline their metallogenetic evolution in association with tectonic and magmatic events (Fig. 1). Accordingly, the Alpine metallogenesis developed during post-Paleozoic times from early aborted intracontinental rifting to spreading areas, subduction-related settings, collision- and post-collision-related settings that are listed below.

Metallgenesis related to intracontinental rifting. The Central East Carpathian Nappes represent a preferential site of Ba-Pb-Zn, Fe, Mo occurrences related to Mesozoic sedimentary and igneous rocks.

At Delniţa dolomitic limestones of Triassic age contain stratabound Fe ores associated in places with barite and base-metal ores. The lens-like bodies are prevailingly sideritic-ankeritic in the east and hematitic in the west.

Along the Ostra-Gemenea-Slătioara alignment barite and witherite accumulations are associated with base-metal ores (Ianovici et al., 1966). At Ostra, N-S striking veins cross the metamorphic basement and the Mesozoic sedimentary cover (Fig. 3). At least two barite generations as well as witherite, pyrite, sphalerite, galena, tetrahedrite were recognized; at depth the penetrated gneisses are impregnated with barite. It seems likely that early barite occurrences were remobilized during post-Jurassic rifting, before major middlecretaceous deformations. The mineraliza-

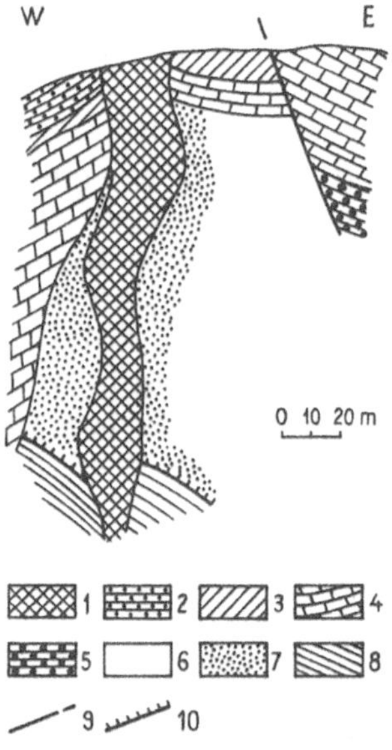

Fig. 3. Cross-section through the Ostra barite deposit (according to Ianovici et al., 1966). 1 = barite vein, 2–5 = Triassic-Jurassic sediments (2 = sandstones, 3 = jasperoid rocks, 4 = dolomitic limestones, 5 = siliceous conglomerates), 6 = Rarău gneisses, 7 = barite impregnated gneiss, 8 = epimetamorphic rocks, 9 = fault, 10 = thrust.

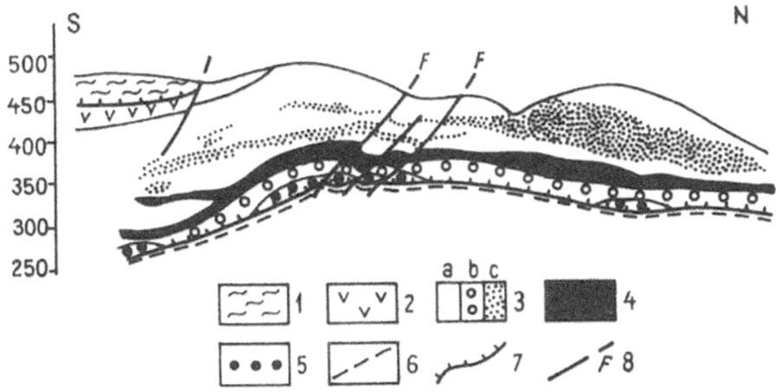

Fig. 4. Section through Cu-pyrite ores from Baia de Aramă (according to CIOFLICA et al., 1980). 1 = Getic Nappe, 2 = serpentinite protrusion, 3 = ophiolite complex (a = upper ophiolites, b = lower ophiolites, c = mineralization), 4 = Cretaceous black argillite, 5 = Cretaceous flysch of the Severin Nappe, 6 = Cretaceous flysch of the Danubian realm, 7 = thrust, 8 = fault.

tion is found discontinuously up to Gemenea and Slătioara where base-metal and barite veins cut the crystalline schists (IANOVICI and BORCOŞ, 1982).

The Jurassic-Lower Cretaceous ring-like intrusion of alkaline character from Ditrău exhibits a zonal structure (ANASTASIU and CONSTANTINESCU, 1980): its inner part consists of foidites, surrounded discontinuously by syenite and monzonite rocks. Hornblendites and diorites as well as granites and alkali granites occur marginally. Numerous lamprophyre, microsyenite, alkali granite and aplite dykes cut the massif. Albitite segregations and carbonatite veins are found in places. Mo-bearing carbonatites occur especially in the north (Jolotca). They are hosted by diorite and hornblendite rocks and contain usually ilmeno-rutile, ilmenite, monazite, tapiolite, columbite and sulphides as well as by alkali syenite rocks when xenothime, sulphides and niobo-tantalates are the common minerals.

Metallogenesis related to ocean-floor spreading areas. The Severin Nappe of the South Carpathians contains ophiolites with Cu-pyrite ores. During early Mesozoic times an elongated basin with oceanic crust acted between the Getic and the Danubian realms (RĂDULESCU and SĂNDULESCU, 1973). The resulting basaltic flows and pyroclastics associated with Lower Cretaceous flysch sediments of the Severin Nappe were obducted eastwards during Laramian compressions when the Getic and the Danubian realms collided. The ophiolite-related mineralization occurs at Baia de Aramă (Fig. 4). CIOFLICA et al. (1981) provided geological and geochemical evidence to characterize these ophiolites as tholeiitic ocean-floor basalts formed in a small ocean setting; the related ore deposits were ascribed to the Joma type of PEARCE and GALE (1977). They occur in basalts as small size stratiform pods of massive chalcopyrite, with subordinate amounts of pyrite and sphalerite in quartzose gangue. The massive ore is commonly underlain by pyrite + chalcopyrite + spalerite stockworks.

Metallogenesis in subduction-related settings. Jurassic-Cretaceous, Laramian and Neogene subduction events promoted ore deposition associated with island arc or Andean type magmatic belts.

The earlier event is represented by tri-stadial magmatism of the South Apuseni Mountains, that is tholeiite series and subsequent calc-alkaline series representing island arc products and final spilitic complex associated with active marginal basin (CIOFLICA et al., 1980; CIOFLICA, NICOLAE, 1981). The related metallogenesis consists of Fe-Ti-V and Ni late magmatic segregations in gabbroic intrusions and Cu-pyrite veins and stockworks in basaltic lavas of the tholeiite series (e. g. CIOFLICA et al., 1980). The calc-alkaline series comprises Mn-volcano-sedimentary ores (e. g. CIOFLICA et al., 1980); such a setting encloses virtual Kuroko-porphyry copper systems too (VLAD, 1983).

The Fe-Ti-V segregations are found as titanomagnetite and ilmenite nests, lenses and grains within layered gabbroic bodies (CĂZĂNEŞTI-CIUNGANI). The Ni ores are confined to the Ciungani gabbroic body wherein pyrrhotite, pentlandite with associated chalcopyrite and magnetite occur as a small size pod.

The Cu-pyrite volcanogenic ores are related to basalts and were considered by CIOFLICA et al. (1981) of Gjersvik type (PEARCE and GALE, 1977). Various veins and stockworks (CĂZĂNEŞTI-CIUNGANI, ALMĂŞEL, ROŞIA NOUĂ) are controlled by brecciated basalts. At Pătîrş the mineralization is located in the upper part of a basaltic unit and contains an inner zone with pyrite + chalcopyrite veinlets and a massive pyrite pod, surrounded by a disseminated pyrite aureole (Fig. 5).

Minor Mn volcano-sedimentary ores associated with jaspers occur in Lower Cretaceous sediments at Soimuş-Buceava, Pîrneşti and Godineşti.

A significant subduction event yielded the Laramian calc-alkaline magmatic belt (banatites). The magmatism is represented by two major stages (CIOFLICA and VLAD, 1973, 1980) that is the senonian volcanicity depleted of metallogenetic interest and the subsequent intrusive stage divided into three phases: early minor diorite bodies; plutons and subvolcanic bodies related commonly to monzodiorite, diorite → granodiorite and granodiorite → granite evolution trends; final acidic dykes and concurrent basic dykes and lamprophyre dykes. The main intrusive event promoted ore deposition inside and around shallow intrusions, plutons or subsequent dykes. Porphyry copper and skarn deposits prevail, whereas vein deposits occur scarcely (CIOFLICA and VLAD, 1980). The monzodiorite, diorite-granodiorite magmatism

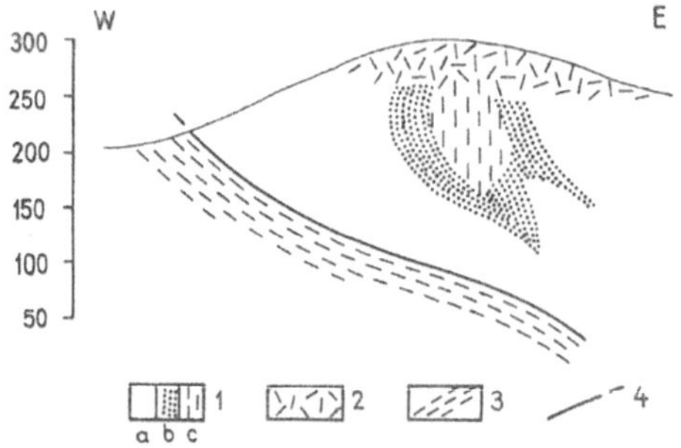

Fig. 5. Section through Cu-pyrite ores from Pătîrş (according to CIOFLICA et al., 1980). 1 = basalts (a = unmineralized, b = pyrite impregnations, c = pyrite + chalcopyrite stockwork), 2 = Gossan, 3 = Cretaceous flysch, 4 = reversed fault.

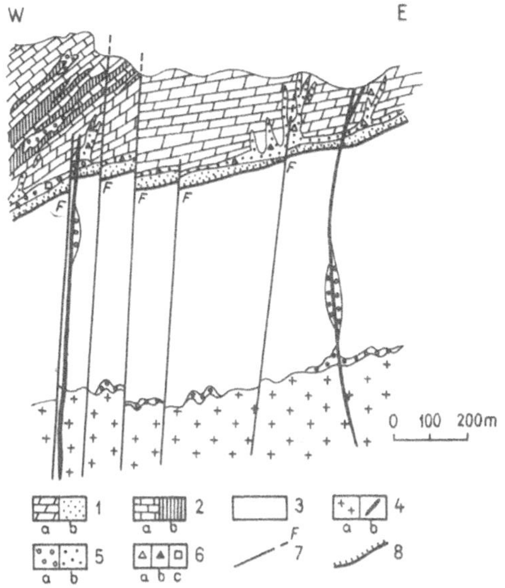

Fig. 6. Cross-section through the Crişul Negru Valley, North Apuseni Mountains (according to Cioflica and Vlad, 1980). 1 = recrystallized Carnian sediments of the Băiţa Unit (a = dolomitic marble, b = apodetrital hornfels), 2 = recrystallized Norian sediments of the Băiţa Unit (a = calcareous marble, b = apodetrital hornfels), 3 = recrystallized Cretaceous limestones of the Bihor Autochthon, 4 = Laramian (Banatitic) igneous rocks (a = granite, b = andesite), 5 = skarns (a = Ca-skarns, b = Mg-skarns), 6 = mineralization (a = Pb-Zn-B, b = Cu-W-Bi, c = Mo-W), 7 = fault, 8 = thrust.

yielded Cu deposits, whereas the granodiorite-granite magmatiam especially base-metal ores.

The North Apuseni Mountains sub-belt is characterized by granodiorite-granite magmatism and widespread base-metal metallogenesis (Ianovici et al., 1977). The ore zoning is expressed by the complex zone of the Bihor-Gilău Mountains (Mo, Bi, W, Cu, Co, Ni, Pb, Zn, B, Fe ores in skarn environment) (Fig. 6), followed landwards by the base-metal zone of the Vlădeasa Massif (hydrothermal Pb-Zn ores).

The South Apuseni Mountains sub-belt is represented by monzonite, diorite-granodiorite magmatism with Cu-impregnated and Fe skarn deposits and granodiorite-granite magmatism with Pb-Zn and Mo vein deposits (Cioflica and Vlad, 1984).

The Banat-Poiana Ruscă Mountains sub-belt is represented by the inner zone (South Banat) with monzodiorite, diorite-granodiorite magmatism and Cu-Mo porphyry copper deposits of Lowell and Guilbert model, stockwork type, monoascendant evolution (e. g. Suvorov, Moldova, Nouă, Fig. 7) as well as Cu-Mo skarn deposits (e. g. Sasca); the North Banat-Poiana Ruscă zone located landwards consists of granodiorite-granite magmatism and Fe, Pb-Zn skarn deposits (e. g. Dognecea, Onca de Fier) or restricted Mo-porphyry occurrences (e. g. Oraviţa). The well-

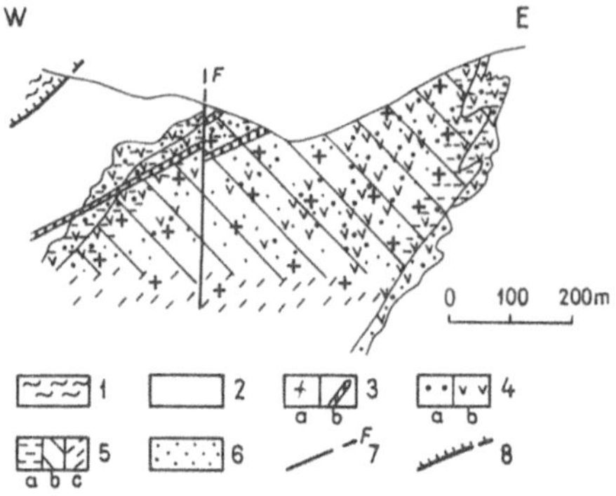

Fig. 7. Cross-section through the Suvorov porphyry copper deposit, Moldova Nouă, South Banat (according to VLAD, 1983). 1 = Supragetic Nappe, 2 = Mesozoic limestones converted into marble, 3 = Laramian (Banatitic) ingenous rocks (a = quartz diorite porphyry, b = lamprophyre), 4 = apocalcic metasomatites (a = grandite skarn, b = propylitic zone), 5 = apoalumosilicate metasomatites (a = argillic zone, b = phyllic zone, c = potassic zone), 6 = copper mineralization, 7 = fault, 8 = thrust.

expressed Cu-Mo-Fe, Pb, Zn transverse zoning and Cu-Pb line that delineates the magmatic-metallogenetic zones are characteristic of the Andean subduction type (VLAD, 1979).

The Neogene subduction gave rise to various volcanics that occurred rhythmically from Helvetian to Pliocene times. Among them the Sarmatian volcanics are commonly of metallogenetic significance. The petrogenetic and metallogenetic characteristics of the Neogene volcanicity were summarized by various authors (e. g. GIUŞCĂ et al., 1968; CIOFLICA et al., 1973; RĂDULESCU et al., 1981; IANOVICI and BORCOŞ, 1982).

The East Carpathian volcanic arc of mainly andesitic composition developed within the Oaş-Gutîi-Ţibleş-Bîrgău-Rodna-Călimań-Gurghiu-Harghita Mountains and contains hydrothermal ore deposits of Au-Ag and base-metal type with subordinate amounts of Hg, exhalative S and hydrometasomatic Fe ores. It is to be suggested that such setting may promote porphyry copper occurrences at depth, near root zones of Au-Ag and base-metal veins. The northern part of the Oaş Mountains contains base-metal veins between Tarna and Bicsad. The southern part of the Gutîi Mountains locates significant Au-Ag and base-metal veins (Fig. 8) along the WNW-ESE Baia Mare metallogenic zone (e. g. Ilba, Săsar, Herja, Baia Sprie, Capnic, Băiuţ). In the Ţibleş-Bîrgău Mountains base-metal veins are related to subvolcanic bodies, whereas in the Rodna Mountains they are connected with breccia pipes or replace marbles. The Călimań-Gurghiu-Harghita Mountains contain minor S (Călimań), siderite (Vlăhiţa), Hg (Sîntimbru), Au-Ag and base-metal ores. The root zones of the volcanic structures may enclose more significant ore occurrences.

Back-arc reactivazion rifting areas trending NW-SE and E-W in the South Apuseni Mountains promoted Au-Ag and base-metal deposits associated with andesite volcanics within the Brad-Săcărîmb, Almaj-Stănija, Roşia Montană-Bucium

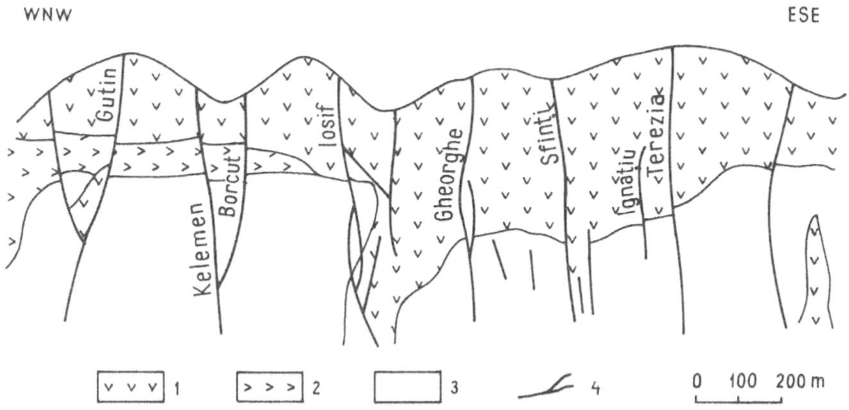

Fig. 8. Cross-section through the Cavnic ore deposit (according to IANOVICI and BORCOŞ, 1982). 1 = Pontian pyroxene andesite, 2 = Pannonian-Pontian quartz andesite, 3 = Tertiary sedimentary rocks, 4 = base-metal veins.

and Baia de Arieş districts. Subvolcanic bodies of andesitic-microdioritic composition which penetrated the volcanic piles contain polyascendant porphyry copper deposits of the diorite model. They contrast with Laramian porphyry-copper systems by lack of skarnified halo which is replaced by peripheral Au-Ag and Pb-Zn veins. The porphyry copper systems show a Cu-Au character at Tarniţa, Rovina, Valea Morii, Musariu and a Cu-Mo character at Deva and Roşia Poieni (IANOVICI et al., 1977). At Baia de Arieş significant base-metal replacement ores occur within marbles in similar setting with the above mentioned Rodna Mountains occurrences (Fig. 9).

Metallogenesis in passive margin-related settings. Restricted Fe occurrences of Paleogene age are found at Căpuş (Gilău Mountains) within epicontinental sediments. The ores are represented by limonitic oolites found in carbonatic, chloritic and glauconitic cement (VINOGRADOV et al., 1964). Such sedimentary Fe

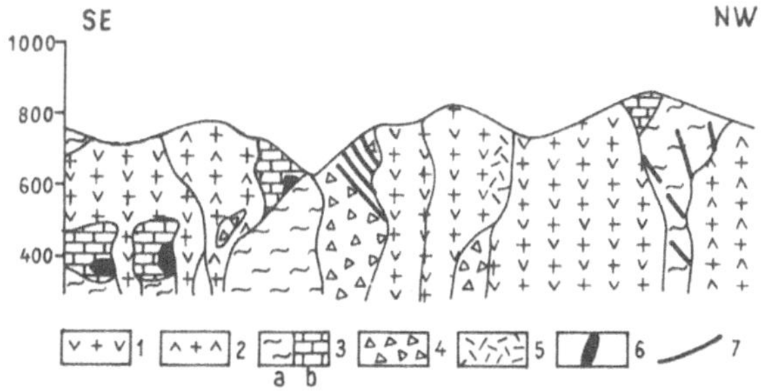

Fig. 9. Cross-section through the Baia de Arieş ore deposit (according to IANOVICI and BORCOŞ, 1982). 1 = Sarmatian amphibole subvolcanic andesites, 2 = Sarmatian biotite-amphibole subvolcanic andesites, 3 = metamorphic rocks of the basement (a = crystalline schists, b = marbles), 4 = brecciated andesitic bodies, 5 = Au-Ag breccia pipes, 6 = Pb-Zn metasomatic bodies, 7 = fault.

ores were considered to be formed on the north-western passive continental margin of the Transylvanian paleo-basin after suture of the South Apuseni island arc to the North Apuseni sialic block (CIOFLICA and VLAD, 1984).

Metallogenesis in the continental collision-related settings. Supragetic and Getic crystalline schists of the East Făgăraş and Leaota Mountains (South Carpathians) are cut by minor veins with Bi minerals, Co-Ni sulphides and arsenides, Cu minerals, galena, sphalerite, pyrite, pyrrhotite, silver and gold. VLAD and DINICĂ (1984) suggested that ore deposition took place during remobilisation connected with major Cretaceous deformations (Dobšina type). On the other hand, geophysical evidence, the NNE-SSW regional distribution of the ores and indications of metallic periplutonic arrangement in both Leaota and East Făgăraş Mountains may suggest that mineralization is provided by a deep-seated continent-continent collision belt of Erzgebirge or Cornwall type (Fig. 10).

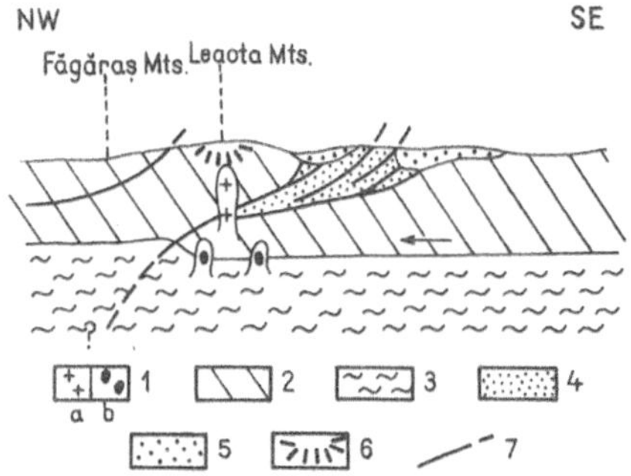

Fig. 10. Presumed continental collision magmatism and related metallogenesis in the eastern South Carpathians (according to VLAD and DINICĂ, 1984). 1 = deep-seated plutons (a = granite-alkali granite, b = ultrabasite, anorthosite), 2 = continental crust, 3 = upper mantle, 4 = Lower Cretaceous flysch ± obducted ophiolites, 5 = molasse, 6 = Co-Ni-Bi-As-Cu ores.

Metallogenesis in the post-collisional-related settings. It seems likely that a post collisional event acting in the South East Carpathians promoted sedimentary base-metal ores in Helvetian sandstones of molasse type at Jitia. On the other hand, late post-collisional rifting yielding Pliocene basalts lacks in metallogenetic importance.

Conclusions

The Carpathian structure was accomplished during the Alpine orogenic cycle. Magmatic as well as other factors intimately controlled the formation of metallic mineral deposits during the Mesozoic-Cainozoic sequence of intra-continental rifting, ocean-floor spreading, subduction, collision and eventual post-collision rifting. The major metallogenetic events are definitely associated with Laramian and Neogene subduction-related settings. It is suggested that the Laramian monzodiorite, diorite-granodiorite magmatism extracted Cu from depth and carried it upwards, whereas sialic contaminated granodiorite-granite magmatism mobilized Mo, Pb, Zn from the continental crust. On the other hand, Cu-Au and Cu-Mo differentiation of the Neogene porphyry copper seems to be explained by Au remobilization from early Mesozoic island arc tholeiites and Mo from the crystalline basement. The spatial superposition of three magmatic-metallogenetic events in the South Apuseni Mountains suggests that the inheritance-reactivization concept is to be taken into account (VLAD, 1980).

Besides the abundant Laramian skarn-porphyry copper deposits and Neogene vein-porphyry copper deposits, relations between plate tectonics and mineralization during Alpine times emphasized that numerous settings are potentially favourable for additional ore bodies. Thus mature island arc structures of the South Apuseni Mountains may contain Kuroko-porphyry copper systems whereas the roots of the East Carpathian volcanic structures porphyry copper mineralization; the geological-geophysical considerations suggest that a deep-seated continental collision-related belt may be found in the eastern part of the South Carpathians and, eventually, geological guides indicate that the Helvetian molasse of the East Carpathians could contain additional occurrences of Jitia type.

References

ANASTASIU, N., CONSTANTINESCU, E. (1980): Structure du massif alcalin de Ditrău. An. Univ. Buc., XXIX, 3–22.

BALINTONI, I., GHEUCA, I., VODĂ, A. (1983): Alpine and Hercynian overthrust nappes from central and southern areas of the East Carpathian Crystalline-Mesozoic Zone. An. Inst. Geol. Geofiz., LX, Lucr. Cgr. XII As. Geol. Carp. Balc., 15–22.

BERZA, T., KRÄUTNER, H., DIMITRESCU, R. (1983): Nappe structure in the Danubian window of the Central South Carpathians. An. Inst. Geol. Geofiz., LX, Lucr. Cgr. XII As. Geol. Carp.-Balc., 1–39.

BLEAHU, M., BOCCALETTI, M., MANETTI, P., PELTZ, S. (1973): The Carpathian arc: a continental arc displaying the features of an "island arc". Journ. Geoph. Res., 78/23, 5025-5032.

BOCCALETTI, M., MANETTI, P., PECERILLO, A., PELTZ, S. (1973): Evolution of the Upper Cretaceous and Cenozoic magmatism in the Carpathian arc: geodynamic significance. Mem. Soc. Geol. Hol., XII, 267–277.

CIOFLICA, G., VLAD, Ş. (1973): The correlation of Laramian metallogenic events belonging to the Carpatho-Balkan area. Rev. roum. Géol. géoph. géogr., sér. géol., 17/2, 207–224.

CIOFLICA, G., SAVU, H., BORCOŞ, M., STEFAN A., ISTRATE, G. (1973): Alpine volcanism and metallogenesis in the Apuseni Mountains. Symp. Volc. Met., Bucharest, Guide Ex. 3AB, 70 p.

CIOFLICA, G., VLAD, Ş. (1980): Copper sulphide deposits related to Laramian magmatism in Romania. European Copper Deposits, Belgrade, 67–72.

CIOFLICA, G., LUPU, M., NICOLAE, I., VLAD, Ş. (1980): Alpine ophiolites of Romain, tectonic setting, magmatism and metallogenesis. An. Inst. geol. geofiz., LVI, 79–95.

CIOFLICA, G., NICOLAE, I. (1981): The origin, evolution and tectonic setting of the Alpine ophiolites from the South Apuseni Mountains. Rev. roum. géol. géoph. géogr., sér. Géol., 25, 19–29.

Cioflica, G., Vlad, Ş., Nicolae, I., Vlad, C., Bratosin, I. (1981): Copper metallogenesis related to Mesozoic ophiolites from Romania. UNESCO Inst. Symp. Maf. Ultram. Compl., Athens, *2,* 156–171.

Cioflica, G., Vlad, Ş. (1984): Alpine metallogeny in Romania. An. Inst. Geol. Geofiz., LXIV, p. 175–184.

Codarcea, Al. (1940): Vues nouvelles sur la tectonique du Banat méridional et du Plateau de Mehedinţi. An. Inst. Geol. Rom., XX, 1–74.

Giuşcă, D., Bleahu, M., Borcoş, M., Dimitrescu, R., Kräutner, H., Savu, H. (1968): Neogene volcanism and ore deposits in the Apuseni Mountains. Int. Geol. Cgr., XXIII Ses., Prague, 51 p.

Ianovici, V., Pitulea, G., Lemne, M. (1966): Contribuţii la studiul zăcămîntului de baritină de la Ostra. St. cerc. geol. geof. geogr., ser. geol., *11,* 331–340.

Ianovici, V., Vlad, Ş., Borcoş, M., Boştinescu, S. (1977): Alpine porphyry copper mineralization of West Romania. Mineral. Deposita, *12,* 307–317.

Ianovici, V., Borcoş, M. (1982): Romania, Mineral deposits of Europe, 2: South-east Europe. Min. Soc., Inst. Min. Met., London, 142 p.

Motaş, C. (1983): Nouvelles données sur les rapports structuraux entre les Carpathes Méridionales et la Dépression Gétique. An. Inst. Geol. Geofiz., LX, Lucr. Cgr. XII Asoc. Geol. Carp.-Balc., 141–146.

Pearce, J. A., Gale, G. H. (1977): Identification of ore deposition environment from trace element geochemistry of associated igneous host rocks. Volc. Proc. Ore genesis, IMM and GSL, Proc. joint Meet. Volc. Stud. Group, 14–24.

Rădulescu, D., Săndulescu, M. (1973): The plate-tectonic concept and the geological structure of the Carpathians. Tectonophysics, *16,* 155–161.

Rădulescu, D., Borcoş, M., Peltz, S., Istrate, G. (1981): Subduction magmatism in Romanian Carpathians. Guide ex. A 3, Carp.-Balk. Geol. Assoc. 12th Cgr., Bucharest, 132 p.

Săndulescu, M. (1983): Le problème de la marge continentale européenne dans l-aréal Carpatho-Balkanique. An. Inst. Geol. Geofiz., LX, Lucr. Cgr. XIII Asoc. Geol. Carp.-Balc., 199–208.

Săndulescu, M., Visarion, M. (1978): Considérations sur la structure du soubassement de la Dépression de la Transylvanie. D. S. Inst. Geol. Geofiz., LXIV, *5,* 153–173.

Vinogradov, C., Barbu, I. Z., Hesselman, A. (1964): Contribution to the study of the iron sedimentary deposit of Căpuş (Cluj region). Rev. roum. géol. géogr., sér. Géol., *8,* 21–32.

Vlad, Ş. (1978): Metalogeneza triasică a zonei Tulcea (Dobrogea de nord). St. cerc. geol. geof. geogr., ser. Geol., *23/2,* 249–258.

Vlad, Ş. (1979): A survey of Banatitic (Laramian) metallogeny in the Banat region. Rev. roum. géol. géoph. géogr., sér. Géol., *23/1,* 39–44.

Vlad, Ş. (1980): Alpine plume-generated triple junction in the Apuseni Mountains and metallogenetic implications. Rev. roum. géol. géoph. géogr., sér. Géol., *24,* 71–81.

Vlad, Ş. (1983): Geologia zăcămintelor porphyry copper. Ed. Acad. Bucureşti, 156 p.

Vlad, Ş., Dinică, I. (1984): Vein mineralization of the Leaota Mountains – preliminary considerations. Rev. roum. géol. géoph. géogr., sér. géol., *28,* p. 35–41.

The Metallogeny of the Eastern Alps in Context with the Circum-Mediterranean Metallogeny

By W. E. PETRASCHECK*)

Zusammenfassung

Trotz der Erkenntnis einer großen Zahl paläozoischer Erzvorkommen in den Ostalpen ist auf die Anzeichen einer jungen (80–100 my) Metamorphose und Mineralneubildung hinzuweisen. Ein Forschungsprojekt „Erzmobilisierung und alpine Metamorphose" wird bearbeitet. Die Ostalpine Metallogenese wird mit der plattentektonischen Entwicklung des Gebirges in Zusammenhang gebracht und eine Sonderstellung der Ostalpen im alpin-mediterranen Orogen für unwahrscheinlich erklärt. Die Besonderheit der alpinen Metallogenese liegt vielmehr im spezifischen Charakter des tektonischen Baus der Alpen begründet: flache Subduktionen, die nur den obersten Mantel ergriffen haben, und eine ungewöhnliche Aufstapelung von Decken innerhalb der kontinentalen Kruste, durch die das Aufdringen kretazisch-tertiärer Magmen und erzbringender Lösungen erschwert wurde.

In a publication 20 years ago, I subdivided the alpine-Mediterranean Metallogenic Belt into three ore provinces: 1. The East-mediterranean province, comprising the Balkan-Peninsula and Anatolia, 2. the central province, consisting of the Western Carpathians and the Alps, and 3. the West-mediterranean province with the Appenines, the Betic Cordillere and the Atlas Chain. Each of these ore provinces is distinguished by its specific style: the Eastern one by the great number of mineral deposits apparently associated with various magmatic rocks of mesozoic and tertiary age, the central one characterized by an abundance of so-called telethermal deposits with almost no visible magmatic affiliation and the Western one, showing again connection with tertiary magmatites, but also deposits of the platform type. The

*) Erdwissenschaftl. Kommissionen. Österr. Akademie d. Wissenschaften, A-1010 Wien, Dr.-Ignaz-Seipel-Platz 2.

whole Metallogenic Belt was considered to be of predominantly alpidic age (PETRASCHECK, 1964).

Since then new observations have been made, new methods applied, and a new philosophy came in vogue, which particularly has changed our view of the ore genesis in the Alps. However by the new knowledge things became less clear, more contradictions appeared, more open questions arised. In order to articulate the problem, three topics have to be dealt with: the age of the deposits, their origin and their relation to the alpine mediterranean plate tectonics.

The question of the age is the most controversial one. The majority of the deposits is now considered to be palaeozoic. About 80% of the 1500 ore occurences registered in O. M. FRIEDRICH's map (1953) are situated in palaeozoic, respectively metamorphic-palaeozoic rocks; they are classified as volcano-sedimentary, formed in Ordovician till Devonian geosynclinal troughs (cf. SCHULZ, 1974 and later, POHL, 1984 and many others). To this group are attributed the scheelite deposit of Mittersill, the lead-zinc-barite layers in the palaeozoic area near Graz, many sulphide beds in the Grauwackenzone, but also the siderite stock of the Erzberg and the copper ore layers and veins of Mühlbach-Mitterberg in Salzburg. Wherever younger epigenetic features are observable or the mineralization is overlapping into post-palaeozoic strata, the phenomena are explained by mobilization. This delicate problem was discussed for different districts and by different authors in 1981 in Leoben (Ore Mobilization in the Alps and in SE-Europe, Schriftenreihe Erdwiss. Komm. Österr. Ak. Wiss., Vol. 6, Vienna 1983). The results of the discussion were ambiguous.

The "Mobilized" ores are always showing a somewhat different chemical composition than the primary ones—which seems quite natural. But the chemistry of the procedure of dissolution and redeposition of many ore minerals is not understood. Tetrahedrite f. g., occuring in Devonian dolomites in Schwaz, as well as in the adjacent Triassic Muschelkalk, is, according to hydrometallurgical experience, practically insoluble in a reducing environment; a dissolution and decomposition of tetrahedrite to its individual metallic components can only be achieved under high oxygen pressure (WÖBKING, 1983, loc. cit.). Recently in a druse of siderite from the Erzberg, supposed to be a remobilisate from sedimentary siderite, crystals of Xenotym and Circon were found (WALTER and POSTLER, 1984). Ankerite crystals rich in iron were formed at a temperature of almost 400° C (BERAN). Thus the remobilizing fluids were not always harmless chlorine formation waters. Particularly the giant stock of iron carbonate in the palaeozoic limestones of the Erzberg with its famous, large scale replacement structures, locally even ingressing into the overlaying Permo-Triassic sandstones, gives the impression of a throughout penetration by iron bearing solutions. Why, in the case of such an almost complete "remobilizing" environment, local thin layers of iron carbonate should have preserved their structure and even their chemical differentiation from grain to grain (BERAN, 1983)? Would it not be possible that these rare sedimentary occurences are belonging to an entirely different ore forming period—may be an older one, and that the bulk of the ore is the product of a later hydrothermal mineralization? Or in contrary, that the sedimentary ores are younger fillings of palaeo-karstic caves within the siderite stock, like recently assumed by MOSTLER (1984)?

In spite of these sceptical objections many convincing examples of ore remobilization in alpine deposits are described in the proceedings of the above mentioned Symposium.

An interesting research project was started a few years ago by H. WENINGER, 1980, but interrupted by his tragic death: it was the investigation of fluid inclusions in minerals of alpine ore veins. It would be worth studying the chemical composition and the homogenization temperatures of the primary and of the remobilized minerals.

The age of the mobilization is already known in a few cases. The important copper ore vein of Mühlbach-Mitterberg, situated in Silurian schists and crosscutting their stratification and also their transversal schistosity, passes over into the overlaying Permian sandstones with reduced thickness and impoverished mineral content; the higher section of the vein could indeed be a remobilized prolongation. These Permian sandstones contain a few small layers of sedimentary uranium ore, but along the salbands of the mobilized ore vein nodules of uraninite, up to one centimeter in diameter, sometimes with native gold in their cracks, were found (SIEGL, 1972); in the deeper levels of the main vein no trace of uranium at all could be detected. Thus, the uraninite nodules were formed by dissolution and redeposition of the sedimentary ore, when the remobilizing metalliferous solutions pervaded the Permian. An absolute age determination of the uraninite nodules, carried out by Prof. KÖPPEL, pointed to 90 ma, thus indicating the age of the remobilization.

The austroalpine Grobgneis of the Eastern Central Alps is in several places traversed by mylonitic zones containing the so called Weisschiefer, a mixture of chlorite, serizite and quartz; compared with the rather homogenous Grobgneis, the Weisschiefer is showing a certain addition of magnesium. The age of the mylonitization and of the Weisschiefer is alpidic. The supply of magnesium reminds of the underlying Penninic ophiolites, from which magnesium could have been set free by serpentinization (MODSTAHEDI and WIESENEDER, 1974). Anyhow, it must be admitted that typical indicator elements for this provenience, like Co and Nö, are missing (PROCHASKA, 1984).

The siderite deposit of Knappenberg-Hüttenberg (Carinthia), by strong geological arguments declared as epigenetic and alpidic (CLAR, 1981), was also recently considered as synsedimentary-palaeozoic. This deposit contains rare accessory uranium ore, which shows an U/Pb-ratio, possibly pointing to about 70 ma. (Determination by Prof. KÖPPEL according to communication by Prof. PAAR.)

A red haematitic pigment, generated by metamorphism of the palaeozoic magnesites of Entachen (Salzburg), has been proved to be Cretaceous by palaeomagnetic measurements (MAURITSCH, 1980).

These few examples are demonstrating various chemical and thermal reactions in ore deposits of the Eastern Alps during a time span between 70 and 120 ma, that is between Upper Cretaceous and Palaeocene. The same period was attributed to a wide spread metamorphism in the Eastern Central Alps recognized by FRANK and collaborators (1983) on the basis of the K/Ar and Rb/Sr-method. Several regions, g. i. on both sides of the Lavanttal in Carinthia and in SW-Tyrol are showing the amphibolite facies, which means a temperature of about 500° C, whereas the greater parts of the Central Alps, including the southernmost marginal parts of the Kalkalpen, have been affected by the greenschist-facies (around 350° C).

From all these findings we may conclude that a regional thermal and in some way also metallogenic event has occured during this epoch and that the "outdated" idea of an Upper Cretaceous-Tertiary metallogeny in the Alps was not so misleading. This event probably depends on the amount of fluids which have been set free by the metamorphism (Frank).

A recent research project with the scope of correlating the alpidic metamorphism and the ore mobilization can perhaps reconsile the theories of the paleozoic and of the alpidic ore formation.

The origin of the ore metals should be discussed in context with the mediterranean plate-tectonic evolution. In a recent publication Pohl (1984) gave a synopsis of the paleozoic ore deposits in the Eastern Alps; he attributed them to different magmatic phases of diabases and keratophyres. Although in Pohl's view, periods of crustal extension and compression were alternating, no convincing evidence for a real large palaeozoic ocean floor in the main territory of the Alps was demonstrated. I myself am of the opinion—different from Stumpfl and El Ageed (1981)—that the ultramafites of Kraubath and Hochgrössen, containing insignificant occurrences of chromite and asbestos, are not remnants of an ocean floor, but local intrusives from the upper mantle into the continental crust—analogous to the palaeozoic serpentinites in the Eastern Rhodopi Mts. in Bulgaria—because they are not associated with the typical ophiolitic "Steinmann Trinity" of gabbros, diabases and particularly cherts. A real oceanic realm between the alpine area and Africa from Upper Ordovician till Lower Carboniferous may have existed only before the fusion to Pangaea (Frisch et al., 1984). The area of the Alps, from the Cambrium till Permian consisted of a mobile continental crust subdivided by intracontinental troughs ("aulacogens"), similar to the area of Central Europe. The crust was comparatively thin, particularly during the geosynclinal phases. Therefore we have the typical bimodal magmatism and the polymetallic ore association of an intracontinental metallognesis (Sawkins, 1982). W and Be-bearing minerals are indicators of a continental provenience.

A striking phenomenon is the scarcity of ore deposits around the Hercynian granites in the Alps—quite different from the granitic massifs in Central and Western Europe. The same phenomenon applies to the granites in the Slovakian Carpathians and the South Bulgarian granites in the Rhodopi Mts. Oulianoff (1943) once has tried to explain this by a kind of destillation process during the alpidic metamorphism. I am not inclined to accept this hypothesis, because the synsedimentary deposits of palaeozoic age have safely survived the alpidic metamorphism. I would prefer the explanation by Ilavsky (1977), given for the Carpathians, namely that the frequency of the deposits decreases with the thickening of the continental crust which has reached a maximum after the carboniferous compression phase. The only minable fruit of the hercanian orogenesis is an important group of spodumen-pegmatite veins within the crystalline schists of the Koralpe in the Eastern Central Alps.

The ore deposits of the Permian, according to Haditsch and Mostler (1974) and Drovenik et al. (1980) are partly related to the subsequent extrusive magmatism (Cu, U), partly to exogenic enrichment by erosion and sedimentation in post-orogenic basins (U and evaporites) (ERKAN, 1977).

The break up of the Euro-African land mass began in Triassic time with the formation of tension fractures for which the name rifts—as sometimes used—seems not quite adequate. These zones are characterized by porphyritic, partly also diabasic volcanism, by an inquiescent and shallow seawater sedimentation and by Fe-Mn-Pb-Zn-Hg deposits of syngenetic, as well as of epigenetic nature. To these zones belong the deposits of the Northern Dobrogea, of the Peleponnes, of Montenegro (Brskovo), Bosnia (Vareš, Suplija Stjena) and Slovenja (Idrija). The original tectonic features of these "rifts" were completely disturbed by the later compression phases.

Into this group the numerous lead-zinc deposits on both sides of the Alpine-Dinaric cicatrice can be encompassed (Bleiberg, Raibl, Mežica, Salafossa etc.). They are products of a mineralization along an incipient rift which much later was revitalized and transformed into the Periadriatic Lineament. The Triassic age of the ores is now without any doubt; syndiagenetic and epigenetic features are occuring together (SCHULZ, 1983). The Pb-isotopes point to a provenience from the Palaeozoic basement, from which the metals probably were leached by saline brines (KÖPPEL). A direct connection with the Triassic diabases can be excluded, but a magmatic heat source is likely. A clear distinction between ore veins, formed per ascensum and veins formed by remobilization (if ever there existed such ones) is not yet possible. The similar deposits in the Northern Kalkalpen in Tyrol and Bavaria can be best explained by later nappe movements from the South.

The opening of the Tethyan ocean started in Upper Jurassic time and continued until the Lower Cretaceous with the formation of the South- and North Penninic troughs. How large these troughs have been, is under dispute; to my opinion they were "microoceans", similar to those in Greece. But there is a remarkable difference regarding the metallogeny: whereas the ophiolitic assembledge in Greece, Albania, Yougoslavia and Turkey is extremely rich in chromium-, copper- and nickel ores, the alpine ophiolites are poorly mineralized. Only small copper-pyrite occurences are existing in the Tauern Window.

The reason for this may be found in the fact that the alpine underthrusting planes were shallow, flat, intracrustal (TRÜMPY). They did not plough deeply into the upper mantle; there were no real subduction zones operating as metallotects, as EVANS stated already in 1975.

The Upper Cretaceous compression periods which lasted until Mid Tertiary were presumably connected with remelting of the deeper parts of the continental crust and with simultaneous metamorphism. Upper Cretaceous granites, so frequent in the SE-Carpathians and Balkan mountain chains, are missing in the Alps. A remarkable link however between the non-magmatic Cretaceous metamorphism and ore mobilization in the Alps and the magmatogenic ore formation in SE-Europe are the small massifs of granite, 110 ma old, in the Gemerske Rudno Horjie of the Slovakian Carpathians. In this district the assembledge of ores—siderite, chalcopyrite, tetrahedrite, magnesite, accessory cinnabar—is almost identical with that of the Grauwackenzone. A siderite vein was even found within one of these Cretaceous granites (oral communication by Mr. BERNARD). Anyhow, the majority of these deposits is now considered to be Palaeozoic. But until now none of the adherents of a Palaeozoic ore formation in the Slovakian Carpathians has discussed the fact in the nearby situated siderite deposit of Rudabanya in Northern Hungary only Triassic Muschelkalk has been mineralized by the same ore association as as known from the Palaeozoic deposits in the Gemerides.

The idea of a combined magmatogenic-metamorphogenic ore formation in the Alps, a long time ago expressed by CLAR, FRIEDRICH and others, was later abandoned when migrating fronts became out of vogue and metamorphism was regarded as isochemical. I would however like to draw the attention to the occasional occurence of fuchsite in white Triassic quarzites and marbles of the Lower Austroalpine sequence, lying above the Penninic ophiolites. This evidence of a migration of chromium may indicate the possibility of other migrating elements (Mg, Fe, Au, Cu?). A study of the regional distribution of these secondary fuchsite occurences with regard to the position of the ophiolites would be of interest. KARAMATA (1983) has shown that Cu, Co and Cr have been incorporated into Tertiary magmas and their associated ores, when having passed through the ophiolites of the Vardar Zone. In the Slovakian Carpathians fuchsithe represents the earliest stage of the mineralization; its chromium is derived from neighbouring Mocks (P. IVAN 1982). So the problem of metamorphism and ore remobilization arises again.

A Tertiary metallogenetic epoch in the Alps is more clearly established than the Cretaceous one. The most manifest examples are the gold veins in the Hohen Tauern, which intersect all the alpidic nappe structures. Probably of Eocene age are the siderite deposits in the Central Alps of Styria and Carynthia, as they are showing a post-tectonic fabric and are locally penetrating into Triassic limestones. Porphyritic dykes have been brought into genetic association with this mineralization (FRIEDRICH). The Tertiary Periadriatic massifs are remarkably sterile, perhaps because they are the products of continental-crustal collision. The long stibnite veins of Schlaining on the most Eastern end of the Central Alps might be related to an ill defined andesite in the neighbourhood. Their hostrocks are cretaceous anyway.

Conclusions

This article certainly gives the impression of the rather personal authors attempt of maintaining his concept about the alpine metallogeny, expressed 10 and 20 years ago, of repeating his arguments and of adapting his theory to the new findings and ideas. This indeed is the case, because the proponents of the unitaristic Palaeozoic and sedimentary metallogeny, inspite of their new and irrefutable observations, almost never have tried to disprove and to explain the opposite arguments. Therefore in this article a few suggestions for research projects have been made with the aim of finding elements to bridge over the two concepts.

My main reason for an alpidic metallogeny in the Eastern Alps is the fact that this mountain chain is a limb of the giant Mediterranean-Himalayan orogenic belt, which has been composed and formed in Mesozoic-Tertiary time with all the phenomena of magmatism usually related to plate tectonic events. Everywhere else this magmatism is associated with metallogenic phases. Promising steps have been made to explain the character of the alpine-mediterranean metallogeny in the frame of the concept of plate tectonics (PETRASCHECK, 1974, EVANS, 1975, JANKOVIĆ, 1977). These publications were generously ignored by FINLOW-BATES and TISCHLER (1981) in their complex synthesis of the alpidic mineralization.

The metallogenic zoning in the whole belt from Gibraltar till Western Asia, the Alps included, is generally the same: predominance of copper in the Northern branch in connection with subduction zones and ophiolites, predominance of lead-zinc in the Southern branch, inherited from the northern margin of the African Plate. Why should the genesis of the ores in the Eastern Alps be a unique and specific Austrian exception?

The pecularity of the alpine metallogeny is conditioned by the specific character of the tectonic structure of the Alps: flat subductions which have affected only the uppermost parts of the mantle and an unusual accumulation of nappes within the continental crust which has prevented the rising of cretaceous-tertiary magmas.

References

CLAR, E., u. MEIXNER, H.: Die grundlegenden Beobachtungen zur Entstehung der Eisenspatlagerstätten von Hüttenberg. Carinthia II 171/191, p. 55–92, Klagenfurt 1981.

ERKAN, E.: Uran- und gipsführendes Perm in den östlichen Ostalpen. – B. u. H. Monatshefte 122. Jg., p. 9–17, Wien 1977.

EVANS, A. M.: Mineralizationin Geosynclines – the Alpine Enigma. – Min. Dep. 10, p. 254–260, Berlin 1975.

FINLOW-BATES, T., and TISCHLER, S. E.: Contols on Alpidic Mineralization Styles. Proc. IV ISMIDA, Heidelberg 1983.

FRANK, W.: Argumente für ein neues Entwicklungsmodell der Ostalpen. Jahresbericht Hochschulschwerpunkt 1982, p. 249–252, Graz 1983.

FRIEDRICH, O. M.: Erzlagerstättenkarte der Ostalpen. Radex Rundschau 7/8, Radenthein 1953.

FRIEDRICH, O. M.: Die Vererzung der Ostalpen, gesehen als Glied des Gebirgsbaues. Archiv f. Lagerstättenforschung 8, S. 1–136, Leoben 1968.

FRISCH, W., NEUBAUER, F., u. SATIR, M.: Concepts of the evolution of the Austroalpine basement complex during the Caledonian-Variscan cycle. – Geol. Rdsch., Vol. 73, p. 47–68, Stuttgart 1984.

HADITSCH, G. H., u. MOSTLER, H.: Mineralisationen im Perm der Ostalpen. – Carinthia II, 84. Jg., S. 63–71, Klagenfurt 1974.

ILAVSKY, J.: Technique globale et metallogenie dans les Carpathes occidentales. – Tschechoslovakie – Metallogeny and Plate Tectonics in the NE-Mediterranean. Ed. S. Janković, Belgrade 1977.

IVAN, P.: Position and significance of fuchsite in evolution of mineralization of the Spiško-Gemerske Rudohorie M^{ts}. Acta geol.-geogr. Univ. Com. 38, Bratislava 1982.

JANKOVIĆ, S.: Major alpine ore deposits and metallogenic units in the NE-Mediterranean and concepts of Plate Tectonics. *In:* Vol. IGCP Proj. 3, Faculty of Mining and Geology, p. 105–171, Belgrade 1977.

Karamata, St.: The Influence of Opholite Complexes on the geochemistry and metallogeny of Tertiary igneous rocks in central parts of the Balkan Peninsula. Schriftenreihe d. Erdwiss. Komm. d. Österr. Akad. Wiss., Bd. 6, p. 93–102, Wien 1983.

Mauritsch, H. J.: Paläomagnetische Untersuchungen an einigen Magnesiten aus der westlichen Grauwackenzone. – Mittlg. Österr. Geol. Ges. 73, p. 1–4, Wien 1980.

Modstahedi, M., u. Wiesenweder, H.: Entstehung u. Zusammensetzung der Leukophillite in den Ostalpen. Archiv f. Lagerstättenforschung, p. 189–214, Leoben 1974.

Mostler, H.: An jungpaläozoischen Karst gebundene Vererzungen mit einem Beitrag zur Genese der Siderite des Steirischen Erzberges. Geol.-paläont. Mittlg., Vol. 13, p. 97–111, Innsbruck 1984.

Oulianoff, N.: Pourquoi les Alpes Suisses sont pauvres en gites metalliferes exploitables. Bull. Lab. Geol. Geogr. Phys. Univ. Lausanne, 1943.

Petrascheck, W. E.: Die alpin-mediterrane Metallogenese. Geol. Rdsch. 53, p. 267–289, Stuttgart 1964.

– Alpine Metallogeny and Plate Tectonics – still a problematic correlation. – IV. Symposium of IAGOD, Vol. II, 504–509, Varina 1974.

– Introduction to: Mineral Deposits of Europe, Vol. 2 South Eastern Europe, ed. F. W. Dunning et al., Inst. Min. & Met., London 1982.

Pohl, W.: Metallogenic evolution of the East Alpine Palaeozoic basement. Geol. Rdsch., 73, p. 131–147, Stuttgart 1984.

Postl, W. u. Walter, F. Xenotym and Monazit vom Steirischen Erzberg. Mittlg. Abt. Min. Landes.-Mus. Joanneum 51, Graz 1983.

Prochlaska, W.: Neue Geochemische Aspekte zur Genese der Talklagerstätte Rabenwald. BHM vgl. 129, p. 457–462, Wien 1984.

Sawkins, F. J.: Metallogenesis in relation to rifting. – Geodynamics Series, Vol. 8, 1982.

Schulz, O.: Metallogenese im Paläozoikum der Ostalpen. Geol. Rdsch. 63, p. 93–104, Stuttgart 1974.

Stumpfl, E. F., u. El Agad, A.: Hochgrößen u. Kraubath, Teil eines paläozoischen Ophiolit-Komplexes. Mitt. Abt. Geol. Bergb. Landesmuseum Joanneum 42, p. 161–169, Graz 1981.

Trümpy, R.: Tectonic problems of the Alpine System. – Veda, Slovac. Ac. Sc. Bratislava 1975.

Weninger, H.: Mikrothermische Untersuchung an ostalpinen Lagerstätten. T.M.P.M. 29, p. 127–144, Wien 1981.

Geotectonic Evolution and Metallogeny of Italy in the Tethys Realm: A Summary of Data and Hypotheses

By the Italian W. G. for I. G. C. P. Project 169
Ed. P. ZUFFARDI*)

With 3 Figures

Zusammenfassung

Der tektonische Aufbau Italiens und die plattentektonische Entwicklung der Apennin-Halbinsel sowie der benachbarten Inseln wird zusammenfassend beschrieben. Fig. 1 stellt die komplexe Zusammensetzung des Gebietes aus Teilen der europäischen und der afrikanischen Platte dar. Das Wechselspiel von permo-triassischem Rifting, jurassischem Seafloor-Spreading und kretazisch-tertiärer Kollisions- und Kompressionstektonik findet Entsprechung in den magmatischen (Fig. 2) und metallogenetischen Produkten (Fig. 3). Die metallogenetische Armut der italienischen Ofiolithe im Gegensatz zum Erzreichtum der ostmediterranen wird hervorgehoben. Die wirtschaftlich wichtigsten Lagerstätten wurden im Jungtertiär und im Quartiär gebildet.

1. Foreword

Many studies have been carried out till now and are carried on at present about the geo-dynamic evolution of Italy and of the Mediterranean region. Recently (1984) a volume on this topic has been published at the Italian Geological Society.

A number of hypotheses have been put forward; of course someones fit in each others, someones are in disagreement; this fact is connected both to the intrinsec difficulties of the area and to the lack of sufficient data.

The working group charged of preparing the various chapters of the Italian report for this session of IGCP 169 (a short resumé of which is presented here) preferred to summarize the most relevant knowledge on the theme, without venturing

*) Dipartimento di Geologia, Università di Milano.

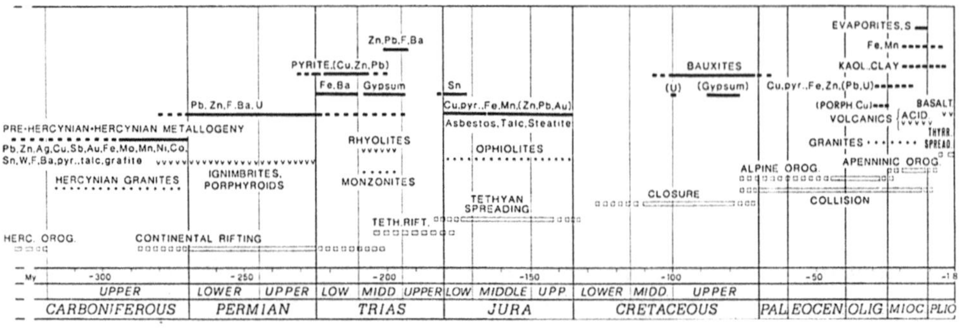

Fig. 2. Schematic chronological sequence of the geo-dynamic events and metallogeny from Middle-Upper Carboniferous to Pliocene.

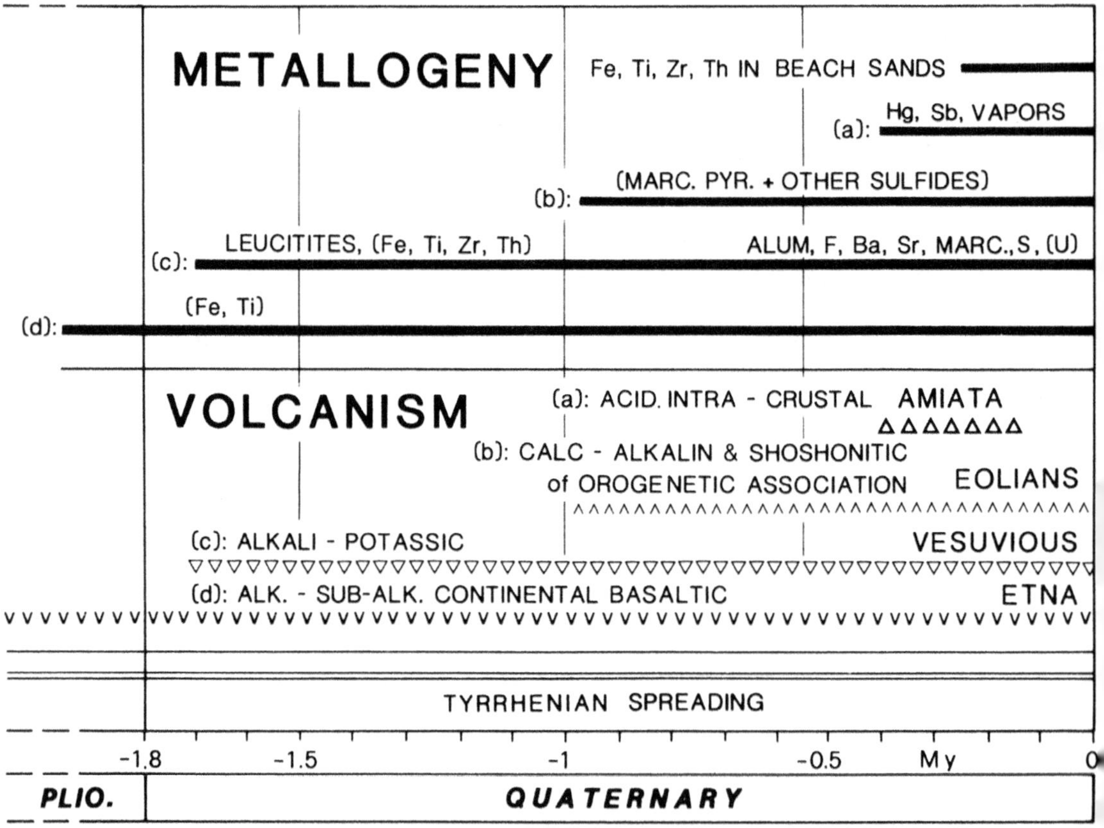

Fig. 3. Schematic chronological sequence of the geo-dynamic events and metallogeny from Pliocene up to Actual times.

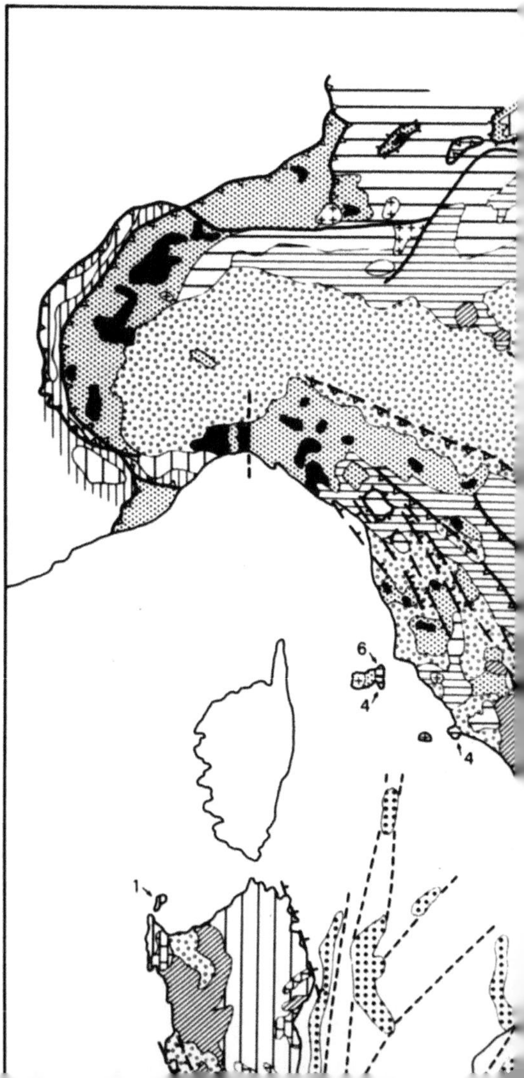
6
4
4
1

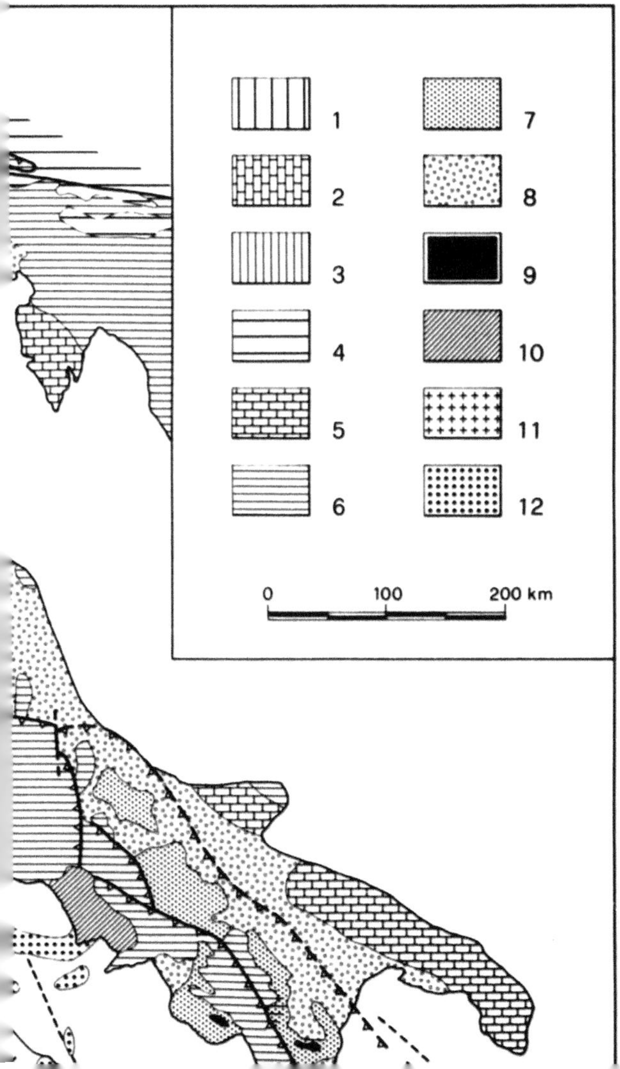
1
2
3
4
5
6
7
8
9
10
11
12
0
100
200 km

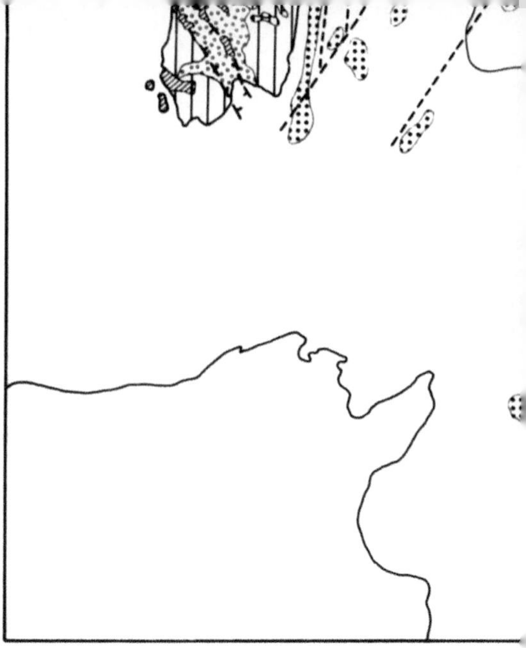

Fig. 1. Geo-tectonic map of Italy (after Parotto &
EUROPEAN PLATE: 1, 2, 3
AFRICAN PLATE: 4, 5, 6
1 and 4 = Hercynian/Pre -Hercynian basements an
or littoral, covers.
2, 3, 5, 6 = Mesozoic (from Middle Triassic) and (p
thrusted and/or folded.
INTERNAL UNITS: 7 = Flyshoid complexes often
bonate platforms. They are called "calc-schists
"Sicilides" (non ophiolitiferous) in the Apennine
the "calc-schists and greenstones". 8 = Molass
MAGMATISM: 9 – Ophiolites.
10 = Tertiary and Quaternary undifferentiated vo
11 = Alpidic granites.
ELEMENTS OF MARINE GEOLOGY: 12 = Sea-
basement, and locally of Mesozoic cover and/

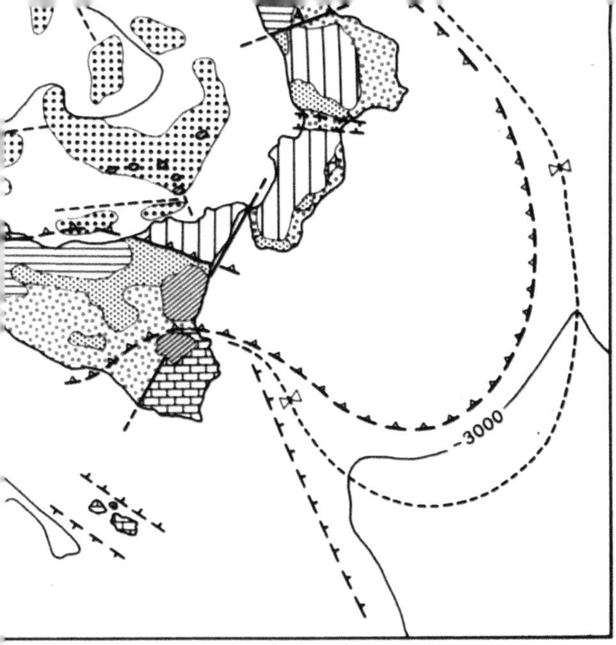

80, simplified).

hyan (Permian to Lower/Middle Triassic), continental and/

ry covers; 2 and 5 = tabular, autochtonous; 3 and 6 = over-

frequently including masses of ophiolites (9), and locally car-
es" in the Alps, "scaly clays", "Ligurides" (ophiolitiferous),
ns of the basement (not reported in the map) outcrop within
rogenic deposits.

ig. 2 and 3).

lly volcanic (mainly tholeiitic), partially made of crystalline
s.

general theories, just elucidating some more or less obvious deductions, and pointing out the doubts that still remain.

The partecipants in the Italian working group are:

A. BOSELLINI, L. BRIGO, T. COCOZZA, F. A. DECANDIA, A. FERRARIO, R. GELATI, A. MARCELLO, P. OMENETTO, A. PRATURLON, S. PRETTI, I. SALVADORI, I. URAS, R. VALERA, E. ZANETTIN-LORENZONI, P. ZUFFARDI.

2. Pre-Tethyan events

We will not enter into details on this point, which is rather outside the aims of this paper. We will just recall that two main tectonic-metamorphic-magmatic-metallogenic pre-Tethian groups of events have to be mentioned; namely:

2.1. Hercynian and pre-Hercynian

Strong effects of the Hercynian tectonics and metamorphism*) are known in the Alps, in Sardinia, in Tuscany, in the Calabrian-Peloritanian Arc. Hercynian granitoid rocks are known in the same region except Tuscany; Pb, Zn, Ag, Cu, Sb, Au, Fe, Mo, Mn, Ni, Co, Sn, W, F, Ba, pyrite, talc, graphite accumulations generated in those times; some of them are very rich in Sardinia (MONTEVECCHIO and MONTEPONI Pb, Zn deposits, for instance), someones are fairly rich in the Alps; they are not so important in the Calabrian-Peloritanian Arc, no deposit of this age is known in Tuscany.

The pre-Hercynian sequences are not the same in the above-mentioned regions; some remarks may be of interest, from the metallogenic point of view, about these terrains. In fact, it has to be noted that the Cambrian is well developed only in Sardinia, where the metallogenically important platform carbonate complex (meaningfully called "Metalliferous") crops out over a vast area, in the South-West section of the Island. Recently Cambro-Ordovician terrains have been recognized in Calabria on the basis of micro-paleontological evidence, but they are terrigenous and barren.

The differences in the distribution of Cambrian (particularly of the Metalliferous and of its huge stock of metals) may be a reason for the different mining importance of the above-mentioned regions. In any case the correlation among the various sections of the Italian Hercynian/pre-Hercynian belt and of their metallogenic districts is much more complex than it appears at a first analysis; considering them simply as various pieces to be shouldered together may became dangerous for prospecting purposes: suffice it to recall the disappointments experienced in Calabria, when one was hoping to find there deposits of Sardinian type.

---

*) The presence of Caledonian effects is controversial: the only undebatable case of pre-Hercynian tectonics is shown by the Cambro-Ordovician unconformity in Sardinia; the latter was not associated with metamorphism and magmatism, but deserves a certain metallogenic interest (karstic Pb, Ba accumulations).

2.2. Post-Hercynian/Pre-Tethyan

A cycle of continental rifting and of acidic volcanism took place after the end of the Hercynian orogenesis (Upper Carboniferous) till Lower-Middle Trias.

It is very developed all along the Alpin chain, where some wide complexes have been generated (such as the so-called "porphyric platform of Adige Valley"); it occurs in two different units, namely: Pennines and Southern Alps. The fact that Tethyan ophiolites are part of the first unit and not in the second, may be considered an evidence for non-correlation between the events under discussion and the Tethyan ones.

Similar cycles are present in South-Eastern Sardinia and in Northern Tuscany; they are not known with certainty elsewhere.

This event is not to be considered as pre-Tethyan rifting, but rather a "tail" of the Hercynian cycle. U, Fe, Zn, Pb, F, Ba, pyrite accumulations generated (in part by remobilization of (pre)-Hercynian pre-concentrations – of Pb, Zn in particular –) in connection to this cycle. Some of these deposits are important in Tuscany (pyrite deposits of GROSSETO area) and in the Alps (U of NOVAZZA, VALVEDELLO); no similar deposit is known in Sardinia.

3. Tethyan events

The evolution of the Italian territory and the adjoining seas was not the same: there are differences between Western and Eastern Alps, between Northern and Southern Apennines, between Sardinia and the other parts of Italy. A schematization to extremes is here described.

3.1. Tethyan rifting

The Tethyan rifting started in Middle Trias, with particular evidence in the Carnian Alps; far less so (or even debatable) in other areas (Liguria, Northern Tuscany, Lucania, Calabria, Sicily).

This rifting lasted till Upper Trias-Lias, as it is proved by the presence of basinal areas originated in Middle Liassic time by "drowning" of large sectors of Liassic-Triassic shallow water domains. Oceanization started in the same moment, since the first oceanic mafites are dated as Liassic.

A scanty magmatic activity is related to this event: the monzonites and the mainly acidic lavas of Eastern Alps are the main instances; minor volcanic complexes of the same age occur in the other above-mentioned regions.

Sedimentary deposits of Zn, Pb, F, Ba, Sn originated in connection to this rifting; they are important in the Central and the Eastern Alps (among them, the Zn, Pb deposits of RAIBL and of SALAFOSSA), not so important, or inexistent, in the other regions.

3.2. Oceanization and ophiolitic volcanism

Oceanization should have started in the Upper Trias, according to the facies of the Norian sediments of the Eastern section of the Southern Alps; the first deposition of mafic rocks – as it was said in the previous chapter – took place in the Lias (the absolute ages being 185–160 My); but the climax of spreading and of the ophiolitic volcanism should have been Middle-Upper Jura; soon after the oceanic closure started.

The intense deformations caused by the subsequent Alpidic-Apeninic orogeneses make the palinspastic reconstruction of the Tethyan domain in Italy difficult. Some characteristies of the ophiolitic sequences can help for this purpose; it has to be remembered that the distribution of the ophiolites in Italy is uneven, so far as both their geographical location and their origin are concerned (and probably their ages).

As a matter of fact, (i) typical oceanic ophiolites occur in Central-Eastern Liguria, in Tuscany and episodically along the Northern border of the Emilian Apennines. The ophiolite bearing sequences are strongly tectonized and displaced as olistolites of various dimension in the last two regions. In any case this group of ophiolites is identical with those of North-Eastern Corsica. (ii) ophiolites, defiled by and/or interbedded with terrigenous sediments, occur in Calabria and along the Alps (in this latter region they make part of the so-called "Calc-schists with greenstones" complex, equivalent to the French "schistes lustrés"). A deduction ensues immediately: the origin of the latter type of ophiolites should have been non-oceanic or at last – should have been closer to a continental margin, than the ones of Liguria and Tuscany. (iii) no evidence of ophiolites or of comparable rocks is known in Sardinia.

These observations suggest that Central-Eastern Liguria (particularly the Bracco zone, according to a more careful examination), should have occupied a middle oceanic ridge type position in the Tethys. This same structure should have been the source of the ophiolite bearing olistostromes of Tuscany and of Emilia.

The "greenstones" of the Alps and the ophiolites of Calabria should outline a belt more or less close to a continental margin (may be at least in part, running in back arc position?).

Sardinia should represent a piece of the (European) continental area in the Tethys domain. As a matter of fact comparative studies of the Mesozoic sediments in the Provence, in the Sardinian-Corsican Massif, in Western Tuscany, suggest that the Massif played a rôle of an emerged craton separated from the Provence by a narrow intracratonic basin with sediments of Germanic facies, and separated from Tuscany by an epi-continental sea, connected with the Triassic Alpine open sea basins.

The tearing of the sialic crust and the Jurassic-Cretaceous marine transgression gave cause to the formation of the "Ligurian-Piedemontese oceanic basin" in the first phase and to the "Paleo-Tyrrhenian" in the second: they represent the Tethys in this area. On the whole it is current opinion that the Tethys, in our area, was not more than a narrow oceanic basin, more or less a strait widening out eastwards, connecting the Atlantic with the eastern paleo-ocean.

A number of deposits of Cu, pyrite, Fe, Mn, (Zn), (Pb), (Au), asbestos, talc and steatite, of various sizes and grades generated in connection with the ophiolitic volcanism, but the only industrially important one is the huge asbestos deposit of BALANGERO (Western Alps); all others (even the historically important Cu ore-body of MONTECATINI included) are modest.

The fact that the Italian "oceanic" ophiolites are lithologically alike the Cr, Ni, Co, Pt bearing ones in the other sections of the Mediterranean, but containing only traces of these metals, represents a puzzling problem from the scientific standpoint and a disappointment from the industrial one.

The only example of a probable remobilization and heredity, connected with the ophiolitic volcanism, is represented by the Au content of some pyrite-Cu deposits and by some (small) Au-pyrite (and minor others sulfides) veinlets occurring in the ophiolitic complexes close to Au bearing sections of the basement: the main localities occur in Valsesia (ALAGNA), in Aosta Valley (BRUSSON) and in Central Liguria (OVADA).

3.3. Closure and collistional stages

As already mentioned, the closure of Tethys started during the Lower Cretaceous collision, initiating the Alpidic-Apenninic orogenesis which began in late Cretaceous-Paleocene and can be considered as terminated in Upper Miocene-Lower Pliocene.

It has to be said that collisional phenomena are yet going on, at present, in the Southern Tyrrhenian, but they have to be considered as a Post-Tethyan event because they are separated from the Tethyan collisional effects by an extension stage (the opening of the Tyrrhenian sea).

Epirogenic emersion of vast carbonatic platforms (Apulia, Central Italy, Southern Alps, Istria, Sardinia, Sicily) characterized the pre-collisional stage, and a number of noteworthy bauxite deposits generated in this time span.

The Alpidic-Apenninic orogenesis deserves particular attention, because the present shape and structure of the whole Italy (Sardinia in a minor scale) are related to it.

The Alpidic and the Apenninic tectonics, metamorphic and magmatic phases, even if belonging to a same event, are slightly displaced in time and affected in various ways the different sections of Italy.

As a matter of fact the Alpidic tectonics are characterized by motions in North-South direction, with intensive overthrusting and recumbent folding; the Apenninic tectonics are instead characterized by motions in latitudinal direction (mainly from West to East), accompained by overthrusting, recumbent folding and collapse structures in particular sites of the displaced masses.

The formation of syntectonic troughs and of insulated evaporitic basins (the so-called "Messinian salinity crisis") are other important phenomena controlled by the Alpidic-Apenninic orogenesis.

The Apenninic cycle started later than the Alpidic one, and reached its climax in Upper Miocene, when the Alpidic orogenesis was practically finished.

The curvature of the Western section of the Alpine chain is a consequence of the overlapping of Apenninic motions on pre-existing Alpidic structures.

On the basis of plate movements it is currently agreed that the Alpidic tectonics were a consequence essentially of frontal approaching of the European and African plates, whilst the Apenninic tectonics were controlled also by longitudinal relative motions of the two plates. The latters are related to the opening of the Northern Atlantic ocean.

It is also agreed that, whilst the Southern front of the European plate was — at those time — fairly smooth and had an East-West trend, the Northern front of the African plate — at the contrary — was uneven and formed a vast nose (the so-called "Adria promontory"), more or less opposite the actual position of Italy. The Apulian and the Sicilian carbonatic platforms are considered to be two major residues of that promontory.

The existence of this structure should explain the involvment of African slabs (the "Austro-Alpin unit") in the Alpine chain, that is mainly made of European and of Tethyan terrains.

The complicated movements, through which the Calabrian-Peloritanian Arc reached its present position (as proved by paleomagnetic studies), could also be explained taking into account that this micro-plate was moving close to a peculiar point of the African plate: the Western root of the Adria promontory.

Meaningful enough is the fact that only collapse structures have been generated in Sardinia by these tectonics, since early Tertiary: this suggests that Sardinia was outside the orogenic Alpidic-Apenninic belt. As a matter of fact, since lower Tertiary a number of extension structures generated in Sardinia; this tensional regime caused, from Upper Oligocene to Lower Miocene, the separation of the Sardo-Corsican massif from the stable Europe and the start of its counterclockwise rotation.

In this same time, intensive, eastwards verging overthrust phenomena, younger and younger in this direction, started in the Apennines.

In Lower-Middle Miocene the counterclockwise rotation stopped and so did the compressive tectonics along the western Apennines; an extensive stage commenced in this area, leading to the opening of the Tyrrhenian sea.

So far as the Calabrian-Peloritanian Arc is concerned, the opinions are divided; the first Authors dealing with this problem proposed that the Arc be simply a microplate coming from the paleo-Europe, such as the Sardo-Corsican massif and the Cabylies.

Subsequent Researchers (see the paper devoted to the Calabrian-Peloritanian Arc in this same volume) proposed a more complex evolution of the Arc.

We shall say here that the problem cannot be considered, at the moment, fully and satisfactory solved, particularly on the metallogenic standpoint, and fig. 1, drawn on the basis of the primitive hypotheses, has to be considered only as a first approximation.

In any case the metallogenic Calabrian-Peloritanian province is different from the Sardinian one, at least on the economic point of view.

The magmatic activity of this stage includes: (i) granitoids rocks: they are widespread in the Alps (the ADAMELLO group is the most important); only small occurrences are known in Tuscany (ELBA Island and GROSSETO district); the small grano-dioritic stock of CALABONA (North-Western Sardinia), that holds the only (small and poor) porphyry copper deposit of Italy, pertain to this group. It is worthwile to quote it, for its possible geodynamic significance. (ii) Volcanic rocks of various composition; they occur: between VICENZA and PADUA, and are mainly made of middle alkaline basalts; in Tuscany: "Selagites" of MONTECATINI, and minor acidic others; in Sardinia: andesites, dacites, rhyolites (in part hyperalkaline: the "commandites" of S. PIETRO and S. ANTIOCO).

The metallogenesis related to the closure/collisional stages and connected magmatisms yielded:

(i) noteworthy bauxite deposits: MONTE ROTONDO (Apulia) and OLMEDO (Sardinia) among others; as already mentioned they generated in pre-collisional stage;

(ii) important Na- K-salts, native S deposits generated during Upper-Miocene (Messinian) in Northern Tuscany (VOLTERRA: Na, alabaster), Eastern Emilia and Marche (gypsum, S) , Campania (S), Calabria (NA, S) and – specially – Sicily (CATTOLICA ERACLEA: Na; PASQUASIA and S. CATALDO: K; TRABONELLA, COZZODISI, GESSOLUNGO: gypsum, anhydrite, S).

(iii) remobilization and (partial) transformation (at places – Calabria – dispersion) of pre-existing metal accumulations in relation to metamorphism and granitization.

The main examples of concentrations are:

(iii – 1): the skarn deposits of Fe_3O_4, Fe S, Fe S_2 of TRAVERSELLA (Western Alps), of U (ADAMELLO district), of high purity Fe S_2 (GAVORRANO, in Tuscany), of Fe_2O_3, Fe S, Fe S_2 of ELBA Island;

(iii – 2): the mixed sulfide veins with quartz gangue of GROSSETO district (BOCCHEGGIANO, CAMPIANO p. p.).

(iv) Ore deposits, with no evident connection to pre-concentrations, and – hence – to be considered "new". They are a few, the mixed sulfide skarn of CAMPIGLIA (Grosseto) is the main example.

(v) porphyry copper deposits: as mentioned before, the only example is CALABONA (North-Western Sardinia). Unfortunately it is of low grade and small tonnage. In the same area the coeval andesites recovering this deposit, hold small, fairly rich Cu, Zn, Fe, Pb sulfide accumulations.

(vi) Industrial deposits, related to (some) volcanic complexes; they include:

(vi –1): Kaolin and Clays (PONZA Island, SARDINIA, South-Eastern Alps between VICENZA and PADUA).

(vi – 2): Mn and Fe Ox (SARDINIA).

4. Post-Tethyan events

The post-collision evolution, encompasses the Neogene-Quaternary events; they are characterized by a new crustal extension, that was responsible both for the opening of the Western Mediterranean basins (the Tyrrhenian in particular), and for several basins in the Apenninic chain and of the compressive stage still going on in the Calabrian-Peloritanian Arc, causing eartquakes, volcanism, quick land upheaval.

The age of the bathyal plain basalts of the Tyrrhenian sea is 7.5 My (Lower-Middle Pliocene), post-dating the beginning of its opening.

The magmatic activity related to this stage includes four types of volcanics; namely:

(i) Alkali-potassic (at places with high K content); it stretches along the Tyrrhenian coast from Northern Latium down to Neaples (VESUVIUS active volcano); also the VULTURE apparatus (Northern Lucania) pertains to the same type.

The age of this volcanism is from 1.7 My to 0 and it is associated with tensional tectonic that followed the Tethys closure.

(ii) acidic intra-crustal: it occurs in Northern Latium and in Tuscany, MOUNT AMIATA being the major apparatus. It is also related to the same above-mentioned structures. The rock age is 0.4 to 0,2 My, but some fumaroles are still active and the LARDERELLO vapor field is connected to it.

(iii) Alkaline/sub-alkaline continental basaltic: important rock masses occur in SARDINIA, in Sicily (MOUNT IBLEI and the active ETNA volcano), in some islets of the Sicilian Seas (PANTELLERIA and others), and of the Tyrrhenian in front of PISA (Isle of CAPRAIA).

Their ages range from 5.4 My to 0. The basaltic deposition in Sardinia followed soon after the volcanic activity quoted at the previous point and is related to the same collapse structures. In Sicily it is thought to be related to tensional structures of the African plate, ensuing from the active convergence of the Calabrian-Peloritanian Arc.

(iv) Calc-alkaline and shoshonitic of orogenetic association: the only example is shown by the Eolian archipelago: its age is very recent (probably from less than 1 My to 0), and is an evidence of the above-mentioned active convergence.

The metallogenesis related to the post-Tethyan event has noteworty industrial interest. One may distinguish:

(i) metallogenesis related to the alkali-potassic volcanism: it includes: huge accumulations of leucitic rocks; huge accumulations of volcano-sedimentary fluorite, at places with admixed baryte, and minor deposits of baryte and celestite; small deposits of volcanic sulfur and marcasite; sub-economic accumulation of U ores; small deposits of alum; traces of Fe, Ti, Zr, Th.

(ii) metallogenesis related to the acidic intra-crustal volcanism: the very important Hg deposits of MOUNT AMIATA and the not negligible Sb deposits of the same region pertain to this group.

Also the vapor fields of LARDERELLO are related to the same volcanism;

(iii) heavy mineral accumulations in beach sands (Fe, Ti, Th, Zr ores) occur in many places of the Tyrrhenian, the Ionian and the Sardic coasts;

(iv) sub-economic sulfide (mainly pyrite) volcano-sedimentary submarine accumulations, related to the Eolian volcanism;

(v) haematite, magnetite, Ti mineral pre-concentrations (giving rise, by erosion, to black sands) occur in some basalts of Sardinia.

Sources of Informations

Boccaletti, M., Cita, M. B., Parotto, M., Sartori, R. (Editors) (1984): „Evoluzione geodinamica del Mediteraneo con particolare riguardo all-Appennino, Sicilia e Sardegna". Mem. Soc. Geol. It., XXIV–1982, 482 p.

Cocozza, T., Decandia, F. A., Marcello, A., Pretti, S., Salvadori, i., Uras. I., Valera, R. (1984): "Geo-tectonic Evolution and Metallogeny of the Sardo-Corsican Massif", in this same volume.

Gelati, R., Praturlon, A., Zuffardi, P. (1984): "The Apenines: geotectonic evolution and metallogeny-an overview", in this same volume.

Ogniben, L., Parotto, M., Praturlon, A. (Ed.) (1973): „Modello Strutturale d'Italia", geological map, scale 1 : 1,000.000 – C. N. R.

Praturlon, A. (1980): „Cadre géologique général de l'Italie. Sa place dans le cadre europeen et méditerraneen". In: „Introduction à la géologie générale d'Italie", XXVI Int. Geol. Congress, 3–14.

Zanettin Lorenzoni, E., Zuffardi, P. (1984): "Geo-tectonic evolution and metallogenesis of the Calabrian-Peloritanian Arc in the Tethys Realm", in this same volume.

The Apennines: Geo-tectonic evolution and metallogeny – An overview

By R. Gelati*), A. Praturlon**), P. Zuffardi*)

With 6 Figures

Zusammenfassung

Der Apennin ist ein relativ einheitliches Orogen zwischen dem Golf von Genua und Sizilien; nur Calabrien und Sardinien-Corsica bilden eine ältere variscische Einheit. Die wechselnden Phasen von Krustenausdehnung und Kompression vom Perm bis in die Gegenwart bestimmten die Metallogenese. Es wird bezweifelt, daß der Ozean zwischen der Adriatischen Platte und Afrika die vielfach angenommene Breite gehabt hat. Herzynische und praeherzynische Vererzung fehlt im Apennin, wenngleich die Isotopenzusammensetzung einiger Erze in der Trias auf ältere Mineralvorkommen im Paläozoikum hindeutet. Dem Öffnungsstadium der Tethys entsprechen die berühmten jurassischen Ofiolith-gebundenen Lagerstätten von Monte Catini. Die kretazischen Bauxite in den Abruzzen und in Apulien sind allochthon und werden von vulkanischen Aschen aus Afrika hergeleitet (Bardossy). Dem miozänen Kollisionsmagmatismus sind die Kontaktlagerstätten von Elba zu verdanken. Hg, Sb und Leuzit sind ökonomisch wichtige Produkte des jüngsten nach-tektonischen Vulkanismus in der Toskana.

1. Definition of the area

On the geographical point of view, the Apennines are the mountain chain stretching all along the Italian Peninsula, in continuation of the Marittime Alps: the geographic boundary between Alps and Apennines is located in the Western

*) Università di Milano, Dipartimento di Geologia, I-Milano.
**) Universita di Roma, Dipartimento di Geologia, I-Roma.

section of the Genua gulf (along a line connecting Noli Head with the Cadibona Pass).

On the geological stand point the boundary Alps-Apennines is displaced 40—80 km eastwards; the first figure (a) refers to the ancient literarure, the second (b) to the most recent authors. In fact, till the 70's it was usual to establish that boundary along the "Sestri-Voltaggio line" (just West of Genua), whilst at present the "Levan-to-Ottone line" (West of Spezia) is supported. These two North-South lineaments enconpass a montainous region, in which the formational-structural characters are (at yet) controversial, so far as they have to be considered dominantly of Alpidic or Apenninic style.

Moreover, on the geological point of view, Sicily can be considered as a continuation of the Apennines, exception made for the most Southern section of the Apennines (from the Sangineto line due South) and the Northern corner of Sicily (from the Taormina line due North), which, togheter, make up the "Calabrian-Peloritanian Arc". The latter is a province having geological, geodynamic and metallogenic characters clearly different from the other sections of the Apennins and Sicily and rather similar to the ones of the Alps and of Sardinia.

As a consequence, the present paper deals with the Apennines plus Sicily, excluding the Calabrian-Peloritanian Arc, which is the topic of another paper in this same volume: from now on the term "Apennines" will concern the above defined area (Fig. 1).

2. General Geological and Metallogenic outline

The Apennines are a part of the Alpidic orogenic belt comprising also the Tunisian Magrebids, Southern Alps, Dinarides, Hellenides. This elongated chain resulted by deformation of a continuous Mesozoic continental margin developed on the Southern part of the spreading Tethys, alongside the African promotory or Adria.

Like other parts of this fold belt sharing the Adriatic stable area as a common foreland, the Apennines area was affected by upper Permian to Middle Triassic intracontinental rifting, by Jurassic to Middle Cretaceous oceanic rifting and spreading and, finally, by closure, collision and post-collision deformations in Late Cretaceous to Recent times.

An outstanding character of the Apenninic geology is the scantiness of Hercynian/Pre-Hercynian outcropping/sub-outcropping terranes, the absence (at least the lack of known evidences) of the Hercynian plutonism and thermo-metamorphism and hydrothermalism. Anticipating what will be said later on in detail, the k n o w n Hercynian/pre-Hercynian stock-metal is practically nil, even if there are some good reasons supporting that it should not be so. This is quite a difference from the Alps, from Sardinia, from the Calabrian-Peloritanian Arc where the Hercynian/pre-Hercynian stock-metal is important.

The main ore/industrial mineral accumulations in the Apennines took place in Permo-Triassic (mainly pyrite), in Jura (Cu, pyrite of the ophiolites), in Cretaceous-

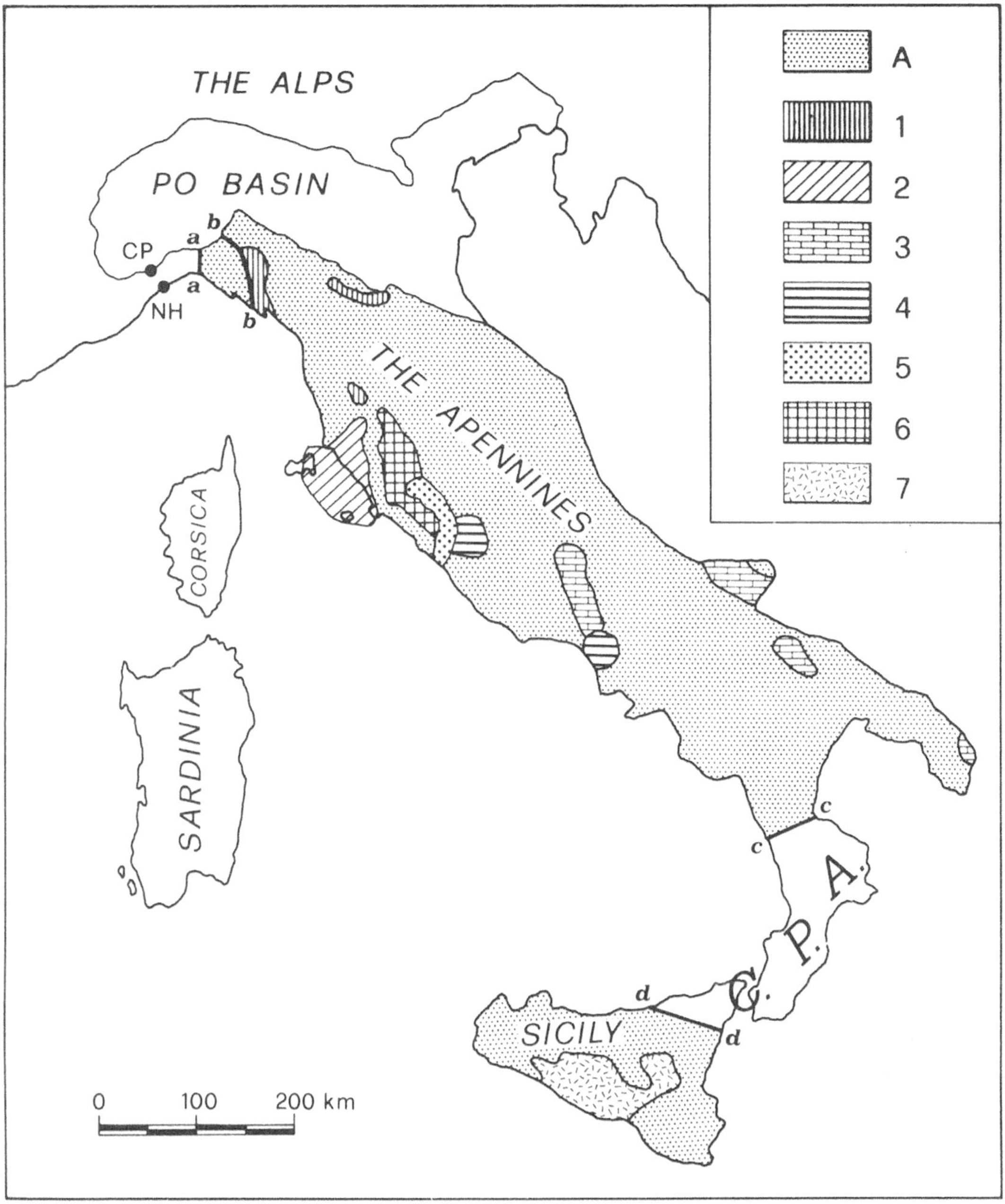

Fig. 1. Location map of the area dealt with in this paper (Apennines plus Sicily excluding the Calabrian-Peloritanian Arc) and of the main mining districts in it. – NH: Noli Head; CP: Cadipona Pass; a–a: Sestri-Voltaggio Line; b–b: Levanto-Ottone Line; c–c: Sangineto Line; d–d: Taormina Line; C. P. A.: Calabrian-Peloritanian Arc; A: Area dealt with in this paper. District n°: 1: Cu, Pyr; 2: Pyr, (Mixed sulfides), Fe_2O_3; 3: Bauxite; 4: Leucite; 5: Hg, Sb; 6: F, Ba; 7: Na, K, S.

Paleogene (bauxite), in Upper Miocene (evaporites and S), in Pleistocene (mainly Hg, Sb, F, leucite).

This metallogenic outline characterizes the Apennines and makes them different from the other Italian geological-metallogenic provinces.

3. Geo-tectonic evolution

3.1. Paleozoic: traces of Hercynian orogenesis

The Paleozoic basement of the Apennines crops out only in Tuscany, where field work and boreholes for mining and/or geothermal investigations evidenced a severely tectonized and partly metamorphosed Ordovician-Devonian sequence affected by the Sudetian and Asturian phases of the Hercynian orogenesis. Large scale thrusting within the basement occurred during the first phase, whilst the Asturian compressional movements were soon followed by crustal extension causing in Tuscany the formation of intramontane troughs where Coalmeasure type sediments (S. Lorenzo Group) deposited (Bagnoli, et al., 1980).

The main effect of the subsequent Saalian phase, at the end of Lower Permian, was the uplift of large crustal blocks. To the same phase has been related also the acidic magmatism of the Red Porphyries which are a part of the "Asciano Group". Both mentioned groups may be interpreted as late-Hercynian molasses, corresponding to the progressive erosion of the chain and to the subsequent post-orogenic volcanism (Cassinis et al., 1980).

Upper Permian was characterized in Tuscany by a vast peneplain and the area was no longer site of accumulation. Exception is made locally for the fine-grained red sandstones of the Castelnuovo Fmt., tentatively correlated with the "Arenarie di Val Gardena" of the Southern Alps.

3.2. Triassic

Northern and Southern Apennines show different evolution during Triassic, in the frame however of a general "intracontinental platform" paleotectonic stage ("intracontinental basin" in the Lagonegro domain).

The early sedimentary assemblages resting on the Late-Hercynian molasse in the Northern Apennines are terrigenous sediments of continental to deltaic and shallow marine environment ("Verrucano"), somewhere with interbedded carbonate bodies and prasinites (Punta Bianca). During Late Triassic, large evaporitic and carbonate platform sedimentary environments spread over the whole area. Sedimentary units of this age are represented by: sulphate rocks interbedded with dolostones (Burano Anhydrite); gypsiferous limestones or dolostones, soon transformed in vuggy complex ("Calcare cavernoso") by leaching of gypsum; dolostones (prevailing in the Grezzoni Fmt. of the Apuane Alps, in the Filettino units of Northern Latium and in other scattered outcrops or boreholes in the Latium-Abruzzi carbonate platform); shales, marls, limestones (in the Rhaetavicula contorta beds and in the Portòro Fmt.).

On the whole, during Norian, it is possible to recognize in Central-Northern Apennines a general paleogeography characterized mainly by a wide evaporitic area and a carbonate belt around it (CIARAPICA & PASSERI, 1981). No evidence of pelagic sediments was found up to now.

In the Southern Apennines quite different sedimentary associations are recognizable. In the Lagonegro Basin sequence, the terrigenous, shallow marine argillites and sandstones with interbedded basalt flows, pillow breccias, somewhere channelled bioclastic limestones, of the M. Facito Fmt. occur at the base. They are followed by pelagic cherty limestones.

Phyllites, with interbedded coralgal carbonates and prasinites („Campotenese-Lungro Phyllites"), are present at the base of the Campania-Lucania carbonate platform sequence. They are covered by back-reef and supratidal dolomites and evaporites.

Upper Triassic carbonates and evaporites show very high rates of sedimentation (up to more than 100 m/m. y.). They were deposited on what had formerly been continental crust of normal thickness (D'ARGENIO & ALVAREZ, 1980) with its surface approximately at sea level; they progressively overlapped all possible source area of terrigenous clastics, so that only cherty limestones and marls could deposit, even in the basins.

This does not explain however the absolute lack of terrigenous clastics along such a wide area from Late Triassic until the deposition of the early Alpine flysch. In fact, passive continental margins, undergoing such strong subsidence, commonly develop thick wedges of clastic sediments transported from the continents interior. In order to have pure carbonate deposition, a continental margin must be really isolated from the influxes of rivers or other sources of clastics. As D'ARGENIO & ALVAREZ pointed out, this is most likely to occur where the rifting, leading progressively to the sea-floor spreading, has blocked out a large sector of continental crust, as it was the case both in the Adriatic promontory and in the Florida-Bahamas region.

Subsidence was very probably an isostatic response to crustal thinning during the rifting and spreading phases. Calculations based on Apennine sequences indicate a thinning of the Italian continental crust for the whole Mesozoic by approximately 40% beneath pelagic basins and 20% beneath carbonate platforms (D'ARGENIO & ALVAREZ, 1980).

Crustal thinning, high subsidence, carbonate or cherty sedimentation everywhere, elimination of any possible cratonic influx in sedimentation and, above all, the early phases of the subsequent Tethyan spreading led to important variations in the paleotectonic regime at the end of Triassic. The Apennine sector of the African continental margin changed in fact at this time, from intracontinental to epioceanic platform/basin stage.

3.3. Jurassic–Lower Cretaceous

The cycle of Triassic rifting ended with deposition of evaporites, shallow water carbonates, deep water cherty limestones. During Early Jurassic time, the Central North Atlantic ocean began to open, a new cycle of rifting began and fragmentation of the Western Tethyan region accelerated, with creation of new rift basins. In Middle Jurassic time, rifting and transform faulting created deep-water corridors, linked the early Atlantic to the Tethyan Ocean and isolated new lithospheric plates in what had formerly been a coherent block (BIJU-DUVAL et al., 1977). According to some authors, the Adria itself became a separate plate. The Troodos ophiolite assemblage in Cyprus has been interpreted as related to an oceanic basin opened between the African and the Adriatic plates. It has however to be said, that against this and similar other hypotheses postulating more or less wide oceanic areas between Africa and Adria, there are a lot of valid arguments (see for a review, CHANNELL et al., 1979).

The age of the earliest appearance of oceanic crust in the new rift system has not been satisfactorily ascertained; the following factual observations are important at this concern: in Sicily, pillow lavas are interlayered in sediments of Late Liassic and Late Bajocian ages. Allochthonous ophiolites of Northern Apennines show radiometric age of 160 to 185 m. y. (Sinemurian to Bathonian; BORTOLOTTI & GIANELLI, 1976).

In the Northern Apennines, basin areas originated in Middle Liassic time by tectonic "drowning" of large sectors of Liassic-Triassic shallow-water carbonate domains, mainly represented by the Calcare Massiccio Fmt. These basins, westwards connected to the newly formed oceanic area of the socalled "Liguride domain", were shaped throughout the Jurassic, by seamounts characterized by very condensed sedimentation. Middle Liassic tensional faulting caused sudden facies variations, neptunian dykes, brecciation, slumping along the slopes.

Liassic platform carbonates pass abruptly upwards to the cherty-marly limestones of the basinal sequence. Limited traces of alkali-olivine basaltic volcanism at this time were detected in Southern Tuscany (BOCCALETTI & MANETTI, 1972). Local emersions, preceding the final drowning, have been recognized in many parts of Tuscany and Umbria.

The above mentioned "Liguride domain" shows the most significant evidences of the Jurassic oceanic rifting/spreading. It is an outstanding paleographic feature in the outline of the Alpidic "geo-syncline", and occurs, fairly continuously, all along the Apenninic chain.

The sedimentary evolution of this sequence is, up to now, controversial, because the terranes, by which it is made up, are known only in allochthonous positions.

Since the 60's (DECANDIA & ELTER, 1969) the Liguride domain is interpreted as an oceanic hiatus connected to a rifting/spreading cycle; in a few words, it is thought to be an oceanic area located between the Paleo-European and the Paleo-African cratons, divided in two (Eastern and Western) sections by a paleo-wrinkle ridge (the so-called Bracco wrinkle) that can be considered as a middle-oceanic ridge.

As a matter of fact the Bracco wrinkle is a vast complex made of mafic/ultramafic crust (peridotites-lherzolites, harzburgites and gabbroes), suffering—in middle to upper Jurassic—from metamorphism of green-schist or amfibolite facies (CORTESOGNO et al., 1975). It was dissected, since the Malm, by normal synsedimentary

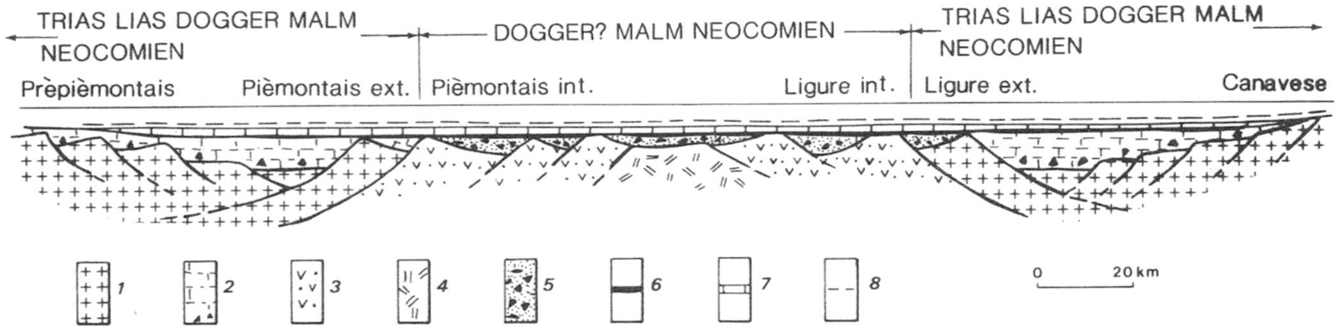

Fig. 2. Palaeogeographic reconstruction of the Liguride Domain in Jura-Lower Cretaceous (from GRANDJACQUET & HACCARD, 1977, simplified). — 1 = Continental Crust; 2 = Trias-Giura; 3 = Lherzolites of the Liguride Domain; 4 = Peridotites and Gabbros of the Liguride Domain; 5 = Pillow lavas and ophiolite breccias; 6 = Radiolarite; 7 = Calpionella Limestone; 8 = Palombini shales.

faults (GRANDJACQUET & HACCARD, 1977), delimitating horst-and-graben systems; the latters, in turn, controlled the depths and localizations of the overlying vulcano-detrital sediments and radiolarites (see Fig. 2 and 3).

The origin of the "ophicalcites" is also debated: according to GRANDJACQUET & HACCARD (1977) they should be related to tectonic accidents, and be generated by carbonate matter filling tension fractures and breccias. FOLK & MCBRIDE (1976), instead, support a pedogenic origin, in connection to paleotopographic highs with subaerial exposition of (ultra)-mafic rocks, prior to radiolarite deposition (Fig. 3). Mainly pelagic facies sediments overlied the radiolarites in Neocomian: namely the Calpionella limestones (Berriasian) and the Palombini shales.

According to ABBATE et al. (1980) the main part of the features of the structural environment in which ophiolites generated in Northern Apennines, can be explained by assuming that most of those ophiolites originated in a narrow ocean, subjected to East-West transcurrent movements, between Africa and Europe, with protrusion of (ultra)-mafic diapirs along the related transform fault zones. A narrow transitional zone, probably characterized by continental crust of reduced thickness, was interposed between the oceanic and the external areas. It corresponds to the Subliguride (or Sicilide) domain, another paleogeographic feature occurring now as allochthonous masses (usually non-ofiolithiferous flysch type) all along the Apennine chain, especially in Sicily.

In Central and Southern Apennines the well-differentiated Triassic sedimentary pattern, already characterized by a carbonate platform-basin system, persisted without important variations. New basins however developed across former carbonate platforms at the expense of pre-existing platform margins. Back-reef

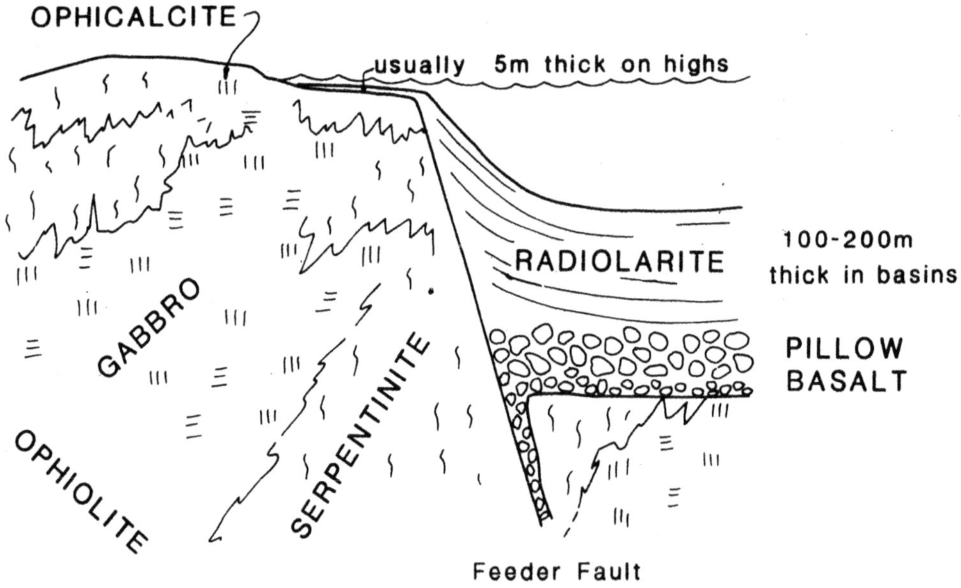

Fig. 3. Geologic relationships of ophicalcite and radiolarite (from FOLK & MCBRIDE, 1976).

lagoon sedimentation of Bahamian type was typical of the persistent carbonate platforms (Latium-Abruzzi pl., Campania-Lucania pl., Gargano-Murge pl. and minor Sicilian representatives), where very thick carbonate sequences accumulated. Intervening basinshaped areas were dominated by siliceous-marly sedimentation, often with thick bioclastic intercalations supplied by the nearby platforms. True radiolarites are sometimes present.

3.4. Middle-Upper Cretaceous

Evidenc of the closing movements occurring at this time in the Tethys are not directly detectable. Subsidence of the platforms was however dramatically reduced to less than 10 m/m. y.; there was an evident reduction of platform domains due to synsedimentary faulting; local basaltic volcanism occurred in the Southern areas; a large part of the carbonate platforms emerged for some m. y. in late Albian–Cenomanian, gaving rise to bauxite formation (BARDOSSY et al., 1977). Only in the innermost zones (Liguride domain) the first flysch facies of the Apennines testify the tectonic crisis occurring in the Tethys.

So far as beginning of the closure is concerned, GRANJACQUET and HACCARD (1977) support that clear symptoms of the first reduction of the Tethyan oceanic area are recognizable in Northern Apennines since the end of Lower Cretaceous and during Middle Cretaceous.

In fact, they postulate the overthrusting of the Liguride domain upon the European craton; intensive syn-sedimentary tectonics consequently took place on the surface: f. i. slumpings and olistostromes; these features are particularly frequent in the terrigenous flysch units, known as "Basal Complexes" in the current literature.

According to ABBATE et al. (1980), the ophiolitic detritus, interbedded in the Basal Complexes, was provided from the paleo-highs located in connection with former transform zones or at the contact between transform zones and normal oceanic crust.

The above mentioned early compressional phase should have played a rôle on the paleogeography of the Liguride domain, controlling in turn, the distribution of the huge turbidite complexes which have been formed from Upper Cretaceous to Paleocene. It is the case of the Helminthoid flysch, calcareous turbidites (M. Antola, M. Cassio, M. Caio, etc.) and of arenaceous turbidites (M. Gottero and M. Casanova); many authors, in the last twenty years, proposed various, often opposite ideas about their correlations: this fact derives (at least in part) from the factual difficulties related to their allochthonous positions.

Some (relatively) ancient studies (ELTER G. et al., 1966; ELTER P. and PERTUSATI, 1973), integrated and/or modified by recent stratigraphic-sedimentologic analyses, deserve still fundamental interest on this topic. According to the synthesis proposed by SAGRI and MARRI (1980), the Apenninic "geo-syncline" has been fed with huge amounts of terrigenous materials from the Sardo-Corsican Massif. The coarser ones deposited close to the continental margin and built up a series of deep-sea fans (Gottero Sandstones, Macinaggio and Elba flysch) (Fig. 4).

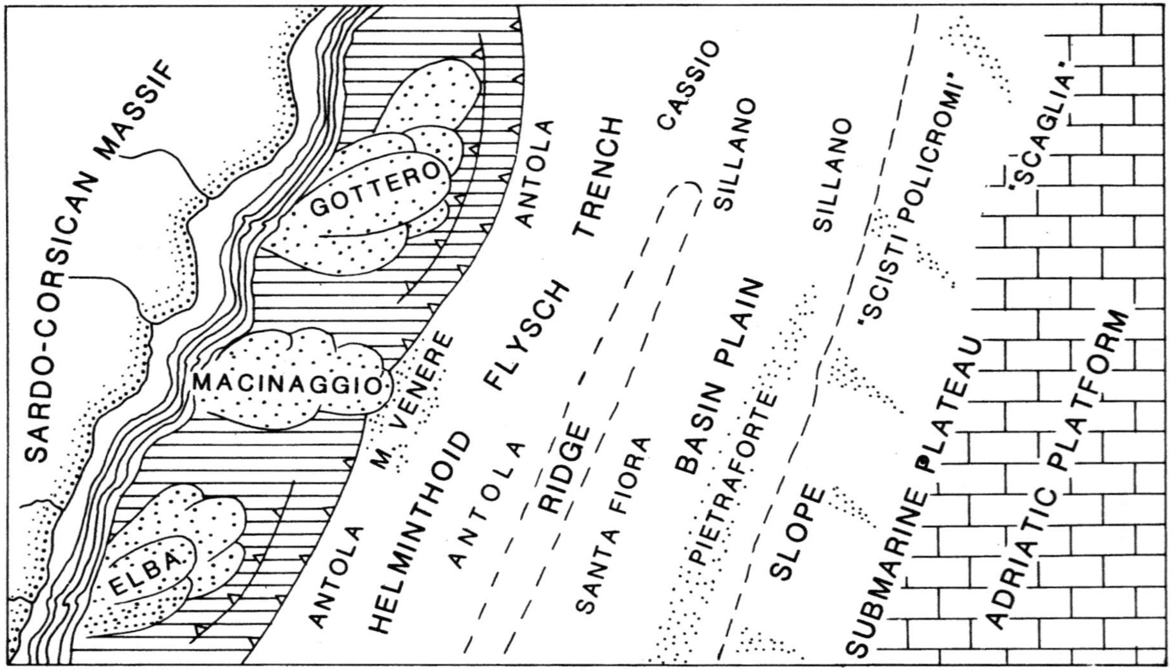

Fig. 4. Palaeogeographic setting of the Liguride Domain in the middle-upper Cretaceous (from SAGRI & MARRI, 1980).

Huge masses of Helminthoid flysch, instead, accumulated eastwards, on a basinal plain, located below the carbonate compensation level; the noteworthy quantities of calcareous muds accumulated in the Helminthoid flysch trough may be related to the high productivity of micro-organisms, that lasted during Cretaceous, in connection with the widespread transgressions which affected the main part of the continental margins (JENKINS, 1980).

Specific studies on the sedimentary facies (ABBATE, personal communication) pointed out that the turbidites of the Gottero sandstones were deposited as a small deep-sea fan in a trench slope basin; they contain foraminifers which indicate deposition at bathial depths and pinch out into and are overlain by, basin plain turbidites.

The Casanova Unit shows peculiar characters in the above defined outline; in fact, according to CASNEDI (1982), the slope and the basin-plain deposits are well represented in this unit, but fans are absent. This is due to the quick activation of a slope along the margin of a major ophiolitic ridge, related to the prosecution of the aforesaid early compressional phase.

At the end of the Cretaceous, all main carbonate platforms emerged. They were affected by a general marine ingression only during Lower-Middle Miocene, whilst sedimentation of the "Scaglia" type (thin bedded varicoloured–dominantly reddish–pelites with marls) persisted continuously in the basins.

In sharp contrast with what happened during Middle Cretaceous, no traces of this long emersion are recognizable. Karst, residual sediments, hard-grounds, erosion, etc. are absolutely absent. Only a very thin stylolitic surface marks the paraconformity between Mesozoic and Miocene.

3.5. Collisional and post-collisional events

During the Cretaceous, the opening of North Atlantic occurred with a greater spreading rate compared with the South Atlantic. This fact, coupled with a strong North-South component in the relative motion between Africa and Europe, squeezed out the Central Tethys, until a continent-continent collision, that mainly occurred in Late Cretaceous-Eocene time, gave rise to the most outstanding part of the Alpine Chain. The Apennines were deformed later, during Mio-Pliocene. Their history was complicated by the opening of small oceanic basins, while orogenetic movements were still going on, in particular the Ligurian and the Tyrrhenian ones, interpreted as back-arc spread marginal basins, with basic magmatism on the floor. The Southern Tyrrhenian should have been related to an arc-trench system with intensive, still active, subduction in corrispondence with the Calabrian-Peloritanian Arc.

The opening of the Tyrrhenian sea started about 7,5 m. y. ago as it is proved by absolute age dating of bathyal plain basalts (BARBERI et al., 1978); the history of the Tyrrhenian area has been connected strictly with folding, faulting, magmatic and metallogenic activity, neotectonics and seismicity of the Apennines.

The Neogene-Quaternary post-collisional evolution has been reinterpreted, in a fairly new way, by BOCCALETTI et al. (1982); these authors suggest a regmatic system consisting of two main conjugated shear trends, according to NE-SW and

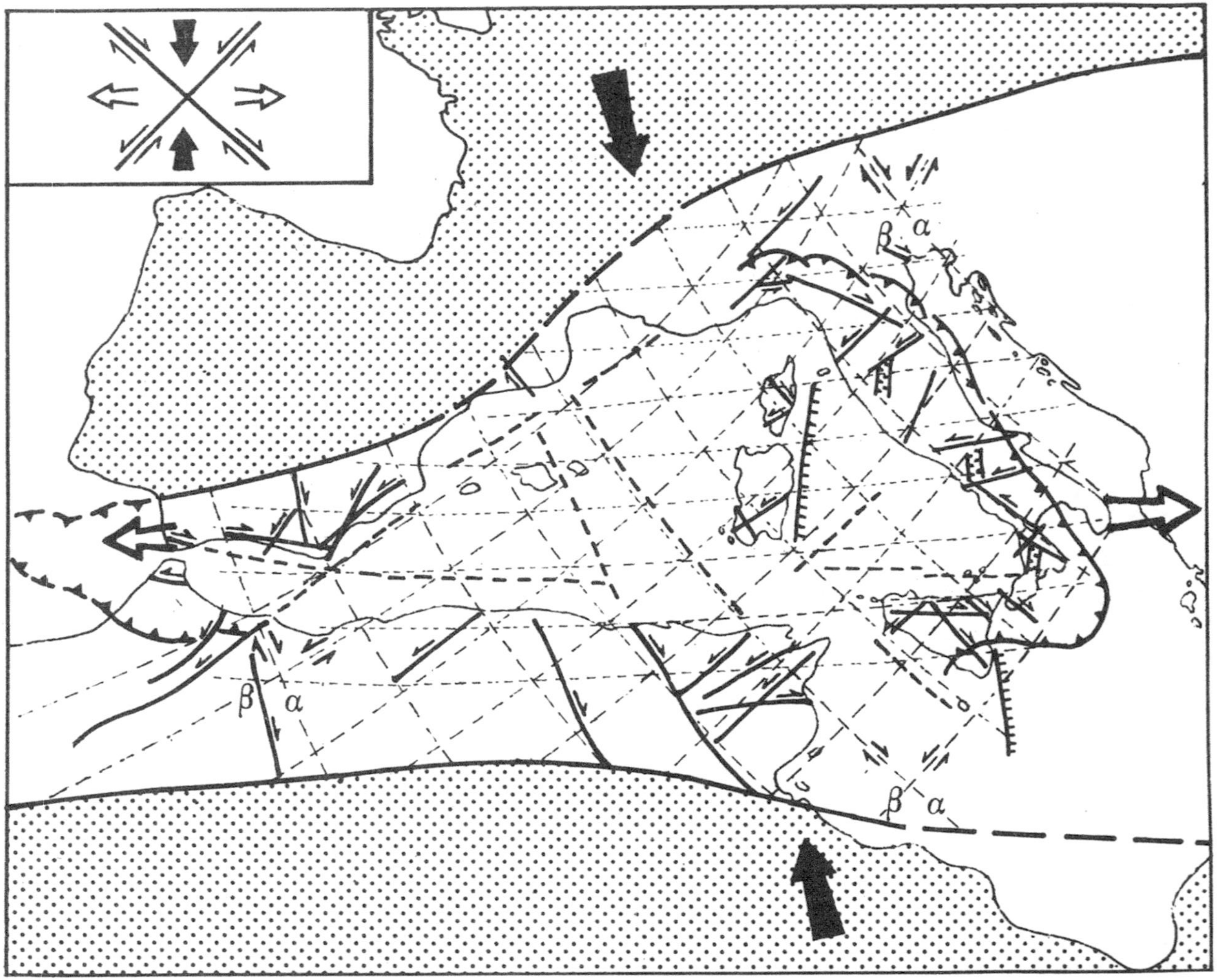

Fig. 5. Actual main shear faults of the Western Mediterranean area compared with the slip-line field developing between the two oblique converging African and European rigid blocks (from Boccaletti et al., 1982).

NW-SE, associated with a mostly shearing E-W trend and a mostly tensional N-S trend (see Fig. 5).

This rather complicated system could explain not only the mechanism of crustal extension, that was responsible for the opening of the Western Mediterranean basin, but also the mechanism of formation of several Neogene—Quaternary basins of the Apenninic chain.

A special significance in this scheme, is attached to some triangular basins occurring in external position relative to the Apennine folded belt: they should represent "pincer" features, generated by converging folds, connected to the two major conjugated shear trends (see Fig. 6).

The so-called "Messinian salinity crisis" (Hsü et al., 1973), has to be located in a rather debated paleogeographic and structural outline, which—so far as the Apenninic domain is considered—should have been controlled by compressional phases in the external part of the chain and by distensional movements in its Tyrrhenian side.

Messinian evaporitic deposits were well known since a long time in a number of outcrops in the coastal Mediterranean countries (such as Sicily, and on both sides of the Apenninic chain). The researches carried out by the Deep Sea Drilling Project, since 1970, found Messinian evaporites both in the rises and in the sub-marine plains. They are underlain by euxinic sediments (including diatomites) and are overlain by Pliocene marine beds; their total thickness is estimated 2000 meters or even more.

Various evaporitic facies are recorded; in fact some deposits have sabkha, ephemeral lake, or not very deep lagoon characters, being made of selenitic gypsum, nodular and/or chickenwire anhydrite, finely laminated gypsum and anhydrite, halite; others are clastic and/or turbiditic, thus suggesting probable sub-acqueous deposition at various depths.

In any case there exist various associated types of facies, with such faunistic contents and such structural characters, as to justify the contrasting theories, that have been proposed for their interpretation.

In short, these theories are based on the contrast between the supporters of the "closed barrier at Gibraltar", with consequent general desiccation of the basin, and the supportes of the "open barrier" hypothesis.

The Messinian evaporites have been involved, in their syn-/ and post-depositional stages, by the compressive and/or distensional tectonics respectively along both the Adriatic-Padanian and the Tyrrhenian sides of the Apenninic chain.

They have been consequently moulded, at places, in complex structures; among them, the diapires, recognized by seismic profiles of the Mediterranean basins (particularly in the Balearic one), deserve special attention.

3.6. The geo-tectonic evolution of the Apennines and of the Tyrrhenian sea, on one hand, and the petrographic composition and localization of the volcanic complexes, on the other hand, are strictly linked.

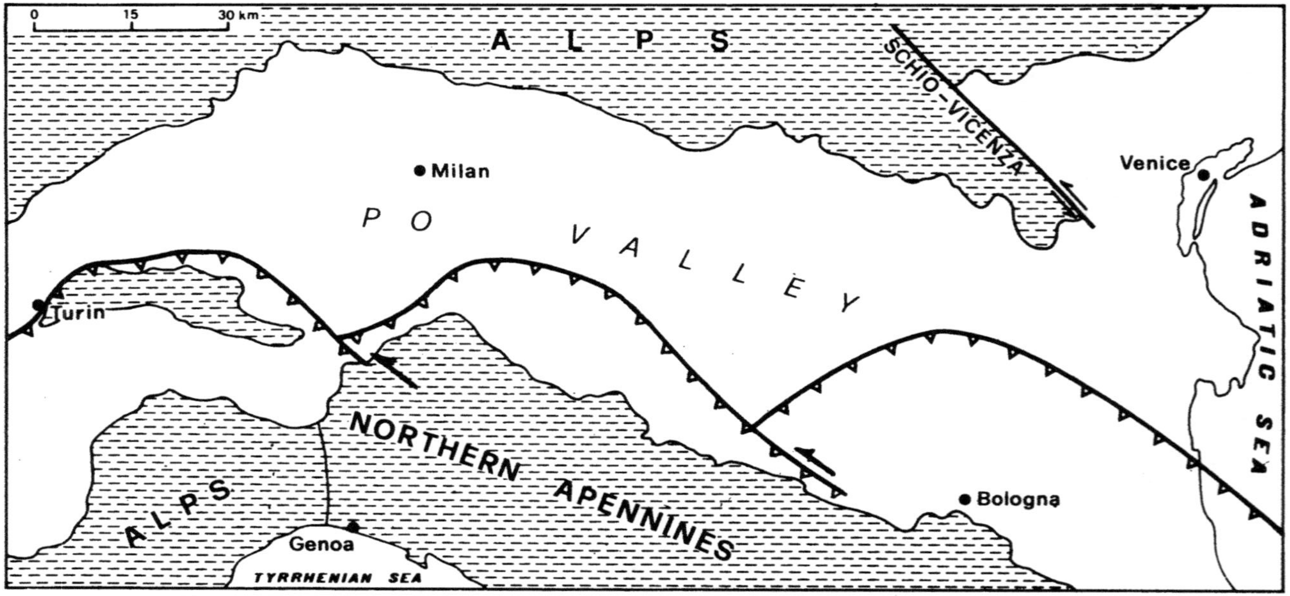

Fig. 6. Scheme of formation of triangular "pincer" basins on the Northern Apennines external front.

As a matter of fact, one may distinguish:

a) calc-alkaline and shoshonitic volcanism: it occurs in the Eolian volcanic arc, its age is fairly young (less than 1 m. y.), and is probably to be connected to the most recent compressional phases;

b) intra-crustal acidic anatectic magmatism: the granites/granodiorites of Tuscany (Gavorrano, Elba Island) pertain to this type and were emplaced in a time span from 8.7—4.9 m. y. Subsequently (from 5 m. y. on), some volcanics of the same composition have been emplaced, in a distensional phase that produced a horst-and-graben structure parallel to the main axis of the Apennines. Among them, the M. Amiata volcanic complex, made of rhyodacitic lavas (quartzlatites) and of minor trachites, played an important metallogenic rôle.

c) Quaternary alkali-potassic volcanims: it stretches from Northern Latium to Neaples, mainly along the Tyrrhenian coast. It is emplaced in the same horst-and-graben structures mentioned above and is interesting from the metallogenic stand-point.

d) Continental basaltic volcanism: the first episodes took place in Upper Miocene, but it became important in Pliocene and in Quaternary; it is still going on at present (M. ETNA).

It is related to local tensional zones connected with the complicated game of plate movements and of the consequent stress distribution in the region.

4. Metallogenesis

Five metallogenic main epochs can be distinguished, in the frame of the present paper; namely:

4.1. Hercynian and pre-Hercynian:

no ore/mineral deposits are known in the Hercynian/pre-Hercynian basement.

This is fairly astonishing, taking into account the metallogenic character of the other parts of Italy (specially of Sardinia), in which the best ore/mineral deposits have been generated in this epoch.

It has to be underlined that the Cambrian "Metalliferous"—that holds rich ore/mineral deposits in Sardinia—is not known in the Apennine basement: this fact is obviously a negative (or, at least, not a positive) point for the metallogenic prospects in the Apenninic basement. On the contrary, Ludlovian (Orthoceras dolomites, Graptolite slates, porphyroids) has been recognized (see BAGNOLI et al., 1980 for a synthesis of the knowledges on the Tuscan Paleozoic); it is well known that the latter complex is ore bearing (mixed sulfides and Sb—W—Hg) in many part of Europe, particularly in Sardinia. This fact should be taken into consideration for new and more researches and explorations.

It has also to be pointed out that the "Boccheggiano phyllites Fmt." should not be considered as Pre-Hercynian, as proposed by some authors (BAGNOLI et al., 1980); we incline to date them Post-Hercynian (Upper Carboniferous? Permian?), according to others authors (VIGHI, 1966, COCOZZA et al., 1978), on the basis of the following facts: the "Boccheggiano phyllites Fmt.", is a sequence made of dominant phyllites, quartzites, occasional chloritic and/or grafitic slates, meta-graywackes, meta-basites; lenses of anhydrite and of evaporitic dolomites are fairly frequent in their upper section. Triassic evaporites ("Vuggy Limestone" type) directly overly them, and have the same isotopic composition (C, O, S) of the above-mentioned lenses interbedded in the BOCCHEGGIANO phyllite Fmt. (CORTECCI et al., 1983, 1984).

Moreover, in many cases (Niccioleta, Elba Island) the same stratiform (volcano-sedimentary?) ores occur both in connection with the evaporites held in the Boccheggiano phyllites and at the base of the Triassic Vuggy Limestone.

All these observations, in our opinion, speak for continuous deposition and metallogenesis between the Boccheggiano phyllites and the Vuggy Limestone, and are against a hiatus between them.

4.2. Post-Hercynian/Pre-Tethyan

The involved time span should be from Upper Carboniferous/lower Permian to the end of Triassic. It is not always easy to date exactly the formations pertaining to this epoch; it is easier and more effective, from the metallogenic stand-point, the use of the formational names, even if with some uncertainties about their very ages.

The formations having main metallogenic interest–from bottom to top are: a) the above-mentioned "Boccheggiano phyllite Fmt." and the "basal slates"–at places called "Farma Fmt."–(a Coal-measure type, slaty, graphytic partially flyschoid sequence, with occasional carbonate lenses); b) the "Verrucano"; c) the evaporitic complexes ("Vuggy Limestones", "Grezzoni", "Burano Anhydrites"; "Small Cell Dolomites")*).

The only ore accumulations known till now in the slates are: i) some small and low grade black shale type, mixed sulfide (mainly: pyrite and chalcopyrite) stratiform occurrences (PONTE SAN PIETRO, West of BOLSENA Lake, described in RIVA, 1983); ii) small and low grade karstic, Pb, Ag, Zn (Sb) occurences (S. ANTONIO of PARI, half way between SIENA and GROSSETO, described in GRIONI, in press).

We spoke yet about the pyrite deposits held in the upper section of the Boccheggiano phyllite Fmt., and will recall them shortly after.

Verrucano–in the Apennines–is normally barren; two groups of ore deposits are known in connection to it; namely: i) inside Verrucano: the only rather puzzling case in the BOTTINO district (East of Pisa, described in CHECCACCI, 1984), where small Pb, Ag, Zn (Cu, Sb, As, Bi) stratiform and/or vein deposits are known. ii) in the transition zone between Verrucano and the overlying evaporites (may be a few meters below, a few meters above, normally along the boundary): it is mainly matter of small baryte (and celestite) stratiform (less frequently vein) deposits; at places haematite occurs with economic interest; pyrite (and other sulfides) is a frequent non-recoverable accessory: only in one case (MONTE ARGENTARIO, South-

---

*) See point 3.1., for details on formation a) and point 3.2. for formations b) and c).

Western Tuscany) the sulfide paragenesis occurs in large—very incompletely explored—quantities. All these deposits have been revised recently by MALANDRINO, 1981; RIVA, 1982; CHECACCI, 1983; DEL FORNO, 1984).

Widespread tourmaline needles and tourmalinites in some Verrucano horizons have to be mentioned, for their possible paleographic significance (see later).

The transition zone between the formations type a) and type c) has paramount importance as metallotect of the major pyrite (with occasional Cu, Zn sulfides) deposits of the Grosseto district(NICCIOLETA, BOCCHEGGIANO p. p., CAMPIANO, p. p.) and the strictly connected haematite-pyrite-pyrrotite Elba Island deposits.

In any case, it is a matter of syn-sedimetary (probably effusive) deposition in evaporitic environment, more or less modified by the Alpidic orogenesis, contact-metamorphism, hydrothermalism (see later on).

The sharp difference in the parageneses and in the industrial importance among the deposits accumulated at the basis of the evaporitic formations, upon the Verrucano or upon the phyllitic sequences cannot be casual: it has probably a paleogeographic significance (distance from the paleo-coast line; basinal energy and related Eh) that can help in the reconstruction of the geological history of the region.

4.3. Tethyan Cycle

For the purpose of the present paper, we shall distinguish the following stages:

4.3.1. Pre-Tethyan rifting:

during this time, the only metallogenic process of a certain interest is the clastic deposition and accumulation of small quantities of cassiterite (MONTE VALERIO, along the Tyrrhenian coast, see VENERANDI-ZUFFARDI, 1983).

The cassiterite source puts problems about the presence at that time of an outcropping basement (the Corso-Sardic massif? A paleo-plate, subsequently sunken in the Tyrrhenian sea?) not so far from the site of accumulation, as it is suggested by the poorly rounded cassiterite grains. This sentence applies also to the tourmaline held in Verrucano described above.

4.3.2. Ophiolitic volcanism:

from the economic stand-point, it had very scanty interest; as a matter of fact the same Cu deposit of MONTECATINI VAL DI CECINA (Tuscany) and the Mn deposits of GAMBATESA (Eastern Liguria), which are the most important ore-bodies of this group, have small tonnages. To complete the list of the deposits related to the ophiolitic Tethyan volcanism, we quote: Libiola (Liguria), Vigonzano, Bisano (Emilia), Rocca Tederighi (Tuscany) for Cu and pyrite; Monte Albareto (Emilia) for talc; Mortale, Terelle (Central Italy) (dubiously) for Mn-oxides; Ziona, Canegreca (Liguria) for (sub-economic) Cr concentrations.

4.3.3. Closure and emersion stages (Middle-Upper Cretaceous):

The deposition of important bauxite masses on the emerging carbonate platform is the major metallogenic event in this stage; the main deposits are widespread in Latium, Abruzzo, Campania, Apulia; the main deposits are S. GIOVANNI ROTONDO and SPINAZZOLA (Apulia).

The studies on these bauxites throw light on the Mediterranean Cretaceous paleogeography: in fact there is general agreement on their allocthonous origin, by alteration of cineritic materials carried by wind from the African plate (BARDOSSY et al., 1977). If this is true, the distances from Africa to the carbonate platforms of the Apennines should have been not so great at those time, probably shorter than now.

4.3.4. Collisional stage:

In the area under investigation (and excluding the Calabrian-Peloritanian Arc) this stage covers the main part of Tertiary and involves the Alpidic orogenesis and plutonism. Bauxite deposition came to end in Paleogene (Southern corner of Apulia) and was followed, at places, by clay accumulation intercalated in flyschoid formations (Neogene of Molise and Apulia).

The Alpidic orogenesis and granitization was characterized, from the metallogenic stand-point, by reworking (metamorphism, recrystallization, selective short distance migration) of pre-existing (pre- and proto-)concentrations.

The following deposits thus originated:

i) high grade, coarse crystalline pyrite (with occasional pyrrhotite) perigranitic masses: GAVORRANO is the main example;

ii) haematite and haematite-magnetite-pyrrhotite-pyrite skarn deposits of Elba Island.

Both i) and ii) accumulations are the result of recrystallization in situ coupled with partial or total oxydation, of pre-existing pyrite masses NICCIOLETA type (see point 4.3.2.).

iii) pyrite and mixed sulfide veins with quartz gangue: BOCCHEGGIANO p. p., CAMPIANO p. p., and others in the GROSSETO district are the main examples.

They are the result of short distance hydrothermal remobilization of pyrite plus mixed sulfide, NICCIOLETA type deposits;

iv) Cu, Zn, Fe, (Pb) skarn deposits: the only example occurs close to CAMPIGLIA (Grosseto district): no evident connection to preconcentrations are known in this case.

The Messinian (Upper Miocene) salinity crisis (see point 3.5.) is the most important metallogenic event (at least from the industrial stand-point) of this time span.

Northern Tuscany (Volterra, Na and alabaster), Eastern Emilia and Marche (gypsum, anhydrite; Perticara, Cabernardi, S); Campania and—specially—Sicily (Cattolica Eraclea, Na; Pasquasia, S. Cataldo, K; Trabonella, Cozzodisi, Gessolungo: gypsum, anhydrite, S) are the interesting areas and the main examples.

4.3.5. Post-collisional stage:

remembering what we pointed out in the first lines of the previous point, this stage starts in Upper Tertiary and includes the not yet terminated post-orogenic volcanism.

Very important ore/mineral accumulations are related to this event, and—particularly—to the volcanics of types b and c, point 3.6.

As a matter of fact, huge Hg deposits (ABBADIA SAN SALVATORE, SIELE, SOLFORATE, etc.) and not negligible Sb deposits (TAFONE, etc.) are, more or less evidently related to the M. Amiata volcanism.

So far as the Quarternary alkali-potassic volcanism is concerned, we may say that enormous potential K—Al reserves are connected with the leucitic lavas; as a matter of fact a few hundreds millions tons of these rocks, holding 20—26% of leucite have been proved; they are waiting for a suitable beneficiation process.

Important fluorite (and baryte) potential reserves, of volcano-sedimetary type, are related to some tuffaceous terranes (Latium: PIANCIANO, etc.); total reserves of about 20 millions tons of run-off, at an average grade of 25% CaF_2 (plus minor quantities of baryte) have been proved. The extreme grain fineness and the intimate mixture with the gangue minerals hinders their economic benefication.

Small deposits of volcanogenic sulfur mixed with marcasite have been partially exploited in the tuffaceous components of these volcanics (Northern Latium).

Small, irregular, generally low grade U occurrences have been recognized, around active or recently extinguished fumaroles of the same area.

Conspicuous high purity clay deposits (bentonites of Ponza Island, etc.) were generated from (mostly hypogene) alterations of perlitic tuffs.

Geo-thermal energy camps are known and exploited in central Tuscany: they are related to a heat dome connected to the M. Amiata volcanic apparatus. Other heat anomalies, at present under active investigation, are located in Latium and in Campania, in connection to the Quaternary alkali-potassic volcanism. Not negligible heavy metal accumulations (Fe, Ti, Zr, Th) are known in the beach sands of the Tyrrhenian coast, in front of the same volcanic apparatus.

Finally it has to be remembered that pyrite deposits, Kuroko type, are generated in the sub-marine areas of the Eolian Arc (HONNOREZ et al., 1970). They have no industrial importance, but represent an interesting metallogenic curiousness.

References

ABBATE E., BORTOLOTTI, V., & PRINCIPI, G. (1980): "Apennine Ophiolite: a peculiar oceanic crust". Ofioliti, Special Issue, Vol. I, pp. 59—96.

BAGNOLI, G., GIANELLI, G., PUXEDDU, M., RAU, A., SQUARCI, P., TONGIORGI, M. (1980): "A tentative stratigraphic reconstruction of the Tuscan Paleozoic Basement". Mem. Soc. Geol. It., Vol. 20 (1979), Roma, pp. 99—116.

BARDOSSY, GY., BONI, M., DALL'AGLIO, M., D'ARGENIO, B., PANTO, GY. (1977): "Bauxites of Peninsular Italy—Composition, Origin and Geotectonic Significance". Monograph Series on Mineral Deposits n. 15, Gebrüder Borntraeger, Berlin-Stuttgart, 61+8 pp.

BIJU-DUVAL, B., DERCOURT, J., LE PICHON, X. (1977): "From the Tethys ocean to the Mediterranean seas: A plate tectonic model of the evolution of the western Alpine system". *In:* BIJU-DUVAL, B., & MONTADERT, L. (Editors), Int. Symp. Structural History of the Mediterranean Basins, Edit. Technip, Paris, pp. 143—164.

Boccaletti, M., Conedera, C., Dainelli, P., & Govec P. (1982): "The recent (Miocene-Quaternay) regmatic system of the Western Mediterranean Region". Journ. Petr. Geol., Vol. V, pp. 31–49.

Boccaletti, M., Manetti, P. (1972): "Traces of lower-middle volcanism in the Crinoidal limestones of the Tuscan sequence in the Montemerano area (Grosseto, Northern Apennines)". Ecl. Geol. Helv., Vol. 65 (1), Basel, pp. 119–129.

Bortolotti, V., Gianelli, G. (1976): „Le rocce gabbroiche dell'Appennino settentrionale: 1. Dati recenti su rapporti primari, posizione stratigrafica ed evoluzione tettonica". Ofioliti, V. 2, Firenze, pp. 99–105.

Casnedi, R.(1982): „Sedimentazione e tettonica delle Unità liguridi nell'Appenino nord-occidentale (Valli Lavagna, Sturla e Aveto)". Atti Ist. Geol. Univ. Pavia, Vol. XXX, pp. 42–66.

Cassinis, G., Elter, G., Rau, A., Tongiorgi, M. (1980): "Verrucano: a tectofacies of the Alpine-Mediterranean southern Europe". Mem. Soc. Geol. It., Vol. 20 (1979), Roma, pp. 135–149.

Channel, J. E. T., D'Argenio, B., Horvath, F. (1979): "Adria, the African Promontory in Mesozoic Mediterranean Paleogeography". Earth Science Reviews, Vol. 15, Amsterdam, pp. 213–292.

Checcacci, M. (1983): „Studio geo-giacimentologico dell'area compresa fra Serravezza, Stazzema, Candella, Pietrasanta (Apuane meridionali)". Graduation Thesis, Dept. of Geo-Sciences, University of Milan, 175 pp.

Ciarapica, G., Passeri, L. (1981): „Tentativo di ricostruzione paleogeografica a livello del Trias nella Toscana Nord dell'Arno e sue implicazioni tettoniche". Mem. Soc. Geol. It., Vol. 21 (1980), Roma, pp. 41–49.

Cocozza, T., & Vai, G. B. (1978): "Paleozoic of Tuscany within the Paleogeographic evolution of Europe". *In:* Tongiorgi, M. (Ed.): Report on the Tuscan Paleozoic Basement. C. N. R. Internal Report of the „Progetto finalizzato Energetica – Sottoprogetto Energia Geotermica". Pisa, pp. 51–58.

Cortecci, G., Klemm, D. D., Lattanzi, P., Tanelli, G., Wagner, J. (1983): "A sulfur isotope study on Pyrite Deposits of Southern Tuscany, Italy". Mineral Deposita, *18*, pp. 285–297.

Cortesogno, L., Gianelli, G., & Piccardo, G. B. (1975): "Preorogenic metamorphic and tectonic evolution of the ophiolite mafic rocks (Northern Apennine and Tuscany)". Boll. Soc. Geol. Ital., Vol. XCIV, pp. 291–327.

D'Argenio, B., Alvarez, W. (1980): "Stratigraphic evidence for crustal thickness changes on the southern Tethyan margin during the Alpine cycle". Geol. Soc. American Bull., Part. I., V. 91, pp. 681–689.

Decandia, F. A., & Elter, P. (1969): „Riflessioni sul problema delle ofioliti nell'Appennino settentrionale (Nota preliminare)". Atti Soc. Toscana. Sc. Nat., Vol. DXXVI, pp. 1–9.

Del Forno, F. (1984): „Studio geo-giacimentologico della zona dei Monti di Capalbio (Toscana Meridionale)". Graduation Thesis, Dept. of. Geo-Sciences, University of Milan, 140 pp.

Elter, G., Elter, P., Sturani, C., & Weidmann, M. (1966): „Sur la prolongation du domaine ligure de l'Apennin dans le Monferrat et les Alpes et sur l'origine de la Nappe de la Simm s. l. des Préalpes romandes et chablaisiennes". Archives des Sc. Genève, Vol. IXX, pp. 279–377.

Elter, P., & Pertusati, P. (1973): „Considerazioni sul limite Alpi-Appennino e sulle sue relazioni con l'arco delle Alpi occidentali". Mem. Soc. Geol. Ital., Vol. XII, pp. 359–375.

Folk, R. L., & McBride, E. F. (1976): "Possible pedogenic origin of Ligurian ophicalcite: A Mesozoic calichified serpentinite". Geology, Vol. IV, pp. 327–332.

GRANDJACQUET, C., & HACCARD, D. (1977): ›Position structurale et rôle paléogéographique de l'unité du Bracco au sein du contexte ophiolitique liguro-piémontais (Apennin-Italie)‹. Bull. Soc. Géol. France, Vol. XIX, pp. 901–908.

GRIONI, F. (1984): „Studio geo-giacimentologico della zona di Pari (Toscana Meridionale)". Graduation Thesis, Dept. of Geo-Sciences, University of Milan, 270 pp.

HSÜ, K. J., CITA, M. B., RYAN, W. B. F. (1973): "The origin of the Mediterranean evaporites". *In:* RYAN, W. B. F., HSÜ et al ii, Initial Reports D. S. D. P., n. 13, Washington, pp. 1203–1231.

MALANDRINO, G. (1982): „Studio geo-giacimentologico della zona di Cetine di Cortorniano-Camporedaldi (Toscana Meridionale)". Graduation Thesis, Dept. of Geo-Sciences University of Milan, 122 pp.

RIVA, A. (1983): „Studio geo-giacimentologico della zona Valle del Tafone-Ponte S. Pietro (Toscana meridionale)". Graduation Thesis, Dept. of Geo-Sciences, University of Milan, 164 pp.

SAGRI, M., & MARRI, C. (1980): „Paleobatimetria e ambienti di desposizione delle unità torbiditiche cretaceo-superiori dell'Appennino settentrionale". Mem. Soc. Geol. It., Vol. XXI, pp. 231–240.

TANELLI, G., LATTANZI, P. (1983): "Pyritic ores of Southern Tuscany, Italy". ICAM 81, Geol. Soc. of South Africa, special pubbl., 7 (in press).

VENERANDI PIRRI, I.,ZUFFARDI, P. (1982): "The Tin Deposit of Monte Valerio (Tuscany, Italy): Pneumotolitic-hydrothermal or sedimentary-remobilization processes?". Ramdhor special volume, Springer-Verlag, Berlin, 57–65.

VIGHI, L. (1966): „Descrizione di alcuni sondaggi che hanno attraversato lenti anidritico-dolomitiche intercalate alle filladi triassiche (Verrucano) nei dintorni di Massa Marittima (Grosseto-Toscana)". Atti del Symposium sul Verrucano. Pisa, settembre 1965. Soc. Tosc. Sci. Nat. pp. 72–95.

Geo-tectonic evolution of the Sardo-Corsican massif and metallogeny of Sardinia

By T. Cocozza*), F. A. Decandia*), A. Marcello***), S. Pretti**), I. Salvadori**), I. Uras**), R. Valera**)

With 2 Figures

Zusammenfassung

Die palaeogeographische Entwicklung des Sardo-Corsischen Massivs wird dargestellt. Während des Paläozoikums war das Massiv mit der Toskana verbunden. Zur Triaszeit setzte Krustendehnung ein und im Jura ereignete sich die Trennung des Massivs von der Toskana, wobei Sardinien-Corsica mit der europäischen Platte verbunden blieb, während in der trennenden Furche der Ozeanboden in Form der ligurischen Ophiolite in Erscheinung trat. Gegen Ende der Kreide wurde der Ligurische Trog geschlossen, weil der Afrikanische Kontinent sich wieder an Europa heranschob. Im Untermiozän begann das Massiv sich abermals von Europa zu trennen. Im Massiv ist der „Metallvorrat" in zwei Hauptgruppen von Vorkommen einzuteilen: in echte Lagerstätten und in Metallgehalte der Gesteine. Zumeist sind die Lagerstätten von den älteren (paläozoischen) Gesteinen abgeleitet; dabei sind exogene Bildungen bedeutender als endogene.

Geological History along the Thetys Evolution

After the end of the Paleozoic up to Messinian time, the Sardo-Corsican Massif was affected by a number of movements, that account for its present position and for the evolution of the Western Mediterranean.

---

*) Dipartimento Scienze della Terra, Via delle Cerchia 3-53100 Siena
**) Istituto Giacimenti Minerari, Facoltà Ingegneria, Piazza d'Armi 09100 Cagliari.
***) Ente Minerario Sardo, Via XXIX Novembre 31-09100 Cagliari

During the Upper Paleozoic Sardinia and Corsica were involved not only in the latest Hercynian collapses, but also in horizontal translations, along the North-Pyrenean transcurrent fault, approximately along the Provence area (ARTHAUD & MATTE, 1977b).

The former movements evidently affected also the basement of Tuscany, whose Paleozoic terrains appear strictly analogous to the Sardinian ones (CARMIGNANI & MECCHERI, in print). Thus some of the present major transverse faults of the Northern Apenines are likely to be related with the same fault system of late Hercynian age.

During this period Sardinia emerged and was characterized by structural highs and lows; a long continental period began, and a wide peneplanation was reached.

Also during the Mesozoic, extensive movements went on and intracratonic sea basins, controlled by different morphology, were formed (GANDIN et al., 1982).

At this time the western margin of the Massif (together with the adjacent north-western areas) wide platforms and shallow, narrow basins have been formed where Germanic facies sediments were being deposited, while its eastern margin represented the western border of the Triassic Alpine Basin.

In opposition to what some authors believe, the Punta Bianca (Tuscany) complex, which includes alkaline volcanics (RICCI & SERRI, 1976), is part of the eastern border of the last basin; this is also in agreement with the paleogeographic and structural situation of the Cape Argentario (DECANDIA & Lazzarotto, 1980; DECANDIA et al., 1980).

A transgression started in Triassic time and, during the Jurassic, extended gradually toward the eastern part of the Massif and on the platform a Provençal facies carbonatic complex was deposited. Confined pelagic episodes testify the presence, inside the platform area, of basins related to the synsedimentary extensive tectonics, that are the early signs of the opening of the Ligurian-Piedmontese oceanic basin.

The tearing of the sialic crust in the intracratonic Triassic Alpine Basin, during the Upper Jurassic, separated the Sardo-Corsican Massif from the Tuscan one (DURAND-DELGA, 1981). In fact an oceanic basin was formed, with simatic crust (ophiolites) and pelagic sediments, which evolved, in the Upper Cretatic, to flysch. After the tearing of the sialic crust, the Sardo-Corsican Massif continued evolving together with the stable side of the European plate up to present, while the Tuscan basement followed the new evolution phenomena of the Apennines continental margin.

The paleographic differentation, which was evident since Cretaceous time with carbonatic platform sediments on the western side and a mainly pelagic deposition on the eastern one (FOURCADE et al., 1977), is consistent with such a structural evolution.

Nearly at the end of the Cretaceous, the closure of the Ligurian-Piedmontese basin commenced and, as a consequence of the consumption of the related simatic crust, the African continent approached again the European one. The collision took place during the Middle Eocene. However, it has to be pointed out that clear evi-

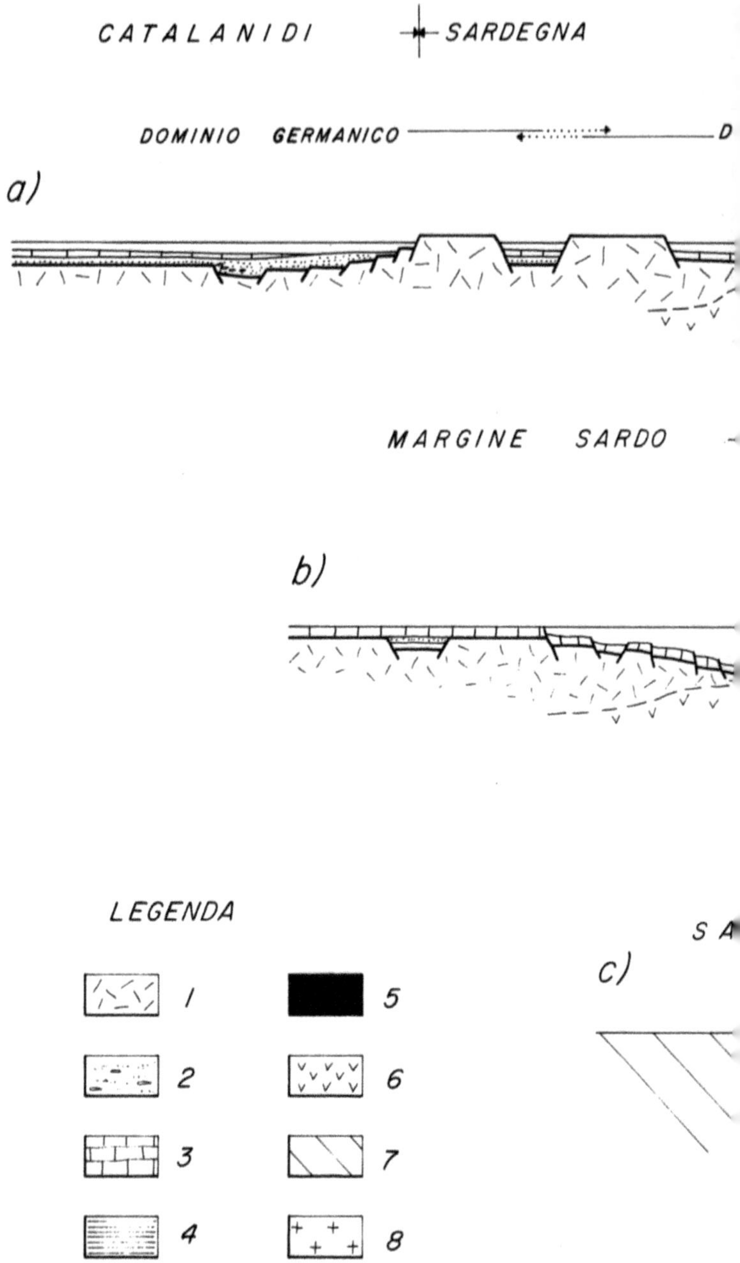

Fig. 1. Interpretative sections (not in scale) relative to the main
BOCCALETTI et al. (1981). – 1 = Pre-triassic basement; 2 = Most
volcanics; 6 = Oceanic lithospher

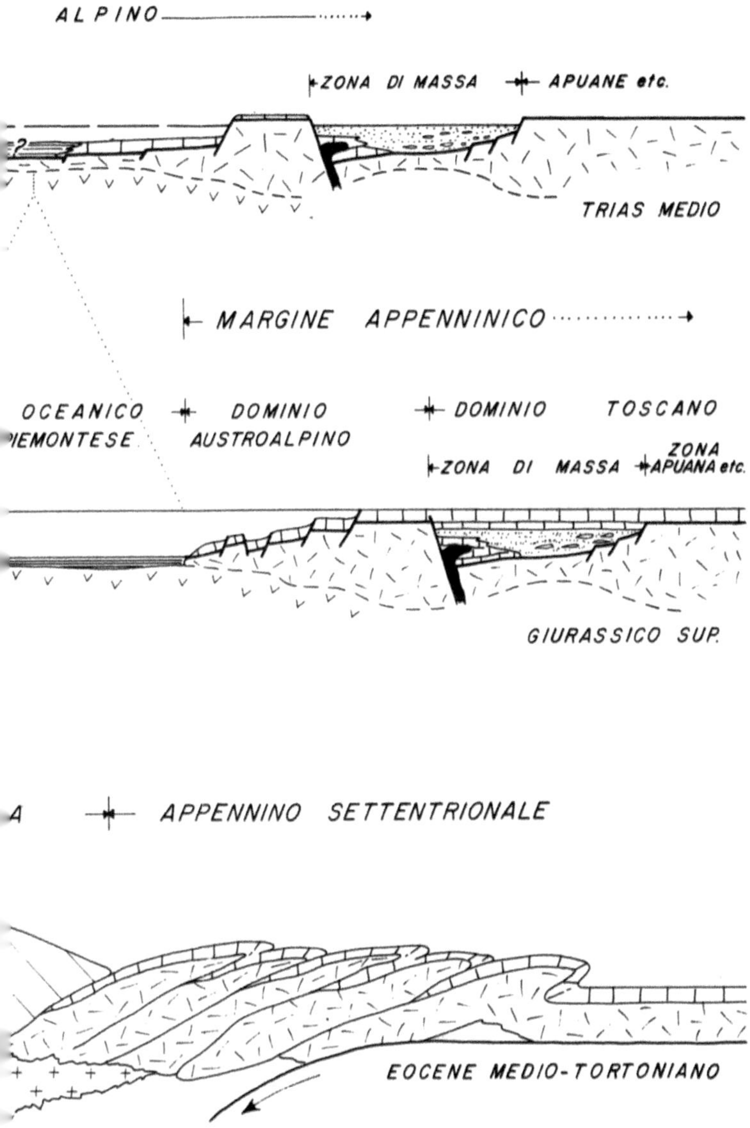

nic-sedimentary evolution of the Western Mediterranean. Section C from
deposits; 3 = Platform marine deposits; 4 = Pelagic deposits; 5 = Alkaline
nt and cover; 8 = Synkinematic granites.

dence of compressive phenomena related to this event are lacking in Sardinia. On the contrary, all its post-Hercynian evolution seems to be characterized by distension phenomena (ALVAREZ & COCOZZA, 1974).

After the collision, the structural evolution of the Apennine chain strictly matches with that of the Sardo-Corsican Massif. During the Upper Eocene, extension structures developed in Sardinia (COCOZZA & SCHÄFFER, 1973); these structures caused, in the Lower Miocene, the separation of the Massif from the stable Europe and then its counterclockwise rotation. At the same time, in the Apennines, the ensialic tectonics started with the deformation of the continental margin, that became shortened by overthrusting phenomena, which are younger and younger eastwards (BOCCALETTI et al., 1980).

Some Sardinian structural lows were covered by epicontinental seas, some formed lagoonal or lacustrine basins.

During the Oligocene continental conditions prevailed, and an intense calc-alkaline magmatic activity, related to the rotation of the Massif, developed.

After this long continental phase, during the Miocene the sea covered the area of the "Sardinian Rift" and separated two emerged areas: the western one and the eastern one. This transgression, characterized by mainly terrigenous deposition, is related to the opening of the Algerian-Provençal basin (BARBERI & CHERCHI, 1980).

In the Upper Miocene almost the whole Sardinia emerged again, only a few, marginal areas being still characterized by lagoonal conditions.

In that moment the counterclockwise rotation stopped, and an alkaline magmatism followed; at the same time the extension phase commenced in the area of the Apennines.

The Metallic Stock in the Sardinian Paleozoic Terrains

As a metallic stock we mean the presence of "metallic" elements of economic interest, apart from their concentrations and mineralogical framework. However, two groups mut be considered:

a) anomalous concentrations of "metallic" elements (ranging from geochemical anomalies up to ore-bodies);

b) presence of "metallic" elements in rock-forming minerals (in normal concentrations, according to rock types).

Taking into account these definitions, we can divide the paleozoic Sardinia into a few main sectors, characterized by the presence of a given set of elements and by their relative abundance.

NW Sector (Nurra). It is the smallest of the Paleozoic sectors, and completely separate from the others.

In it a few ore bodies (group a) are known: Pb-Zn veins of some importance, Sb veins of lesser importance, and Fe lens-shaped bodies exploited in the past; in granites and metamorphic rocks (group b) Fe and Al are occuring.

NE Sector (Gallura-Logudoro-Nuorese). In this sector, mainly granitic, many ore-bodies (group a), mainly vein-shaped, are known, but their importance is low; they contain Pb-Zn ores, barite and fluorspar. Cu, Mo (and minor W) are known in a few localities as low-grade disseminated and stockwork mineralization or scattered veinlets, in intrusive rocks. Also a few lens-shaped Fe-bearing bodies are known in roof-pendants. The group b includes Al, Fe and heavy minerals (especially monazite) in granitic and metamorphic rocks. Also U as geochemical anomalies (group a) in microgranites has been reported.

SW Sector (Arburese-Iglesiente-Sulcis). Many and important ore-deposits of several types (group a) are known in this sector. They can be summarized as follows: Pb-Zn in vein-shaped and stratabound bodies, numerous and often important; a few veins of minor importance bear minerals of Sn, Ni, Co and As; F and Ba are widespread in vein-shaped and stratabound bodies, sometimes important; mixed sulphides (mostly Pb-Zn and minor Cu) in numerous lens-shaped bodies; also widespread, but seemingly less important, Mo and minor W in greisen-type mineralization, and disseminations and quartzose veinlets in granites; a few lens-shaped bodies contain Fe and less W and F; yet, for the group a, geochemical U anomalies in microgranites are known. For the group b, still Al, Fe and heavy minerals in granites and metamorphic rocks are worth mentioning.

SE Sector (Barbagia-Ogliastra-Sarrabus-Gerrei). Also in this sector several ore-bodies (group a) are known. Veins containing F-Ba and minor Pb-Zn-Ag are widespread and sometimes important. Frequent, but less important, are mixed sulphide ores (Pb, Zn, Cu, Fe, As) in lens-shaped bodies.

Scarce, but exploited in the past, stratabound Sb-ores with minor amounts of W are existing.

In the intrusive rocks, poor disseminations of Cu and less frequently Mo, as well as a few veins containing Mo (-W-Sn) minerals are known. Always Fe, Al and heavy minerals represent the group b in granitic and metamorphic rocks.

Metallogenic Process in Sardinia during the Tethyian Evolution

The previous outline of the geological history shows that Sardinia, in the Thetian domain, represented a continental fragment, episodically covered by shallow seas. Thus its metallogenic evolution is dominated by the following phenomena:

i) the long continental period after the Hercynian uplift;
ii) the recurrent marine transgressions;
iii) the intense volcanic activity in Tertiary times, with possible intrusive episodes;
iv) the final general uplift.

On the basis of these elements, two main groups of metallogenic events must be distinguisched:

A—Exogenous phenomena: they include all the processes of erosion and leaching in subaerial environments, those of deposition in continental and marine basins, and phenomena of mobilization and remobilization. In all these processes almost only "metals" inherited from Paleozoic rocks are involved.

B—Endogenetic phenomena: they are related to the phases of magmatic activity. A Paleozoic heritage is likely in some cases, while in other cases a supply of "new" elements is more probable.

Group A. The occurrences belonging to this group are numerous and sometimes important. On the basis of the already described main characters of the evolution of Sardinia, they can be further subdivided as follows.

1. Mobilization and remobilization phenomena: They started in connection with the erosional cycle subsequent to the Hercynian uplift in the Upper Paleozoic. During the Mesozoic the landscape reached the stage of senility, and consequently these processes became less and less active. The rejuvenation related to the Alpine uplift activated similar phenomena, that are often superimposed on the older ones, so that it is difficult to make a distinction between the two phases.

From the economic point of view, the mobilization processes were effective only in already mineralized carbonatic rocks, i. e. in the Cambrian districts of the south-western part of the island (Fig. 2, site a). The ore-bodies are mainly the result of the filling of karst cavities (Padalino et al., 1973). The main ore minerals in these bodies are: barite (Marcello et al., 1983), galena (sometimes very rich in silver) and oxidized lead minerals, oxidized zinc minerals. Important lead and zinc occurrences are known, and exploited, near Iglesias.

In rocks other than carbonatic, these phenomena, even though frequent, and displaying a wider range of minerals (at least Ag, Ba, Cu, F, Fe, Pb, U and Zn minerals are known), never gave birth to economic occurrences. The most remarkable example consists of thin fracture fillings and veinlets of oxides and phosphates of uranium, in thermometamorphic rocks at the north-eastern edge of the Sulcis granite (Fig. 2, site b).

All the "metals" involved in these processes were present in the Paleozoic rocks mostly as both true ore bodies and geochemical anomalies; only Fe was present also as a normal component of the rocks. Thus we can infer the origin of the most part of these ores from a stock of type (a).

2. Sedimentation in lakes, lagoons, coastal basins: Many of the known deposits are related to Mesozoic sediments. In other cases their age is uncertain, ranging from Permian to upper Tertiary or even Quaternary times.

They can be classified in four main types.

i) Thin-bedded barite-galena-silica layers (Brusca et al., 1971) (Fig. 2, site a). The known occurrences, associated with sandstones, are always related to the Cambrian carbonatic rocks, in a senile morphological framework (post-Hercynian peneplain). Althoug they are not yet dated, a relation between them and the Sulcis Basin of Eocene age is likely. Their rich content in $BaSO_4$ and the proximity to the main remobilized barite-bearing bodies allow their economic exploitation, in spite of their generally small size. Both barite and galena of these beds belong to type (a) stock.

ii) Pb-Zn (and minor Cu) sulphides in carbonatic depositions. Only one, very small example is known in central-eastern Sardinia (Salvadori & Zuffardi, 1966) (Fig. 2, site c).

Its stock is always of type (a). A few other occurrences are but geochemical anomalies.

iii) U oxides and phosphates in fluvial-lacustrine clays, associated with carbonized wood. Only one occurrence has been discovered so far; it is a small basin in a granitic area (northern Sardinia) possibly of late Tertiary-Quaternary age, as its morphological characteristics allow to infer (Fig. 2, site d). The uranium was supplied by the surrounding granites, where it occurs in geochemically high values [type (a) stock].

iv) Fe oxides, associated with clays and/or carbonatic rocks (MARINI, 1983) of the basal Mesozoic in central Sardinia (Fig. 2, site e).

The ores occur both as impregnation of the Paleozoic footwall rocks (thus belonging rather to the next subgroup) and as thin beds in the Mesozoic.

These occurrences are widespread and remarkably extensive, but never reach a satisfying thickness. The stock of the iron is likely to be both of type (a) and of type (b).

3. Residual concentration: Two types of deposits developed from the Mesozoic onwards. The first one is related to the long lasting continental conditions in the Cambrian area; it consists of mainly clastic accumulations of barite with minor galena (Fig. 2, site a). One important occurrence is known, and exploited near Iglesias; others are less explored, and some might be important. The stock is obviously of type (a); for the younger occurrences, related to the last rejuvenation, at least a part of the ore is likely to stem from post-Paleozoic bodies belonging to the subgroups 1 and 2.

The second type is related to an infra-Cretaceous continental phase in the Mesozoic basin of north-western Sardinia (Nurra) (Fig. 2, site f). It is a fairly continuous bauxite bed, resting on carbonatic rocks; the interest in them is growing more and more as the exploration goes on.

The stock of the aluminium is obviously of type (b).

4. Alluvial and eluvial accumulation of Tertiary and Quaternary age. Barite, with minor galena, is known in scarcely explored, perhaps important occurrences, always in the Cambrian districts (MARCELLO et al., 1983) (Fig. 2, site a). The stock is of type (a), possibly derived from both Paleozoic and post-Paleozoic pre-existing bodies.

Interesting placer indications of other heavy minerals (mostly monazite, cassiterite and Fe-Ti minerals) have been discovered in recent times (MARCELLO et al., 1978).

Their stock is mostly of type (a), but some occurences of Fe-Ti minerals, that have been known for several tens of years along the north-western coast, emanate from Tertiary volcanic rocks, thus a post-Paleozoic stock of type (b).

Group B. The endogenetic ore-forming phenomena did not give rise to orebodies comparable, in size and importance, to those of the former group.

A further division of this group is possible, as follows.

1. Hydrothermal veins. Several, but worthless veinlets occur in Tertiary volcanic rocks of the western half of Sardinia, mainly barite-bearing (MARCELLO et al.,

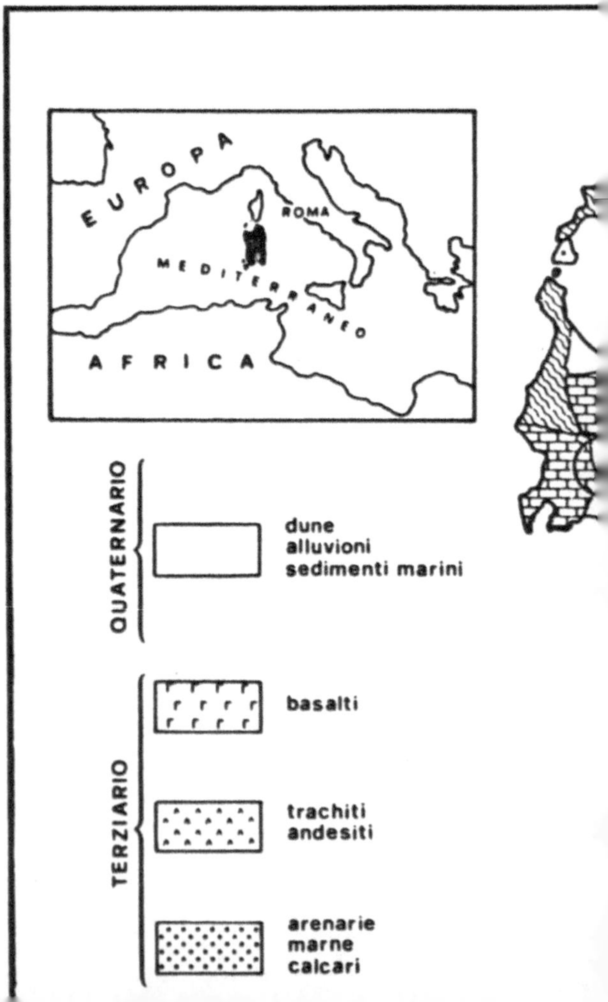

EUROPA
ROMA
MEDITERRANEO
AFRICA
QUATERNARIO
dune
alluvioni
sedimenti marini
TERZIARIO
basalti
trachiti
andesiti
arenarie
marne
calcari

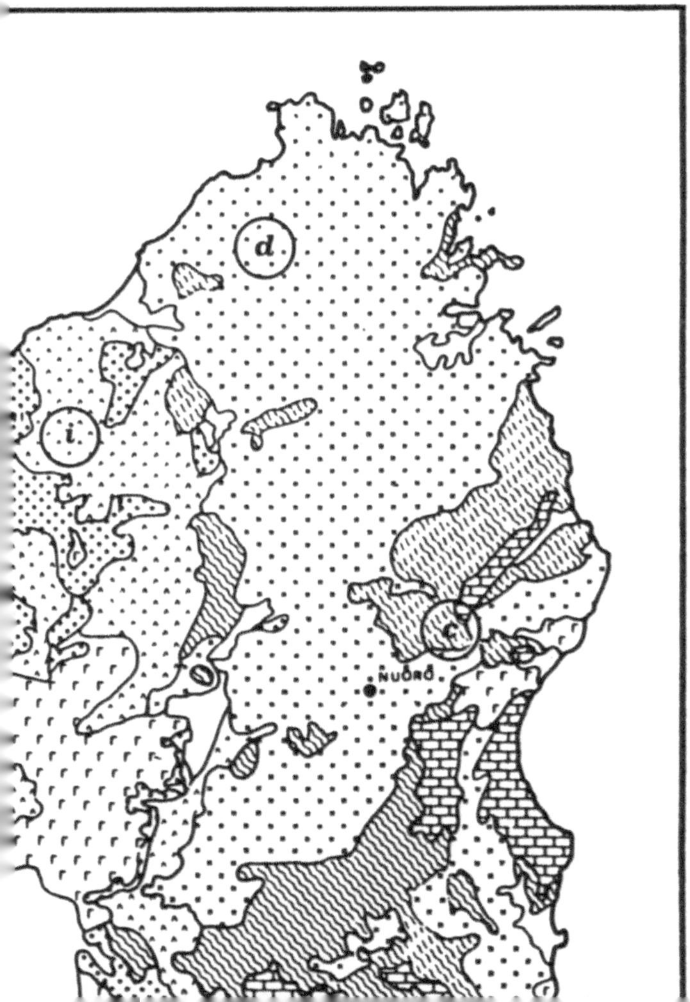
d
i
c
NUORO

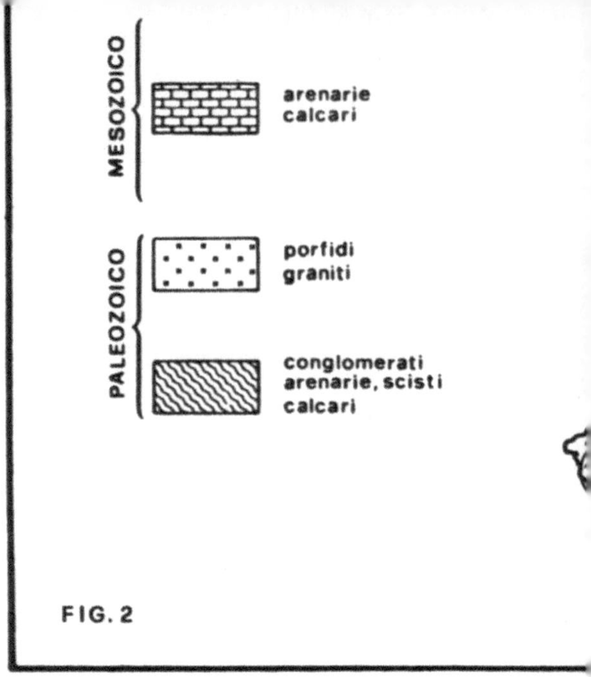

Fig. 2. Ore occurrenc

a – Ba-Pb-Zn in bodies of various origins: mobilized an lized, sedimentary (clastic to chemical), residual and often important occurrences).

b – U oxides and phosphates mobilized in veninlets (sir rence).

c – Pb-Zn sulphides in carbonatic rocks (one occurrence indications).

d – U oxides in continental clays, associated to orga (single occurrence).

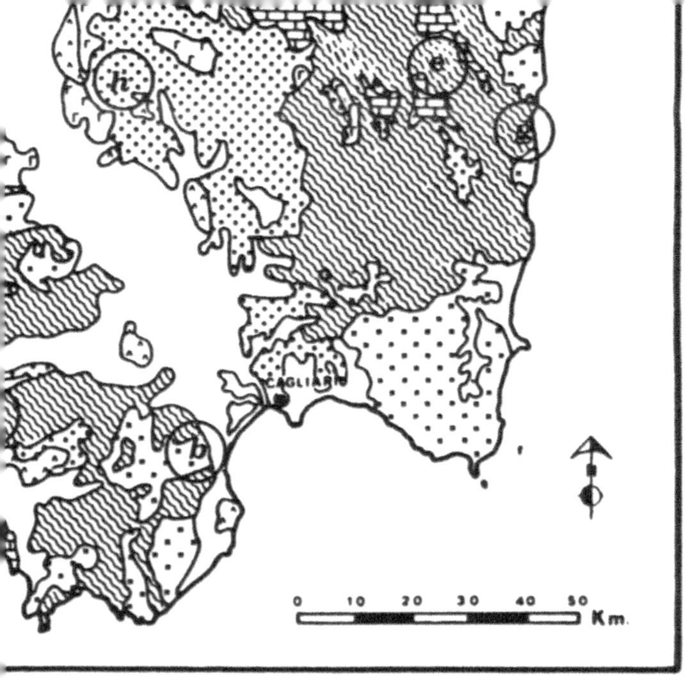

ons in post-Paleozoic terrains.

– Fe oxides in thin beds and impregnations (several occurrences).

– Bauxite interbedded in carbonatic rocks (broad occurrence)

g – Barite in veins (elongated vein system).

a – Fluorspar in veins, associated to volcanic rocks (one occurrence, a few indications).

–l – Mn in stockworks and lenses, and disseminated Cu-Fe sulphides, in volcanic rocks (several occurrences and indications).

n – Mn in beds, stockworks, and veinlets (several occurrences).

1983). The only noteworthy barite occurrence is a NS-striking vein system, in the south-western sector, hosted in Paleozoic terrains (Fig. 2, site g).

Isotopic studies on these barites commonly show heritage from Paleozoic rocks.

Also very small are a few fluorite occurrences, both in volcanic and sedimentary Tertiary rocks (Cavinato, 1954); among them, only one vein-shaped occurrence in central Sardinia was an object of prospecting (Fig. 2, site h). Also in this case, a Paleozoic heritage is likely. A last occurrence, an iron-bearing vein in tertiary volcanics in central-western Sardinia, was also explored, and abandoned. Its iron ore may or may not be inherited.

2. Volcano-sedimentary Mn bodies (Fig. 2, sites i, l, and m). They occur in the western side of Sardinia, strictly related to Tertiary volcanic rocks (Uras, 1965). These occurrences, both stratiform and stockworks, were explored and partially exploited in several places in the past; to-day they are under re-examination.

While Mn looks to be a "new" metal in Sardinia, in some occurrences of the northern districts also W was detected in several assays; this last metal might be inherited from Paleozoic rocks.

3. Stockworks and disseminated occurrences of mainly Cu-Fe sulphides in Tertiary calc-alkaline volcanics (Fig. 2, sites i and l). Several indications of this type are known, but exaustive investigations have not yet been performed. Even though copper-bearing mineralization is known in Paleozoic terrains, a porphyry-copper-type phenomenon of formation of these bodies should exclude any important derivation from occurrences of this kind.

References

Alvarez, W. (1972): Rotation of the Corsica-Sardinia Microplate. Nature Phys. Sc., 235, 553–559.

Alvarez, W., & Cocozza, T. (1974): The Tectonics of Central-Eastern Sardinia- and the Possible Continuation of the Alpine Chain to the South of Corsica. In "Paleogeografia del Terziario Sardo nell'ambito del Mediterraneo occidentale". Cagliari 1973, Sem. Fac. Sc. Univ. di Cagliari, suppl. vol. 43 (1973), 5–34.

Arthaud, F., & Matte, P. (1977 a): Synthèse provisoire sur l'évolution téctonique et les raccords entre les segments hercyniens situés autour du bassin nord-baléare (Sud de la France, Espagne, bloc Corso-Sarde). Coll. Intern. CNRS, n. 243, 497–513.

Arthaud, F., & Matte, P. (1977 b): Determination de la position initiale de la Corse et de la Sardaigne à la fin de l'orogenèse hercynienne grace aux marqueurs géologiques anté-mésozoiques. Bull. Soc. Géol. France, Ser. 7, 19, 833–840.

Barberi, F., & Cherchi. A. (1980): Excursion sur le Mésozoique et le Tertiaire de la Sardaigne occidentale. Livret-guide C. N. R., Prog. Finaliz. Geodinamica, Pubbl. n. 345.

Boccaletti, M., Coli, M., Decandia, F. A., Giannini, E., & Lazzarotto, A. (1981): Evoluzione dell'Appennino settentrionale secondo un nuovo modello strutturale. Mem. Soc. Geol. It., *21* (1980), 359–373.

Brusca, C., Pretti, S., Tamburrini, D., & Tocco, S. (1971): Su alcune mineralizzazioni a barite e galena insediate nei „karst" del penepiano ercinico (Iglesiente meridionale – Sardegna). Res. Ass. Min. Sarda, LXXVI, *5*, 25 pp.

Carmignani, L., & Meccheri, M. (1984): Contributo alle conoscenze stratigrafiche del basamento delle Alpi Apuane. Soc. Tosc. Sc. Nat., Mem., Ser. A. (in print).

Cavinato, A. (1954): La fluorite nel Terziario in Sardegna. Res. A. M. S., LVIII, *4,* 11 pp.

Cocozza, T., & Schaffer K. (1973): Cenozoic Graben tectonics in Sardegna. *In:* „Paleogeografia del Terziario sardo nell'ambito del Mediterraneo occidentale", Cagliari 1973, Rend. Sem. Fac. Sc. Univ. Cagliari, suppl. V, *43,* 145–162.

Cocozza, T., Jacobacci, A., Nardi, R., & Salvadori, I. (1974): Schema stratografico-strutturale del Massiccio sardo-corso e minerogenesi della Sardegna. Mem. Soc. Geol. It., *13,* 85–186.

Decandia, F. A., & Lazzarotto, A. (1981): Le unità tettoniche del Monte Argentario (Toscana meridionale). Mem. Soc. Geol. It., *21* (1980), 385–393.

Decandia, F. A., Giannini, E., & Lazzarotto, A. (1981): Evoluzione paleogeografica del margine appenninico nella Toscana a Sud dell'Arno. Mem. Soc. Geol. It., *21* (1980), 375–383.

Durand-Delga, M. (1981): Ouvertures océaniques de la Méditerranée occidentale et dislocations des chaînes alpines. *In:* Wezel, F. C. (ed.), Sedimentary basins of Mediterranean margins, Bologna, 417–431.

Elter, P., Giglia, G., Tongiorgi, M., & Trevisan, L. (1975): Tensional and compressional areas in the recent (Tortonian to present) evolution of the Northern Apennines. Boll. Geof. Teor. Appl., Trieste, *17,* 3–18.

Fourcade, E., Azema, J., Chabrier, G., Chauve, P., Foucault, A., Rangheard, Y. (1977): Liaisons paléogéographiques au Mésozoiques entre les zones externes Bétiques, Baléares, Corso-Sardes et Alpines. Rev. Géogr. Phys. et de Géol. Dynam., *19* (4), 377–388.

Gandin, A., Tongiorgi, M., Rau, A., & Virgili, C. (1982): Some examples of the Middle-Triassic Marine Transgression in South-Western Mediterranean Europe. Geol. Rund., *71* (3), 881–894.

Giglia, G. (1973): L'insieme Corsica-Sardegna e i suoi rapporti con l'Appennino settentrionale: rassegna di dati cronologici e strutturali. *In:* „Paleogeografia del Terziario sardo nell'ambito del Mediterraneo occidentale", Cagliari 1973, Rend. Sem. Fac. Sc. Univ. Cagliari, suppl. V, *43,* 245–276.

Marcello, A., Pretti, S., & Salvadori, I. (1978): Le prospezioni geominerarie in Sardegna: la prospezione geochimica strategica. Boll. Serv. Geol. It., XCIX, 277–310.

Marcello, A., Pretti, S., & Uras, I. (1983): La barite in Sardegna. L'Industria Mineraria, *5,* 1–32.

Marini, C. (1983): Le concentrazioni residuali post-erciniche di Fe nell'Ogliastra (Sardegna orientale). Contesto geologico e dati mineralogici. Cagliari, 20 pp.

Padalino, G., Pretti, S., Tamburrini, D., Tocco, S., Uras, I., Violo, M., & Zuffardi, P. (1973): Ore Deposition in Karst Formations with Examples from Sardinia. *In:* Amstutz, G. C., & Bernard, A. J. (ed.) „Ores in Sediments", Springer-Verlag, 209–220.

Reutter, K. J., Günther, K., Groscurth, J. (1978): An Approach to the Geodynamics of the Corsica-Northern Apennines Double Orogene. *In:* Alps, Apennines Hellenides. Inter-Union Commission on Geodynamics Scientific Report no 38, 299–311.

Ricci, C. A., & Serri, G. (1976): Evidenze geochimiche sulla diversa affinità petrogenetica delle rocce basiche comprese nella serie a facies toscana. Boll. Soc. Geol. It., *94* (1975), 1187–1198.

Salvadori, I., & Zuffardi, P. (1966): Indizi di mineralizzazioni metallifere nel Mesozoico sardo. Res. A. M. S., LXXI, *6,* 27–53.

Uras, I. (1965): Depositi manganesiferi in Sardegna. Parte I: generalità, studi sul terreno. Res. A. M. S., LXX, *3,* 32 pp.

Geo-Tectonic Evolution and Metallogenesis of the Calabrian-Peloritanian Arc in the Tethys Realm

By E. Zanettin Lorenzoni*), P. Zuffardi**)

With 4 Figures

Zusammenfassung

Der Calabrisch-Peloritanische Bogen, die Südspitze Italiens und Nord-Sizilien umfassend, gehört zum Südteil der Europäischen Platte. Er ist vorwiegend aus praeherzynischen und herzynischen Serien aufgebaut. Die Entwicklung seiner Tektonik wird eingehend beschrieben. Die nicht sonderlich reiche Vererzung ist zumeist paläozoisch; die spärliche alpidische Metallogenese ist auf die Abwesenheit triassischen Riftings und tertiärer saurer Magmatite zurückzuführen.

Summary

The Calabrian-Peloritanian Arc is mainly made of Hercynian/pre-Hercynian terranes, moulded in complicated structures, by both the Hercynian and the Alpidic tectogeneses.

The ore/mineral accumulations are prevailingly related the Hercynian/pre-Hercynian metallogeny and have been partially dispersed and/or altered in connection with subsequent orogeneses and metamorphism.

The interpretation of the evolution of the Arc is controversial: the differences in opinion concern the domains to which the complexes belong, the origin of the ophiolites, the position of the Tethys, the age of tectogenesis, the dynamics of the orogenesis, the vergence of the orogene.

To solve these problems, new contributions particularly on the basis of the metamorphic and magmatic history of the area, are presented in this paper.

---

*) Department of Earth Sciences, University of Bari.
**) Department of Earth Sciences, University of Milano.

1. Foreword

The Calabrian-Peloritanian Arc represents a well defined geological and metallogenic province, covering almost the whole Calabria and the Northern corner of Sicily; it is delimited by two tectonic lineaments: the "Sangineto line" in the North and the "Taormina line" in the South (Fig. 1).

From the geological and metallogenic stand-points, the Arc is undoubtedly different from the other parts of the Apennines and of Sicily, and it is rather similar to Sardinia and to the Alps. As a matter of fact the Arc, like Sardinia and large sections

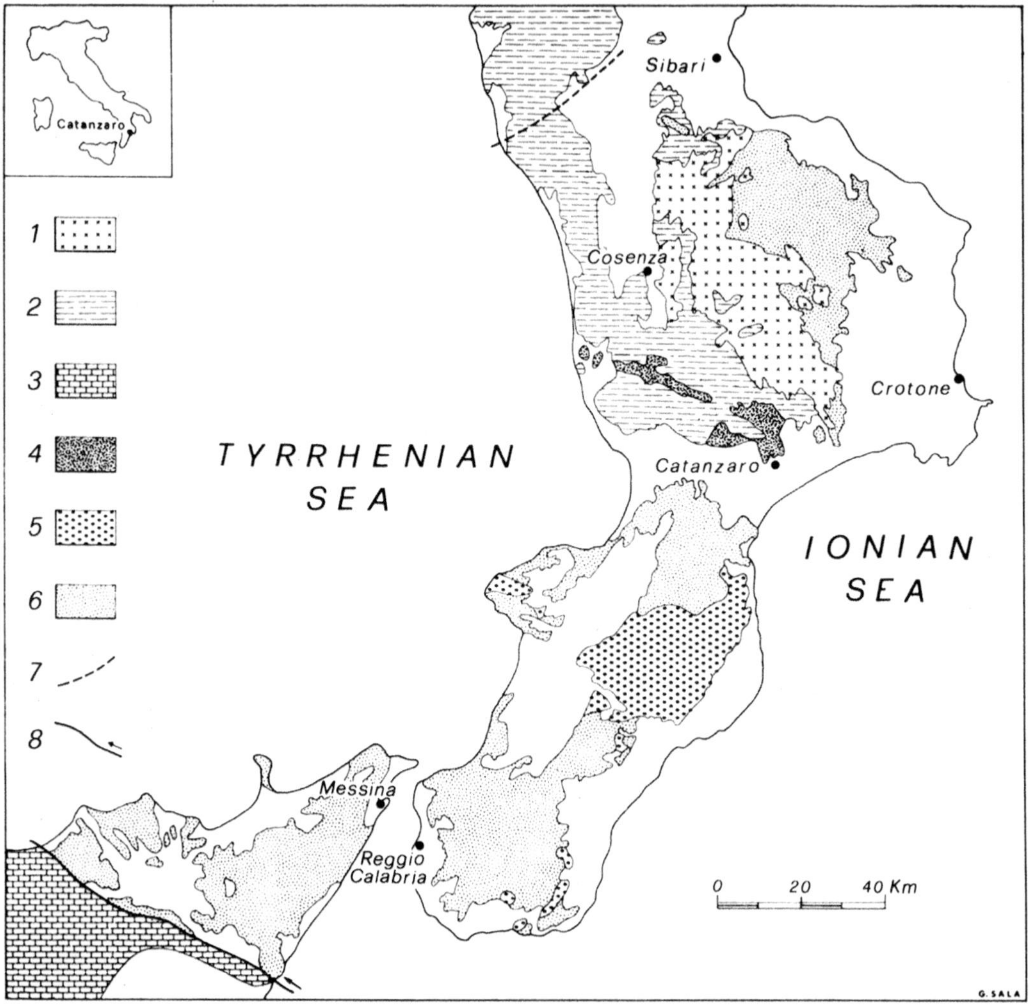

Fig. 1. Tectonic sketch of the Calabrian-Peloritanian Arc. 1 = Monte Gariglione Unit; 2 = Calabrian Apennine Range; 3 = Sicilian Maghrebids; 4 = Tiriolo Unit; 5 = Stilo Unit; 6 = Calabrian-Peloritanian Hercynian Range; 7 = Sangineto Line; 8 = Taormina Line.

of the Alps, is mainly composed of pre-Hercynian terranes, whilst post-Hercynian ones are dominant in the Apennines and in Sicily, at least at the surface and till the depths reached by drilling and mine works.

These differences and similarities inspired some conclusions and recommendations, that frequently turned out deceiving: first of all, the suggested probability of finding rich ore deposits in Calabria, like those that are well known in Sardinia.

The geo-tectonic evolution of the Calabrian-Peloritanian Arc is still controversial; some new data and the most recent hypotheses will be dealt with in the next chapters, that has been entirely compiled by E. ZANETTIN LORENZONI. The other chapters of the present paper are the results of debates between the two co-authors.

2. Geo-Tectonic Evolution of the Calabrian-Peloritanian Arc

The Calabrian-Peloritanian Arc is key zone for understanding the geological evolution of the Mediterranean and, as such, it has been investigated by many workers. Among these, interpretations of a synthetic character have been put forward recently by OGNIBEN (1973), DUBOIS (1976), AMODIO MORELLI et al. (1976), GÖRLER & GIESE (1978), GRANDJAQUET & MASCLE (1978), LORENZONI & ZANETTIN LORENZONI (1980, 1980a), SCANDONE (1979, 1982), WEZEL (1980), and ZANETTIN LORENZONI (1982). The interpretations differ both concerning the geo-tectonic structure of the Arc and its history. The differences stem not only from the basic complexity of the problems but also from the fact that often the authors have considered only a few of the many involved aspects, completely ignoring or undervaluing, in particular, the information provided by the metamorphic and magmatic history of the area. The differences in opinion concern such fundamental facts as the domains to which the terranes belong, whether or not the ophiolites are oceanic, the position of the Tethys, the age of the tectogenesis, the dynamics of the orogenesis and the vergence of the orogene.

The aim of this paper is to help solve the geological and tectonic problems of the Calabrian-Peloritanian Arc on the basis of knowledge acquired to date (December 1983).

The geotectonic structure of the Arc has been discussed in previous papers (ZANETTIN LORENZONI, 1982; LORENZONI & ZANETTIN LORENZONI, 1983; ACQUAFREDDA et al., 1983; FERLA et al., 1983). The interpretation put forward here (Figs. 1 and 2, Table 1) is that the Calabrian-Peloritanian Arc is formed of two ranges which came together and were partly overthrusted (the Calabrian-Peloritanian Hercynian Range mainly of Hercynian tectongenesis and the Calabrian Apennine Range mainly of Alpine tectogenesis) and of three minor tectonic elements of Alpine age (the Stilo Unit, the Tiriolo Unit and the Monte Gariglione Unit).

The Calabrian-Peloritanian Hercynian Range is the frontal, outermost unit of the Arc (or more precisely the emergent part thereof). During the Alpine orogeny the Range was subject to essentially fracture deformations, while metamorphic effects were only local, weak and of limited extent. The main phases of Alpine compression are Eocene-Oligocene.

Table 1: Main tectonic elements of the Calabrian-Peloritanian Arc. C = Cainozoic; M = Mesozoic; P = Paleozoic; pD = pre-Devonian; dotted lines = unconformities; unbroken lines = Alpine contacts; wavy lines = late-Hercynian contacts; dashed lines = Hercynian contacts.

MONTE GARIGLIONE UNIT

pD	medium-grade gneisses and metabasites	late-Hercynian magmatites
- - -	- - - (Hercynian contact)	
pD	high-grade gneisses and metabasites	
———	Middle-Upper Miocene	———

CALABRO-PELORITAN HERCYNIAN RANGE

pD	S. Sofia d'Epiro Unit: medium-high grade gneisses and metabasites
- - -	- - - (Hercynian contact)
pD	Castagna Unit: dynamic metamorphites of "granites" and gneisses
- - -	- - - (Hercynian contact)
pD	Bagni Unit: metapelites, metapsammites, metavolcanites, marbles
———	Upper Cretaceous
M	Reventino-Gimigliano Unit: metabasites, serpentinites, phyllites, marbles
———	Lower Miocene
MC	Verbicaro Unit: slightly metamorphic limestones and dolomites, locally mafic lavas
———	Lower Miocene
MC	San Donato Unit: phyllites, metabasites, marbles
———	Middle Miocene

TIRIOLO UNIT

C	conglomerates, sandstones	
······	······ (unconformity)	······
M	limestones, quartzose conglomerates	
······	······ (unconformity)	······
P	metapelites, metalimestones	late-Hercynian magmatites
- - -	- - - (Hercynian contact)	
pD	medium-grade gneisses	(granites-tonalites)
———	Oligocene	———
pD	high-grade gneisses and metabasites	
———	Middle-Miocene	———

STILO UNIT

M	limestones, quartzose conglomerates	
······	······ (unconformity)	······
P	metapelites, metalimestones	late-Hercynian magmatites
- - -	- - - (Hercynian contact)	
pD	medium-grade gneisses	(granites-tonalites)
———	Oligocene	———

CALABRIAN APENNINE RANGE

MC	conglomerates, sandstones, limestones, marls, clays	
······	······ (unconformity)	······
pD	numerous pre-alpine tectonic units constituted by gneisses, micaschistes, metabasites, porphyroids, marbles, phyllites	late-Hercynian magmatites (granites-tonalites)
- - -	- - - (Hercynian contact)	
P	metapelites, metapsammites, metavolcanites, metalimestones	

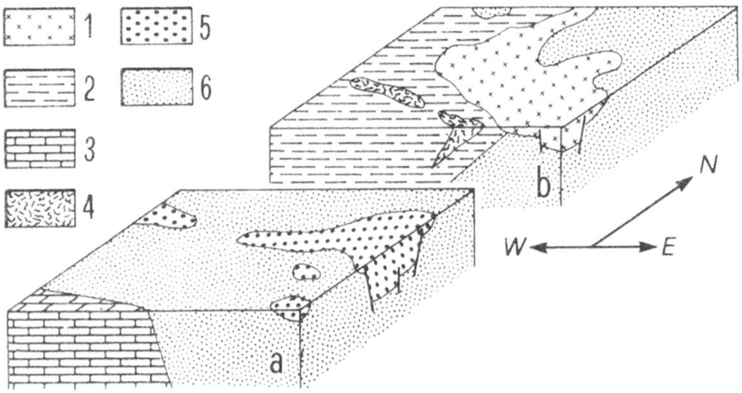

Fig. 2. Tectonic block-diagram of the Calabrian-Peloritanian Arc (a = southern stretch, b = northern stretch). 1 = Monte Gariglione Unit; 2 = Calabrian Apennine Range; 3 = Sicilian Maghrebids; 4 = Tiriolo Units; 5 = Stilo Unit; 6 = Calabrian-Peloritanian Hercynian Range.

The Calabrian Apennine Range is a more internal element. It can be attributed to an original domain with thin continental crust and Triassic-Miocene basins. The Alpine deformations were accompanied by metamorphism. The main compressive phases are Lower Cretaceous and Lower Miocene.

During the Middle Miocene the two ranges were juxtaposed with partial overthrusting of the Calabrian Apennine Range on the Calabrian-Peloritanian Hercynian Range (on which the Stilo Unit had already been overthrusted).

The Stilo Unit is an Alpine tectonic element in which we find evidence of Hercynian and pre-Hercynian sedimentary, metamorphic, tectonic and magmatic events. It was overthrusted on the Hercynian Range coming from more internal regions during the Oligocene compressive phases without metamorphism.

The Tiriolo Unit is a fragment that broke away from the Calabrian-Peloritanian Hercynian Range + Stilo Unit in the Middle Miocene during the overthrusting of the Calabrian Apennine Range; there are no Alpine metamorphic effects.

The Monte Gariglione Unit is the innermost element of the Arc and the last to be emplaced in the Middle-Upper Miocene by overthrusting on the Calabrian Apennine Range and the Calabrian-Peloritanian Hercynian Range; there are no Alpine metamorphic effects.

Before going on to discuss the Alpine evolution of the Calabrian-Peloritanian Arc, it will be usefull to clarify some particularly significant geological aspects of the matter.

The first aspect concerns the attribution of all the pre-Triassic formations of the Calabrian-Peloritanian Arc to the same Hercynian orogenic domain (a hypothesis advanced by LORENZONI & ZANETTIN LORENZONI, 1983). This attribution was proposed on the basis of analogies regarding the tectonic, metamorphic-deformational and magmatic history of the pre-Triassic rocks of the Arc now represented in the Calabrian-Peloritanian Hercynian Range, the Stilo Unit, the Tiriolo Unit, the upper part of the Calabrian Apennine Range and the

Monte Gariglione Unit. All these elements of diverse Alpine evolution, in fact, have the following points in common:

– The tectonic superimposition—at the end of the Hercynian—of units consisting of higher-grade metamorphic rocks on those of a lower grade (it is very likely that the upper units of the Calabrian Apennine Range are an exception to this case).

– The tectonically highest Hercynian elements all consist of complexes that are essentially polymetamorphic, the metamorphism being Hercynian and pre-Hercynian.

– The rocks of the Paleozoic basins with only Hercynian metamorphism always occur in the lower Hercynian tectonic positions.

– Hercynian metamorphism is characterized by a dynamic phase followed by thermal phases (only the rocks of the Stilo Unit seem not to have been affected by the latter).

– The emplacement of the Hercynian tectonic units was later than the thermal metamorphism phases.

– The late-Hercynian intrusive magmatism is characterized by an initial cycle due to deep anatectic magmas followed by a second cycle due to moderate-depth anatectic magmas (there are only very rare magmatites attributable to the second cycle in the Calabrian Apennine Range).

– Alpine metamorphic effects are absent, or weak and local, except in the upper part of the Calabrian Apennine Range where the pre-Triassic rocks are affected by weak but widespread high-pressure/low-temperature effects.

An interpretation of the Hercynian evolution, recently advanced for the Calabrian-Peloritanian Range by FERLA et al. (1983), postulates that the Range was formed on the southern edge of paleo-Europe due to subduction of the southern plate with a crust that was oceanic at least in part.

According to the interpretation proposed by the authors (Fig. 3a)—which of course takes account of the Alpine evolution and the present geometry, as well as of the Hercynian history—the domains of the Calabrian-Peloritanian Hercynian Range and of the Stilo Unit occurred in the relatively more external parts of the Hercynian orogen, while the domains of the three upper units of the Calabrian Apennines Range and the Monte Gariglione Unit belonged to the relatively internal parts. Indeed, terranes of the Paleozoic basins on which the pre-Hercynian basement was overturned, and magmas intruded under conditions of greater tension, are represented in the Hercynian Range and in the Stilo Unit. The position attributed to the Monte Gariglione Unit, instead, is in accordance with the syn-late-tectonic and relatively deep-crust character of the magmatic intrusions (GURRIERI et al., 1982), thus implying also the absence of Paleozoic basin complexes. The position attributed to the upper units of the Calabrian Apennine Range is explained by the great scarcity of late-Hercynian intrusive magmatites, as well as by the lack of Paleozoic basin terrains.

Mention should be made here to the correlation noted by DAL PIAZ et al. (1983) between the rocks of Monte Flavio Gioia in the Tyrrenian basin and those of the Calabrian-Peloritanian Hercynian Range. The Authors consider that the Monte Flavio Gioia phyllites, with their Alpine tectonic-metamorphic overtones, fit in well

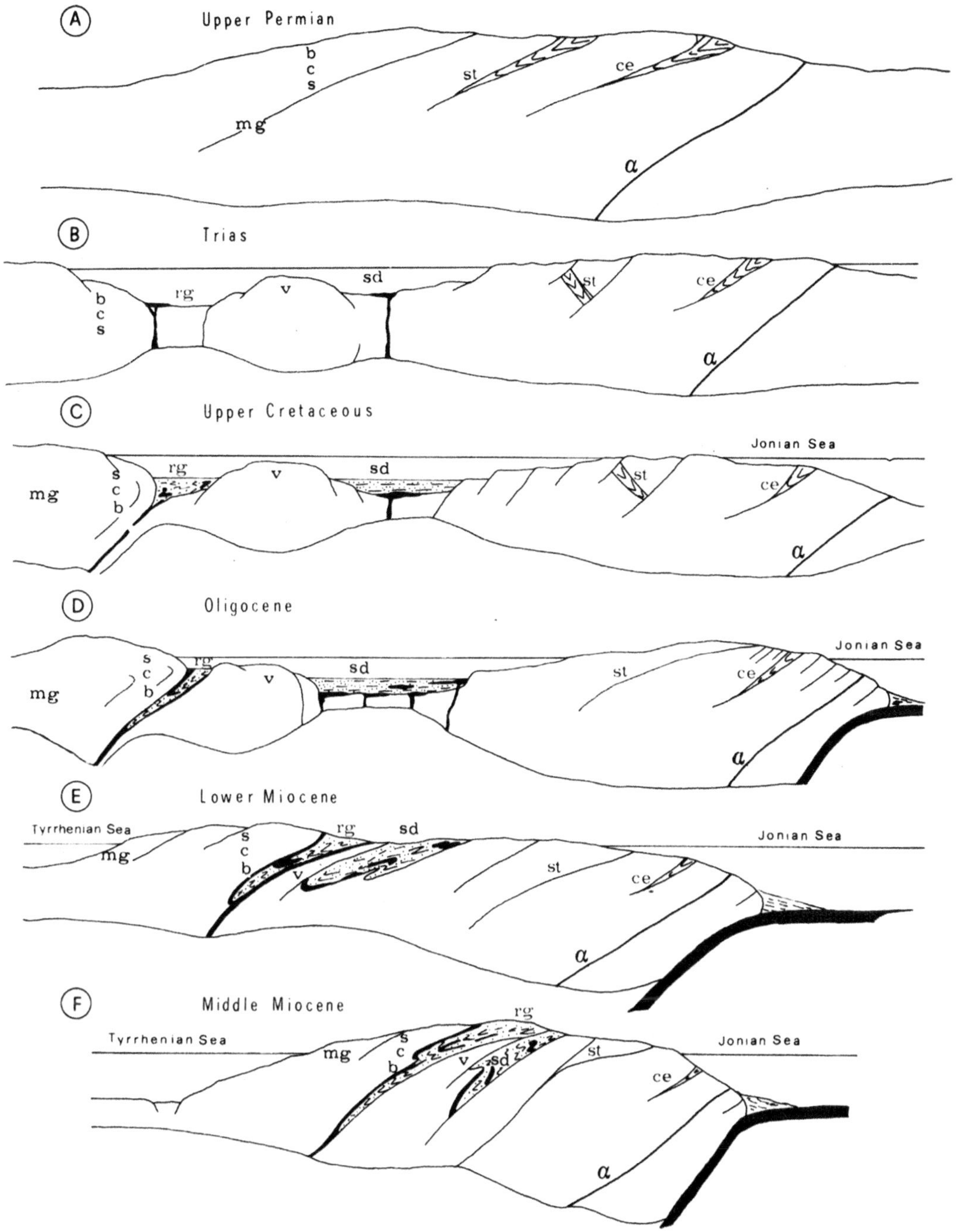

Fig. 3. Geo-tectonic evolution of the Calabrian-Peloritanian Arc from Upper Permian to Middle Miocene. mg = Monte Gariglione Unit; s = S. Sofia d'Epiro Unit; c = Castagna Unit; b = Bagni Unit; rg = Reventino-Gimigliano Unit; v = Verbicaro Unit; sd = San Donato Unit; st = Stilo Unit; ce = Calabrian-Peloritanian Hercynian Range; a = scar of the Hercynian subduction.

with this correlation, and not merely because of their marked mineralogical and structural resemblance to the pre-Hercynian phyllites of the Hercynian Range. The presence of Alpine metamorphic effects in domains that on the whole exhibit no such effects, is also characteristic of the Hercynian domains outcropping in Calabria, as will be made clear ahead.

A second aspect of fundamental importance in the interpretation of the Alpine evolution of the Calabrian-Peloritanian Arc concerns the possibility that the "ophiolitic" terrains may belong to a thin-crust continental basin (Fig. 3b). It should be made quite clear here that the authors consider that the ophiolitic terrains of the Calabrian-Peloritanian Arc belong to a single tectonic unit (the Reventino-Gimigliano Unit; the reservations expressed in LORENZONI & ZANETTIN LORENZONI, 1983, on the Malvito Unit have now been withdrawn). The possibility of the ophiolitic areas belonging to a thin mainly continental crust basin is supported by various considerations.

The first significant factor is that all the "ophiolites" are associated with flysch-like, mainly clastic formations. Of these, suffices to mention here the minute metaconglomerates (common throughout the "ophiolite" outcrop area) with clasts deriving almost certainly from rocks of the upper units of the Calabrian Apennine Range. Also of great significance is the existence of metabasite-gneiss associations in southern Lucania (SPADEA, 1982)—attributable to the Reventino-Gimigliano Unit and S. Sofia d'Epiro Unit, respectively—their characters indicate a common geological history from the time of the basalt outpourings onto the pre-Triassic rocks of the Calabrian Apennine Range (LORENZONI & ZANETTIN LORENZONI, 1983). According to this interpretation, there was a sedimentation basin in which the basement and the feed area included the rocks which subsequently have formed the Calabrian Apennine Range. Further support for this thesis comes from considerations on the tectonic and metamorphic situation as a whole. The ophiolitic unit lies tectonically above a unit formed of continental basin or shelf rocks (Varbicaro and San Donato Units), and below units formed of pre-Triassic metamorphites. In the former, tectogenesis—and metamorphism—started in the Lower Miocene, in the latter in the Lower Cretaceous. Again as regards the latter, unlike the former, overturning occurred, as it did in the case of many of the rocks of the ophiolitic basin. It is clear, therefore, that the domain of the pre-Triassic units of the Apennine Range could not be on the same side of the basin as that where the Verbicaro and San Donato domains were located. Their position is thus "northerly".

A second point concerns the existence of continental crust subduction which is demonstrated by the high-pressure metamorphism encountered in the Reventino-Gimigliano Unit and the overlying units. Additional evidence is provided by the fact that the terrains nearest the "southern" edge of the ophiolitic basin now visible are of continental shelf type.

In addition to all the statements made so far, it must be accepted that subduction occurred in a very short space of time, entirely within the Albian, according to SPADEA (1982), and that there is no proof of any involvement of the oceanic crust in the subduction; so the most convincing idea is that the basin is continental. In the line of this idea are the analogies of lithological associations between the Reventino-Gimigliano Unit and the San Donato Unit, which are certainly those of a continental basin. The Authors' interpretation of this fact also fits in with the reconstruction made on a paleomagnetic basis by MANZONI & VANDERBERG (1982), who consider Sardinia, the southern Apennines, the Calabrian-Peloritanian Arc and Sicily as forming part of Africa at least up to the start of the Tertiary.

The third fundamental aspect to be considered is the vergence of the Calabrian-Peloritanian Arc. It is held that all the tectonic elements, Hercynian and Alpine, were emplaced when moving in the "African" direction. As regards the Calabrian-Peloritanian Hercynian Range, FERLA et al. (1983) postulate its southern Hercynian vergence on the basis of its geometry and the fact that it belongs to the southern margin of the paleo-European continent. The vergence effect of the Alpine compressions that generated a slice structure with major movements in the Eocene-Oligocene was also "African". Where the Stilo Unit is concerned, its assignment to a more "northerly" paleogeographic position in the Hercynian Range (FERLA et al., 1983; ACQUAFREDDA et al., 1983) at the end of the Hercynian orogeny establishes the southerly direction of the movement in the Oligocene. The vergence of the San Donato and Verbicaro Units commonly considered as the continuation of the southern Apennines in Calabria, is evidently in keeping with that of the Apennines themselves and hence is southerly. In the case of the upper units of the Calabrian Apennine Range, the Authors consider their vergence to be African and not European (as held by AMODIO MORELLI et al., 1976; FAURE, 1978; SCANDONNE, 1983), since the domain of the three upper units (BAGNI, CASTAGNA and S. SOFIA D'EPIRO) was on the northern edge of the ophiolitic basin. The nature of the vergence is, of course, independent of the fact as to whether the basin is oceanic or not, while attribution of the domain of the units in question to Europe or Africa would vary. There are no special problems with the Tiriolo Unit if the interpretation of ZANETTIN LORENZONI (1982) is correct, namely that this is a fragment which broke away when the Calabrian Apennine Range collided with the Calabrian-Peloritanian Hercynian Range + Stilo Unit. In fact, in this case the unit is the effect of movement in a "southerly" direction. In the Alpine, the Monte Garigline Unit (attributed, in the reconstruction at the end of the Hercynian, to a deeper part of the crust than the S. Sofia d'Epiro Unit) is placed farther from the continental margin, because it exhibits non-Alpine metamorphic effects. The Monte Garigione Unit, the last to be emplaced, is thus considered to be of more internal origin and it was evidently moved in a southern direction. The interpretations made by the authors – which await confirmation or rebuttal in particular by structural analysis – are in accordance with the interpretations (on paleomagnetic basis of MANZONI & VIGLIOTTI, 1983) which exclude the existence in the Sila of opposite vergences for the different tectonic elements.

The Alpine geo-tectonic history of the Calabrian-Peloritanian Arc can now be reconstructed on the basis of present knowledge and the interpretations presented above.

At the end of the Paleozoic (Fig. 3a) the Hercynian orogen—which was subsequently dismembered by Alpine dynamic action, with the fragments going to form various structural elements—had already been highly eroded.

Tension stresses began to occur in the Triassic (Fig. 3b). These led to the creation of basins where basalt lavas spread out and where deposits were laid down; they sub-sequently formed the San Donato and the Reventino-Gimigliano Units (this basin can be dated Triassic according to the discovery of fossils, probably of Anisian-Ladinian age, in the lower parts of the Gimigliano sequence; see COLONNA & ZANETTIN LORENZONI, 1972). The existence of tension was probably also the cause of the partial overturning of that sector of the crust which subsequently formed the Stilo Unit; this overturning occurred after the intrusion of the late-Hercynian magmas and before the start of deposition of the Mesozoic cover (which would, however, appear to be Liassic rather than Triassic). The tension phases were interrupted by compression phases which became more marked towards the Lower Cretaceous; in the "ophiolite" basin (future Reventino-Gimigliano Unit) the build up of basalts and volcanic products gradually declined, while terrigenous flysch-type sedimentation started to prevail.

In the Cretaceous (Albian, of SPADEA, 1982) (Fig. 3c), with the opening up of the Ionian (GÖRLER & GIESE, 1978), there was subduction of the thin continental crust (and perhaps also of limited areas of oceanic crust) and of the basic vulcanites and the flysch, beneath the "northern" edge of the basin itself. The basement and the mesozoic formations that lay towards the edge of the basin were overturned and high-pressure, low-temperature metamorphism occurred, but only in the plate undergoing subduction, and in the trench area.

It is held that in the Cretaceous the sector including the Calabrian-Peloritanian Hercynian Range now outcropping in the Sila, was rotated about 60 ° counter-clockwise with respect to southern Calabria (Fig. 4). In fact, considering the position of the Paleozoic basins (after having brought the Peloritanian Mountains ideally to their original position, namely the southern edge of the Aspromonte: see FERLA et al., 1983) and also the trend of the late-Hercynian tension lines which governed the intrusion of the magmatic bodies, the Sila sector of the Hercynian Range appears to be roteted about 90 ° counter-clockwise as regards the Aspromonte-Peloritanians sector. The direction of vergence of the Hercynian slices in the two sectors, instead, forms and angle of not more than 30 °. Hence there was rotation of about 60 ° between the Permian and the Eocene. A similar type of rotation is reported by MANZONI & VANDERBERGER (1982) on the basis of paleomagnetic measurements; this is dated as post-Liassic. Taking account of all these data, it would seem that rotation most probably occurred in the Cretaceous, in connection with the subduction movements which occurred at differential speed from present north to present south. It should be observed in this respect that the "ophiolitic" basin probably did not extend much beyond what can be seen in outcrop nowadays: to the south of the Catanzaro isthmus there are no traces of rocks that can be ascribed to that basin nor even of Cretaceous dynamic movements or Alpine metamorphism. At this point, mention should be made of the traces of metamorphism encountered on a geochronological basis by SCHENK (1980) in the northwestern Serre, namely in the postulated closure zone of the "ophiolitic" basin.

Subduction ceased in the Upper Cretaceous and the Lower-Middle Eocene, but the compression regime remained, while tension increased in the Ionian. The most marked effects are the stratigraphic gaps in the cover formations (especially in the domains of the Stilo Unit and the Calabrian-Peloritanian Hercynian Range) and the deposition of the Eocene breccias on the latter.

Subduction of the Ionian bathyal area beneath the Calabrian-Peloritanian Arc started in the Oligocene (Fig. 3d) (GÖRLER & GIESE, 1978) while, at the same time (DURAN DELGA, 1980) or at the latest in the Lower Miocene (FABBRI et al., 1980), the Tyrrhenian opened. A regime involving intense compression occurred in the frontal part of the Calabrian-Peloritanian Arc which gave rise to the slice structure with "African" vergence of the Calabrian-Peloritanian Hercynian Range (perhaps this structure had already started to form in the Upper Eocene) and caused the Stilo Unit to be overthrusted on the Hercynian Range. The Oligocene deformations were brittle and occurred without metamorphism.

In the Lower Miocene (Fig. 3e) the compressions shifted towards the inner part of the Arc and affected only what is today its northern sector, namely that of the thin crust basins. The basin complexes were deformed, and the terranes of the more internal carbonate shelf (Verbicaro Unit), the already metamorphosed formations of the ophiolitic basin (Reventino-Gimigliano Unit) and the basement of their northern margin (Bagni, Castagna and S. Sofia d'Epiro Units) were shifted over them. The miocene deformations were accompanied by metamorphism, and the authors consider that it was caused basically by high tectonic pressures. In fact, on the Calabrian-Lucanian boundary the metamorphic restructuring is intense in the San Donato Unit (the lowest), extremely weak in the Verbicaro Unit (overlying) and absent–relative to post-Cretaceous metamorphic effects–in the Reventino-Gimigliano Unit (again overlying). In this latter unit, instead, metamorphic effects of normal pressure are evident in southern direction, with a sharp gradient a little to the south of Sybaris Plain where the tectonic units of the Calabrian Apennine Range are thickening rapidly. Yet another effect of the Lower Miocene compressions is postulated: the further anticlockwise rotation of the Sila "basement" by about 30° as regards southern Calabria (Fig. 4).

In the Middle Miocene (Fig. 3f), the Calabrian Apennine Range which had already been formed, moved in the "African" direction, coming alongside and being partially overthrusted on the Calabrian-Peloritanian Hercynian Range (on which the Stilo Unit had already been overthrusted): at the same time a fragment of the "southern" element broke away and became wedged in the upper part of the Calabrian Apennine Range to form the Tiriolo Unit. The compressive phases were followed by tension phases with formation of fault systems, some of which were reactivated subsequently (the Crati and Catanzaro grabens). The compressing started again. In the Middle and Upper Miocene this caused the overthrusting of the Monte Gariglione Unit on the Calabrian Apennine Range and on the Calabrian-Peloritanian Hercynian Range. All the deformations of the Middle and Middle-Upper Miocene occurred without any metamorphic effects.

In the sector which now corresponds to southern Calabria and the Peloritanian Mountains, tectogenesis had already ceased by the beginning of the Miocene and no compression is encountered here after the start of deposition of the Capo d'Orlando Flysch (as the authors have no direct knowledge of the processes of the emplacement of the "Argille Scagliose Varicolori", nothing will be said on this matter). Ten-

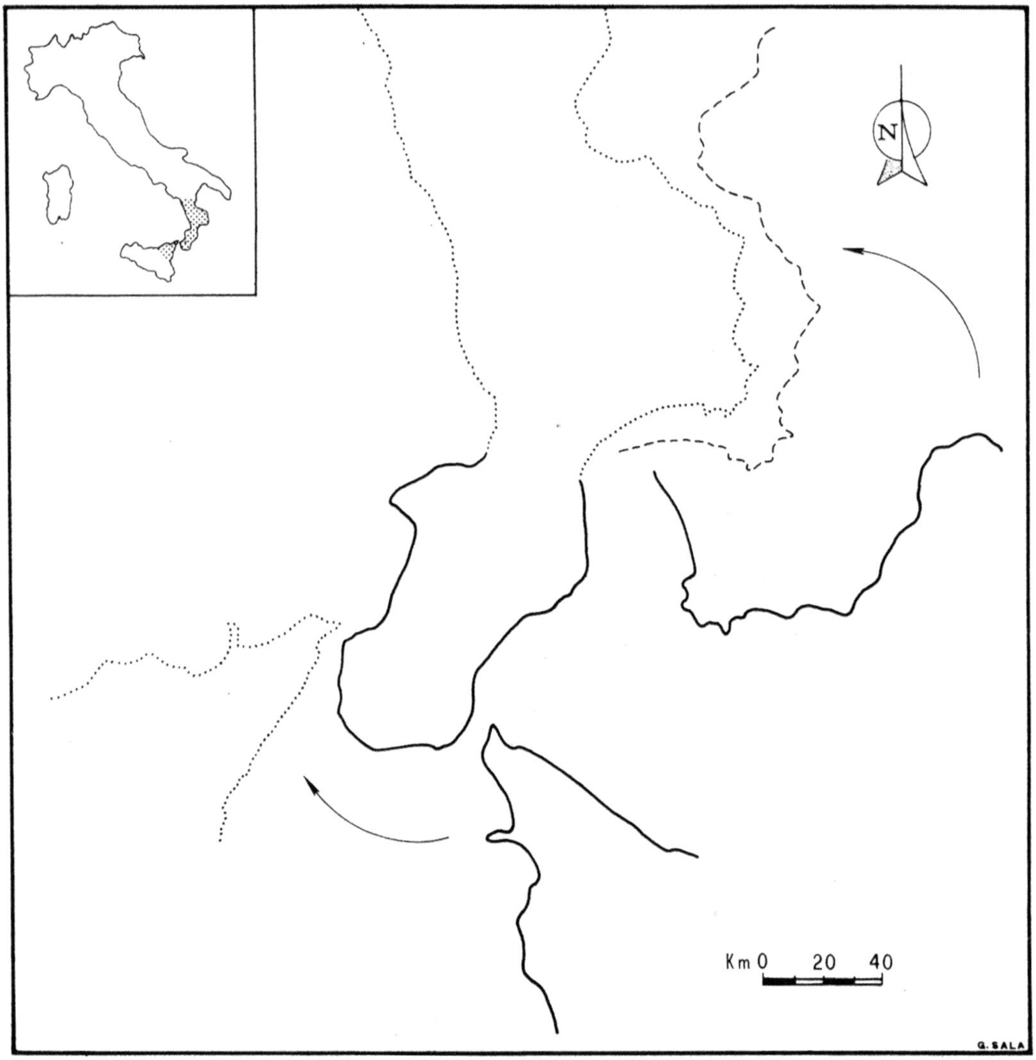

Fig. 4. Comparative position of southern Calabria, Sila, Peloritani in Upper Jurassic (unbroken line) and Oligocene (dashed line).

sion again occurred in the Miocene, with dismemberment and movement (and/or rotation) of the fragments. The breakaway from the Aspromonte of the sector that now corresponds to the Peloritanian Mountains occurred in particular in the Middle Miocene with a 90 ° clockwise rotation of the Peloritanians which were juxtaposed—along the "Taormina line"—with the Sicilian Maghrebids (Fig. 4). The suggested interpretation is that the Taormina line is thus a left transurrent fault, whose significance is quite different from that of the "Sangineto line" (commonly considered as the northern limit of the Arc) which the authors hold to be a fault zone within one and the same domain.

3. The Calabrian-Peloritanian Metallogenesis

If one takes into consideration deposits/districts having a size more than 1 of the conventional scale adopted in the metallogenic maps of the Mediterranean Region, the Calabrian-Peloritanian Arc has to be considered a barren region.

As a matter of fact ore/mineral occurrences in that area are numerous and various, but they are of small (size 1) or even they can be classified as simple indications (size 0).

According to the most recent authors (LORENZONI et al., 1983; BONARDI et al., 1982; CASTALDO and STAMPANONI (eds.), 1975) about a hundred ore/mineral occurrences are known in the Calabrian-Peloritanian Arc (excluding coal, feldspar and quartz); they may be grouped, in the optics of the present paper, in 11 paragenetic types and in five metallogenetic epochs, as it is shown in Table 2. For more details, the above-mentioned papers should be consulted.

A first factual observation stands out: the post-Hercynian pre-Tethyan and pre-rifting volcanism, which is well known all along the Alps (WINTERER and BOSELLINI, 1981) and is responsible of interesting ore/mineral depositions of U, Fe, Ba (CADEL, G. et al., 1984), is absent (at least: not yet recognized) in the Calabrian-Peloritanian Arc.

The second remark is that the largest majority of Calabrian-Peloritanian ore/mineral deposits are held in, and have been generated by, pre-Hercynian/Hercynian events, involving (according to LORENZONI et al., 1983) deposition in connection to various Hercynian/pre-Hercynian metallogenic processes, and remobilization, reshaping, partial dispersion by the Hercynian and by the Alpine metamorphism and magmatism.

Coming to the post-Hercynian events, we can distinguish four stages, namely: i) post-Hercynian pre-Alpine events, not related to the ophiolitic volcanism; ii) the ophiolitic volcanism—whatever its domain (oceanic or non-oceanic); iii) Alpine; iv) post-Alpine orogenesis.

The following factual observations can be made:

i) ore/mineral deposition related to the first above-mentioned stage are neither so big nor so rich; as a matter of fact the only not negligible ore accumulations are the residual Fe Oxides (with minor sulfides at places) of Pazzano-Stilo on the post-Hercynian peneplane and below to the Mesozoic (Jurassic?) carbonate platform. They are quite similar to the coeval deposits of western Sardinia.

Table 2. Ore/mineral occurrences in the Calabrian-Peloritanian Arc

	Metallogenic Epoch				
Main Paragenesis	(pre)-Hercynian	post-Hercynian pre-Alpine non-ophiolite	Ophiolite	Alpine	late- and post-Alpine
Mixed Sulf.	A	C	C		
Pb-Zn-Ag	A			No new contributions, only (partial) remobilization, displacement, dispersion of pre-existing ores/minerals	
F-Ba	A	C			
Pyr-Cu	B				
U-Th-REE	C				C
Mo-Cu	A				
Fe ox, Ti		A			B
Mn-Fe ox			A		C
Hg		B			
Sb-W-Hg	B				
Na-K-S					A

Relative importance (in the Calabrian-Peloritanian metallogenesis)

A: frequent and someones of not negligible tonnages
B: not so frequent and often small
C: occasional and small

For details: see LORENZONI et al., 1983.

Small accumulations of Ba, F, pyrite, Zn, Pb, Cu, Hg are known in a section of the Northern part of the Western Coastal Range (North-North-East of Cetraro). If these occurrences can be considered equivalents to the rich, paragenetically similar, Triassic deposits of Central-Western Alps (but this hypothesis needs confirmation), the difference between them is stricking and has to be explained.

ii) very few and not so rich ore/mineral deposits generated in connection to the ophiolitic magmatism. As it was said at point 2 the ophiolite sequences occur in not so large outcrops, scattered along the Northern section of the Western Coastal Range (from the Sangineto Line to the Lamezia-Catanzaro trench). The related ore concentrations consist of some poor pyrite-chalcopyrite occurrences.

This is a real difference from the, generally rich, deposits related to the ophiolitic sequences of the other parts of Mediterranean.

iii) no new deposit have ben generated in connection to the Alpine orogenesis and metamorphism; we do not speak of Alpine magmatism, because there is no evidence of it in the Calabrian-Peloritanian Arc.

The Alpine orogenesis and metamorphism are responsible of displacements, remobilization, dispersion of pre-existing ore/mineral deposits, as it has been fully demonstrated by LORENZONI et al., 1983.

These events cover large areas of the region under investigation, particularly its North-Western part, from the Sangineto line down to the Stilo area. Important sections of the Hercynian/pre-Hercynian regions, and the related mineral accumulations, have been (partially) involved.

iv) The metallogenic importance of late- and post-Alpine events consists in Messinian evaporite and biogenic S deposition of considerable tonnages, even of size no. 1.

Some beach sands, enriched in heavy minerals, and some lacustrine U occurrences, have been also deposited in these times.

The only evidence of recent volcanism and related ores, which are so largely present in Central-Southern Italy, is the submarine accumulation of sulfides (mainly pyrite) in the Eolian basin.

4. Conclusive Remarks

As a conclusion we may say that the (very scarce) ore/mineral wealth of the Calabrian-Peloritanian Arc has been built up mainly in connection to Hercynian/pre-Hercynian metallogenic events; it was partially reshaped or dispersed and only very poorly enriched in connection to the post-Hercynian events.

The absence of the Cambrian carbonatic platform formations*) and—consequently—of their rich metal stock, so well represented in Sardinia (where they are meaningfully called "Metalliferous"), may be one of the causes of the poverty of the metal reserves.

"One of the causes"—we said—but not the only one: other factors had probably played a role; without entering into details, and sticking to the metallogenic Thetyan events, the following characters have to be taken into consideration:

i) the (probable) absence of the Permian-Lower Triassic rifting and related Permian volcanism, well known in the Alps, where it is associated to interesting, U, Fe, Ba accumulations.

ii) the scanty development of Middle-Upper Triassic and its difference from the (richly mineralized) South-Alpine Triassic.

iii) the probable different nature of the Calabrian ophiolites in comparison with the other mediterranean ophiolitic complexes, as suggested by LORENZONI and ZANETTIN LORENZONI, 1980, and confirmed in the previous point.

The geological and metallogenic outline of the Calabrian-Peloritanian Arc, as it comes out from this paper, even if it is not so detailed, suggests that this region has peculiar characters, partly similar and partly clearly different from the other Mediterranean regions.

As a matter of fact the likeness of some Paleozoic formations in the Alpine Arc, in Sardinia and in the Calabrian-Peloritanian Arc speaks for a possible paleogeographic connections among them.

---

*) Cambro-Ordovician terrains have been recently discovered in Calabria (BOUILLIN et al., in print), but they have not carbonatic platform facies.

The presence of the Middle-Paleozoic Sb-W-Hg complex in all three regions is another supporting evidence; the occurrences of similar residual Fe Oxide concentrations on the Hercynian peneplane, at the base of the carbonatic Mesozoic formations, in West Sardinia and in Central Calabria, may be considered as another evidence of a similar early post-Hercynian climatic-erosional evolution.

The regional distribution of some ore/mineral deposits in Southern Europe, the Sardinian-Corsic massif and in Calabria—as pointed out by one of us (ZUFFARDI, 1978)—appears consistent with the paleogeographic reconstruction of the Alpine orogenic belt proposed by ALVAREZ et al., 1974, according which the Calabrian-Peloritanian Arc should be considered a microplate of that belt.

On the other hand, many other factual observations discussed in point 2 of this paper, demand for going deeper into this problem and for seeking explanation of not negligible details.

In particular, taking into account the possible relationship between Sardinia and the Calabrian-Peloritanian Arc (which are most interesting on the metallogenic point of view), it has to be noted that there are substantial differences between the distributions of the whole (pre-Hercynian+Hercynian+Alpine) metamorphic-magmatic-tectonic history and in the structural setting of these two regions: to quote the most striking example—in the aim of Tethyan investigations—we point out the decidely different effects of the Alpine orogenesis in these regions (absent or at most rigid in Sardinia, intensive and mainly plastic in the Arc).

Acknowledgments

The Authors gratefully acknowledge to Prof. R. FUNICELLO and M. PAROTTO (University of Roma) and to Prof. S. LORENZONI (University of Bari) for critical review of the manuscript and for suggestions.

References

ACQUAFREDDA, P., LORENZONI, S., ZANETTIN LORENZONI, E. (1983): „Guida all'escursione sull'Unità di Stilo (Calabria)". Gruppo Paleozoico C. N. R., 3–9 ottobre 1983. Ist. Min. Petr. Univ. Bari (inedited).

ALVAREZ, W., COCOZZA, T., WEZEL, C. (1974): "Fragmentation of the Alpine orogenic belt by microplate dispersal". Nature 248, March 22, 309–314.

AMODIO MORELLI, L., BONARDI, G., COLONNA, V., DIETRICH, D., GIUNTA, G., IPPOLITO, F., LIGUORI, V., LORENZONI, S., PAGLIONICO, A., PERRONE, V., PICCARETTA, G., RUSSO, M., SCANDONE, P., ZANETTIN LORENZONI, E., ZUPPETTA, A. (1976): „L'arco calabro-peloritano nell'orogene Appenninico-Maghrebide". Mem. Soc. Geol. It., XVII, 1–60.

BONARDI, G., DE VIVO, B., GIUNTA, G., LIMA, A., PERRONE, V., ZUPPETTA, A. (1982): „Mineralizzazioni dell'Arco Calabro-Peloritano. Ipotesi Genetiche e quadro evolutivo". Boll. Soc. Geol. It., *101,* 141–158.

BOUILLIN, J. P., BAUDELOT, S., MAJESTE'-MENJOULAS, C. (1984): ›Mise en évidence du Cambro-Ordovicien en Calabre central (Italie). Affinité paleogéographiques et conséquences structurales‹. Bull. Soc. Geol. de France (in print).

CADEL, G., D'AGNOLO, M., MENEGHEL, L., ZUFFARDI, P. (1984): "Polygenetism in strati-form/strata-bound Uranium Deposits of the Italian Alps". 27th Int. Geol. Congr. 1984, Special volume, no. 12, 142–160.

CASTALDO, G., STAMPANONI, G. (eds.) (1975): „Memoria illustrativa della Carta Mineraria d'Italia, scala 1 : 1,000.000", Serv. Geol. It. XIV, p. 213.

CHANNEL, J. E. T., CATALANO, R., D'ARGENIO B. (1980): "Paleomagnetism and deformation of the Mesozoic continental margin in Sicily". Tectono-physics, *61,* 391–407.

COLONNA, V., ZANETTIN LORENZONI, E. (1972): „Gli scisti cristallini della Sila Piccola. 2°: I rapporti fra la formazione delle filladi e la formazione delle pietre verdi nella zona di Gimigliano". Mem. Soc. Geol. It., *11,* 261–292.

DAL PIAZ, G. V., DEL MORO, A., DI SABATINO, B., SARTORI, R., SAVELLI, C. (1983): „Geologia del Monte Flavio Gioia (Tirreno centrale)". Mem. Soc. Geol., XXXV, 429–452.

DUBOIS, R. (1976): ›La suture calabro-apenninique crétacé-eocéne et l'ouverture Tyrrhenienne neogene; etude petrographique et structurale de la Calabre central‹. Thèse Univ. P. et M. Curie, Paris.

DURAND DELGA, M. (1980): ›Ouvertures océaniques de Méditerranée occidentale et dislocation des chaines alpines‹. *In:* "Sedimentary Basins of Mediterranean Margins", CNR, Italian Project of Oceanography, Urbino, 417–432.

FABBRI, A., GALLIGNANI, P., ZITELLINI, N. (1980): "Geologic evolution of the peri-Tyrrhenian sedimentary basins". *In:* "Sedimentary Basins of Mediterranean Margins", CNR, Italian Project of Oceanography, Urbino, 101–126.

FAURE, M. (1978): ›Microtectonique et charriage Est-Ouest des nappes Alpines profondes de Sila (Calabre-Italie méridionale)‹. Rev. Geol. Dyn. Géogr. Phys., *22,* 135–146.

FERLA, P., LORENZONI, S., ZANETTIN LORENZONI, E. (1983): "Geological constitution and evolution of the Calabro-Peloritan Hercynian Range". Rend. Soc. It. Miner. Petr. (in print).

GÖRLER, K., GIESE, P. (1978): "Aspects of the Evolution of the Calabrian Arc". *In:* "Alps, Apennines, Hellenides", Schweitzrbartsche Verlag, Stuttgart, 374–388.

GRANDJAQUET, C., MASCLE, G. (1978): "The structure of the Ionian Sea, Sicily and Calabria-Lucania". *In:* "The Ocean basins and margins", Plenum Press New York, *46,* 257–329.

GURRIERI, S., LORENZONI, S., STAGNO, F., ZANETTIN LORENZONI, E. (1982): „Le magmatiti dell'Unità di Monte Gariglione (Sila, Calabria)". Mem. Soc. Geol., XXXV, 69–90.

LORENZONI, S., ORSI, G., and ZANETTIN LORENZONI, E. (1983): "Metallogenesis in the Tectonic Units and Lithogenetic Environments of Calabria (Southern Italy)". Mem. Soc. Geol., XXXV, 411–428.

LORENZONI, S., ZANETTIN LORENZONI, E. (1980): »Cadre structural. Calabre et Peloritains«. *In:* »Géologie d'Italie«, XXVI Congr. Int. Géol. Paris, 45–48.

LORENZONI, S., ZANETTIN LORENZONI, E. (1980a): »Orogènéses, magmatismes et metamorphismes en Calabre et dans les Péloritains«. *In:* »Geologie d'Italie«, XXVI Congr. Géol. Paris, 91–93.

LORENZONI, S., ZANETTIN LORENZONI, E. (1983): „Note illustrative della Carta geologica della Sila alla scala 1 : 200.000". Mem. Sc. Geol. (in print).

MANZONI, M., VANDERBERG, J. (1982): "Peri-Tyrrhenian Paleomagnetic Data and the Setting of the Calabrian Arc". Earth Ev. Sc., *3,* 181–186.

MANZONI, M., VIGLIOTTI, L. (1983): "Further paleomagnetic data from Northern Calabria. Their bearing on directions of emplacement of the Calabrian Nappes". Boll. Geofis. Tecnica Appl., XXV, *97,* 27–43.

OGNIBEN, L. (1973): „Schema geologico della Calabria in base ai dati odierni". Geol. Romana, *12,* 243–585.

Scandone, P. (1979): "Origin of the Tyrrhenian Sea and Calabrian Arc". Boll. Soc. geol. It., *98,* 27–34.

Scandone, P. (1982): "Structure and evolution of the Calabrian Arc". Earth Ev. Sc., *3,* 178–180.

Schenk, V. (1980): "U-Pb and Rb-Sr Radiometric Dates and their Correlation with Metamorphic Events in the Granulite Facies Basement of the Serre, Southern Calabria (Italy)". Contrib. Mineral. Petrol., *73,* 23–38.

Spadea, P. (1982): "Continental crust rocks associated with ophiolites in Lucanian Apennines (Southern Italy)". "Ofioliti", *2/3;* 501–522.

Wezel, F. P. (1980): "The structure of the Calabro-Sicilian Arc: krikogenesis rather than subduction". *In:* "Sedimentary basins of Mediterranean Margins", CNR, Italian Project of Oceanography, Urbino, 485–488.

Winterer, E. L., Bosellini, A. (1981): "Subsidence and Sedimentation on Jurassic Passive Continental Margin, Southern Alps, Italy". Amer. Ass. Petr. Geologists, *65,* 394–421.

Zanettin Lorenzoni, E. (1982): "Relationships of main structural elements of Calabria (Southern Italy)". N. Jb. Geol. Paläont., *7,* 403–418.

Zuffardi, P. (1978): "Plate Tectonics and Ore/mineral Provinces in Western European Mediterranean Region". *In:* „Ergebnisse der österreichischen Projekte des I. G. C. P. bis 1976, Springer-Verlag, Wien, 61–76.

The Distribution of Alpidic ore Districts in the Western Part of the Mediterranean Sea in Relation to Plate Tectonic

By Yves A. Fuchs*)

Résumé

On envisage l'étude des gîtes de la Méditerranée occidentale du point de vue de l'évolution géodynamique de cette zone. L'importance des phénomènes magmatiques et hydrothermaux du Carbonifère supérieur et du Permien est miseen évidence et leur rôle quant à un apport important en métaux est souligné. Cette période est caractérisée comme celle d'une phase initiale du cycle alpin plutôt que comme un phénomène tardif du cycle hercynien.

On insiste, en ce qui concerne le Tertiaire, sur les différences de comportement des bordures des plaques eurasiennes et africaines. L'existence sur la marge de la plaque africaine de phénomènes de coulissement (dilacération de la microplaque d'Alboran) et de zones de subduction crustales paraissent responsables de la présence d'un magmatisme calco-alcalin et de la formation de gîtes hydrothermaux (veines polymétalliques et/ou gîtes volcanogéniques) en Italie et au Maghreb.

Zusammenfassung

Die Bildung der Erzlagerstätten und Erzvorkommen im westlichen Mittelmeerraum steht in enger Beziehung zu den verschiedenen Entwicklungsphasen der Plattentektonik. Wichtig sind die magmatischen und hydrothermalen Ereignisse während des Obersten Karbon und des Perm. Sie spielen eine besondere Rolle für die Metallzufuhr. Dieser Zeitraum stellt mehr eine initiale Phase des Alpidischen Zyklus als eine späte Phase des Variszikum dar.

*) Laboratoire de Géochimie & Metallogénie – Université Paris VI. Tour 16–5e E., 4, place Jussieu – 75230 Paris Cédex 05.

Während des Tertiärs spielen die Ränder der Afrikanischen und Europäischen Platten sehr verschiedene Rollen. Am Rand der Afrikanischen Platte sind es die lateralen Bewegungen, die zum Zerreißen der Alboran Mikroplatte führen; die aktiven Subduktionszonen sind verantwortlich für einen kalk-alkalinen Magmatismus sowie für die Bildung hydrothermaler Lagerstätten (polymetallischer Gänge und/ oder vulkanogener Lagerstätten) in Italien und in Nordafrika.

Numerous ore occurrences or economic deposits of the Western part of the Mediterranean Sea show an evident relationship with particular events of the Alpine Orogenesis. The geochemistry as well as the morphology of these deposits are of quite different types. However it seems that the most important periods for the formation of ore deposits were related to some major events in the evolution of plate tectonics in the western part of the Mediterranean Sea.

The Initial Period

The rifting process began very early, probably during the late Carboniferous and Lower Permian period. Oceanization phenomena were not present during this time, but a distensive tectonic movement was an important fore-runner of the opening of the Téthys. In the French Massif Central during Stephanian time, the main tectonic trend was of compressive type. Important shearing zones with a N 20 E direction were active. In some areas, pull-apart phenomena induced the formation of intermontane subsiding coal basins. These areas were restricted. Therefore, some crustal anatexis, related to these basins, and a silico-aluminous potassium rich magmatism originated. It belongs to the type of "topaz rhyolites" described in the western part of the USA by D. Burt and Sheridan (1980), or to the type of hypovolcanics called ongonites by Kovalenko (1976).

Uranium, tungsten (as scheelite in tuffs), fluorite, tin, . . . occurences are associated to this type of volcanism (Badia et al., 1984). Later during Lower Permian, the distensive phenomenon became the most important. Large graben type basins were developping in the western part of Europe. A dominantly calc-alkaline magmatism was related to this geodynamic evolution (Badia et al., 1984). This magmatism is important all over western Europe (North Italy, Maures, Estérel, Alps, Massif Central and Vosges in France, Sarre-Nahe Basin and Northern Germany in FRG, North Sea, . . .). Its characteristics (petrological, geochemical and isotopes) show that this could be the result of an evolution in a continental plate, particularly along deep seated fault zones, where mantle influence can also be observed. This distensive geodynamic evolution is dominant during the Permian all over Western Europe whereas in some places it began during Stephanian, as the age of 285 My for the ignimbrites of Novazza has shown (S. Philippe, 1984, U-Pb method of zircons). It must be emphasized that the development of hydrothermal convective systems seems related to this magmatic activity. Ore deposits like Novazza in North Italy (uranium with some zinc and molybdenum in ignimbrites), Val Vedello in the same region (uranium in the boundary fault of a graben basin) and minor occurrences like

Pomayrols in France (uranium in dacitic dyke) or Ellweiler in Germany (uranium with some mercury and copper in a permian volcanic complex), seem to be directly related to hydrothermal circulation. The circulating fluids were particularly rich in magnesium and sodium penetrating into a hydrothermally altered zone, and leading to the formation of minerals, such as magnesian chlorite and dravite (BADIA et al., 1985). J. LEROY (1978) in his study on the uranium district of Limousin (France) has noticed that the granites show an age of 350—360 My (Rb-Sr), but that towards 285 My, a lamprophyric magmatism related to a fault tectonism took place. A heat flow was associated to this magmatism. It caused a heating up of the cold water which impregnated the granites and then formed convective circulating hydrothermal systems. The fluids caused the episyenitization of the granite and this phenomen was followed by the deposition of pitchblende and pyrite (275 My) (LEROY, 1978). Not only uranium is affected by these hydrothermal processes. BREVANT et al. (1982) studied the isotopic ratios of lead in galena of the southern part of the French Massif Central. They have shown (Fig. 4) that the isotopes of lead in galena were distributed in two groups, one with $17.76 < 206\ Pb/204\ Pb < 17.94$ and the other one with $18.25 < 206\ Pb/204\ Pb < 18.63$. They concluded that, after a mantle evolution until 3 By the lead was incorporated in a part of the crust. Afterwards the less radiogenic group was associated to the carbonate rocks of the Cambrian. The second group seems to be related to the hercynian granitoids or more exactly to the hydrothermal systems of the late evolution of the hercynian period.

It can be discussed whether these major hydrothermal events were related to the end of the Hercynian orogenesis, or to the beginning development of the Alpine system. Following H. WOPFNER (1984) who investigated the Permian deposits between Val Camonica and the Karawanken, one can assume that the development of a complex rift system during the Permian initiated the alpidic sedimentation. Equally, this tectonic and magmatic environment was favourable for an important metal supply during this initial phase of the alpidic evolution.

Part of this metal supply is directly connected with the transportation by hydrotherms. An other part in the continental and epicontinental domain is provided by the weathering of the old basement and another important part by the weathering of recently vented volcanics. ZIELINSKY (1982) has shown the importance of the weathering of volcanic glasses as a source for metal supply in the supergene cycle. It can be a major source for stratiform deposits and occurrences of uranium located in strata of Permian age (France, Italy, . . .) but also extending up to the Triassic in Austria (PETRASCHECK et al., 1974).

During Triassic, the opening is continuing. In the Western Mediterranean Sea, one can clearly characterize two domains. In basins with Germanic facies, lead-zinc deposits (Maubach and Mechernich in FRG, Saint Avold and Largentières in France, J. Tebagha in Tunisia) are hosted in sandstones. In the other domain of sedimentation, important economic deposits (Bleiberg in Austria, Raibl, Salafossa, Gorno in Italy and in S-W Spain) are related to carbonate rocks deposited on an epicontinental platform. The palynspastic map (Fig. 1) shows that the most important deposits are located in the vicinity of deep seated structures of major importance. It is impossible to conclude that these structures represented the source of the metal

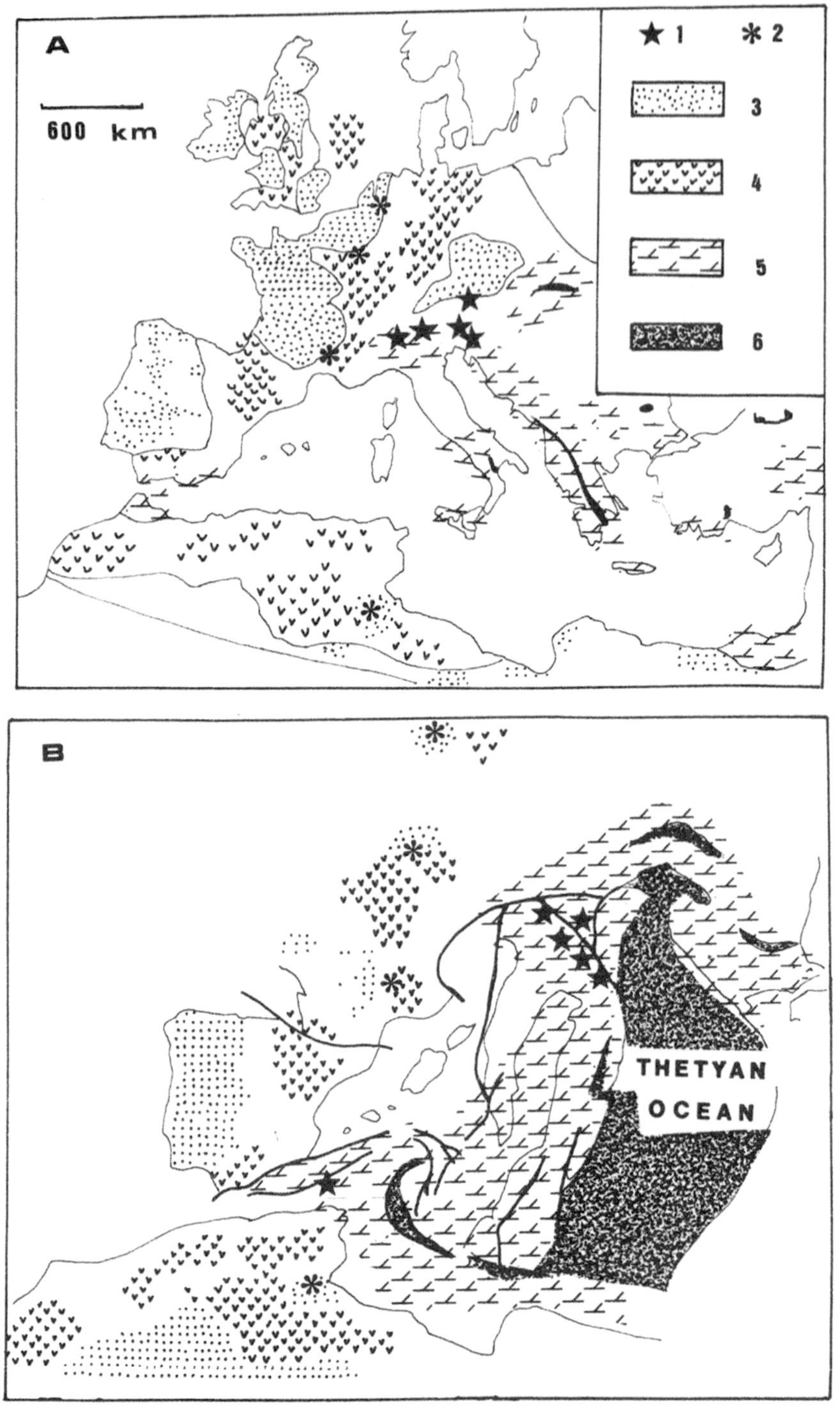

Fig. 1. Repartition map (A) and palynspastic map (B) of Triassic sediments and of related Pb-Zn deposits. 1 = carbonate hosted Pb-Zn deposits; 2 = sandstone hosted Pb-Zn deposits; 3 = sandstone facies; 4 = evaporitic facies; 5 = carbonate platform facies; 6 = oceanic crust.

supply. The discussion remains open. Many geological observations agree with an exogenous source of supply. Nevertheless, these structures induced unstable sedimentation conditions in the neighbouring areas and induced the formation of sedimentary traps (emersion surfaces, karst, unconformities, reef facies, lagoonal facies, slumping, . . .).

Regarding the stratiform pyritic deposits of Southern Tuscany, the presence of evaporites and basic volcanics (TANELLI et LATTANZI, 1983) indicates an environment with volcanogenic supply and a trend to rifting or initial ocean floor spreading. This point of view matches with the palynspastic map (Fig. 1).

A great number of diverse ore deposits is hosted in the sedimentary rocks of Jurassic age in the Western Mediterranean area. These carbonate rocks belong to the epicontinental platform type. The ore deposits consist of lead-zinc ores, but also fluorite and barite (Les Malines in France, Touissit in Morocco, El Abed in Algeria are lead-zinc deposits; Pierre-Perthuis in France is a fluorite deposit; Chaillac and Pessens in France are barite deposits). Bou Grine, Jebel Ajered in Tunisia, Kharzet Youssef in Algeria, Reocin in Spain are lead-zinc deposits hosted in platform carbonate rocks of Lower Cretaceous age.

The controlling features of these ore deposits are mainly related to sedimentary or diagenetic structures. The supply of metals to the platform epicontinental environment is connected mainly with leaching out of weathered sections on merged areas. The stratigraphical control of the occurrences and deposits was a function of epirogenic movements, causing the leaching of pedogenetic concentrations, formed on the continent during periods of stability. The main controlling factors of these types of deposits are of climatological and sedimentological type, but the permanent instability controls the localization of the ores in time (supply) and space (palaeogeographical traps). On the other hand it is possible that some heath flows occured and induced remobilization of ore (SUREAU et al., 1981).

The Orogenic Events and the Closing of the Tethys

The first closing event occured in the internal part of the Alpine system during the Lower Cretaceous (Albian mainly). The evolution of this phenomenon is confusing. Many geological features were overprinted by the later development of the orogenesis. A predominant dynamic metamorphism (except in the Pyrenées) was active from about 110 My (Schistes lustrés from Liguria, Piémont and Calabria) (DURAND-DELGA et FONTBOTE, 1980). The apparent age of the pitchblende from Val Vedello (Bergamasc Alps, Italy) 115 My (S. PHILIPPE, 1984) could be related to a remobilization and rejuvenation connected to this event.

Nevertheless, the most important geodynamic events occured during the Tertiary. At the beginning of the Miocene, the Mediterranean Sea was formed by the collapse of continental blocks. At the same time the crust thinned out and the Alboran Sea developed. An other basin, the Basin of Provence was formed through a 30 to 55 ° rotation of the Corsican-Sardinian microcontinent. The Algerian basin opened through the movement of the Kabylian Massif above a subduction zone, dipping

North during the Miocene which is now situated behind these massifs. The formation of the Tyrrhenian basin is related to a subduction zone situated outside of the Sicilian-Calabrian-Apenninic Arc. This evolution was still proceeding as a dextral movement by tearing the microplate of Alboran into two pieces. (J. ANDRIEUX et al., 1971). Again during the same period a shortening happened between the Eurasian and the African Plate. This shortening is related to the activity of two crustal subduction zones symmetrically dipping on both sides of the mesomediterranean plate. In the Maghrebides the geological units were overthrusted towards the South, in the Betides, towards the North. In the Apennin, it is assumed that a southwest dipping subduction zone could have induced the forward movement of the internal Ligurian zones above the external ones. This subduction zone could be the prolongation in this direction of the Maghrebide subduction zone.

Only is this Southern and Eastern part of the Western Mediterranean area (Betide, Maghrebide, Apennins), a calc-alkaline magmatism developped during this period (Tuskany, Tunisia, Algeria) and some hydrothermal events were associated with this magmatism. The difference in the geodynamic and magmatic evolution on the Eurasian and African plate in the Western Mediterranean Sea induces some differences in the domain of genesis of ore deposits. During Tertiary time, the ore deposits which were formed on the Eurasian plate, were related to supergene phenomena. So are uranium deposits in the sedimentary cover like Coutras in Gironde (France), Saint Pierre du Cantal (France) or the roll type deposits of the Lauraguais (France). On the other hand, on the African plate and at its boundaries on the Southern and Eastern banks of the Western Mediterranean Sea, some ore deposits are connected with magmatism and the associated heat flow and hydrothermal systems.

In Tuscany, the autochtonous basement consists of units from Paleozoïc up to Oligocene. These units are overlain by allochtonous liguride and subliguride units. The whole geological succession up to the Calcare Massicio was intruded by quartz-monzonites (5.7 My old; BORSI et al., 1967). A mostly thermic metamorphism (500° C and < kb) was associated with these intrusions. Then, a volcanic activity (about 4.7 My) began. It showed vented volcanics and porphyry dykes. The dykes are intensively transformed through a potassium rich hydrothermal alteration. Copper-lead-zinc sulfide ores (Campiglia Maritima) are associated with this magmatism. Recent publications (particularly by F. CORSINI et al., 1980; G. CORTECCI et al., 1983) on the base of sulfur isotope data, conclude that the formation took place at low depth and temperatures in the range of 250 to 400° C. The involved hydrothermal fluids could have removed a metal content wich existed in the autochtonous units. An other type of deposit like the Gavorrano pyritic ore body have resulted from contact metamorphism of intrusives with a preexisting deposit (BENCINI et al., 1980).

Ore deposits or occurrences associated with Tertiary magmatism are also important in the Maghrebides. Some of them are directly related to volcanic activity like Oued el Kebir (TOUAHRI et FUCHS, this vol.). In Oued el Kebir area, near Jijel, the first unit consists of vented volcanics forming trachytic domes, breccias and tuffs. After deposition of these effusives, the center of the area collapsed and a caldera was formed. Within the caldera, marine sediments were deposited. This very

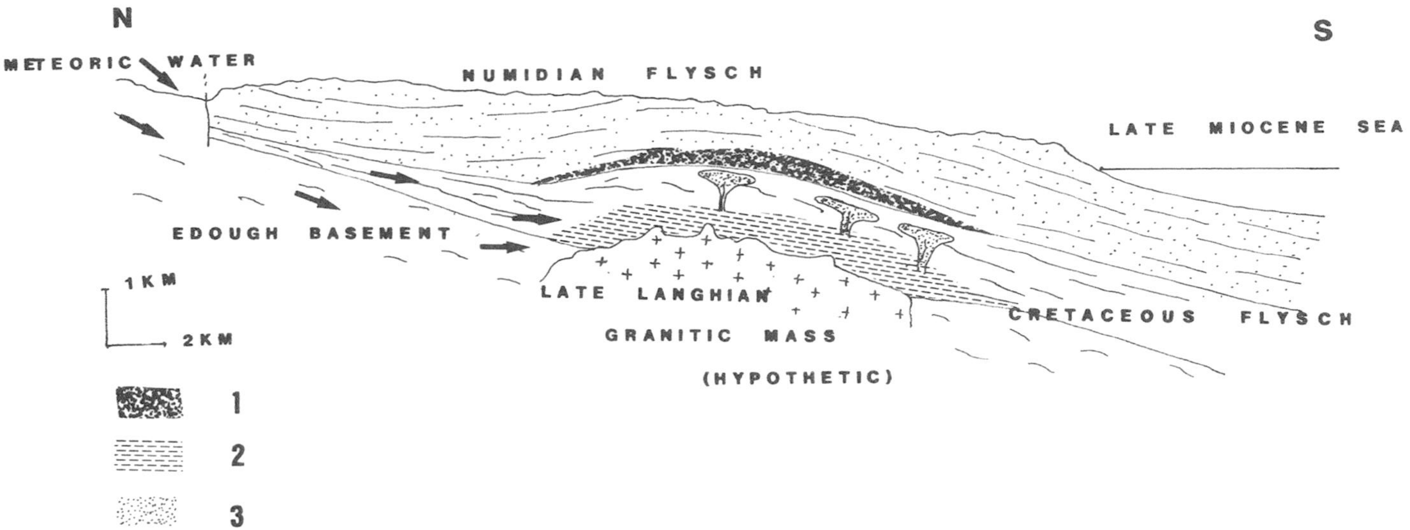

Fig. 2. A possible cross-section of the Aïn Barbar geothermal field at the end of Langhian times (in Marignac, 1983). 1 = silicifications in the Numidian cap rock; 2 = hornfelses with adinolite/solite/ plagioclase; 3 = Ca-Al Silicates in the Maestrichtian aquifer.

short period during the Langhian was followed by a new collapse which induced the formation of an extended unit of breccias. Afterwards, the volcanic activity continued with basaltic and andesitic flows which covered all of the area. In a later stage, monzonitic plugs intruded into the outer part of the complex. The ore deposits (Pb, Zn, Cu, Ba, Ag, Au) include stratiform ore bodies at the bottom of a limited sedimentary sequence and stockwerks in the lower volcanic unit, seated beneath the stratiform ore bodies. Both types of mineralization are characterized by an important pervasive hydrothermal alteration of wall rocks.

An other type of ore deposit of Tertiary age in the Maghrebides is a polymetallic vein type. At Aīn Barbar (Algeria), C. MARIGNAC (1983) the evidence of Tertiary thermal domes has been shown. A paleogeothermal system was active towards the end of Langhian time. An allochtonous cap of Numidian flysch induced the formation of a convective hydrothermal system in the underlying Cretaceous flysch unit (Fig. 2). A zonation of wall rock alteration can be observed (Fig. 3). Paragenetic associations indicate that the temperature was less than 320—330° C and the pressure about 100—130 bars (10—1 M. P. A.).

The boundary of the African Plate from the Betides to the Apennins was subject to an important lateral displacement associated with crustal subduction. For these reasons the magmatic and hydrothermal activities were very well developed, particularly during Miocene. Lead isotope ratios are in full agreement with this assumptions. On the border of the European Plate, lead isotope ratios show the existence of two groups: one related to the Cambrian carbonate rocks, the other one to late Hercynian hydrothermal activity (BREVANT et al., 1982, Fig. 4). These two groups were recognized even in the deposits of the post Permian sedimentary units. On the other hand, P. RAJLICH et al. (1983), in their studies of Moroccon deposits Fig. 5 and V. KÖPPEL (1983), in his studies on deposits of the Eastern and

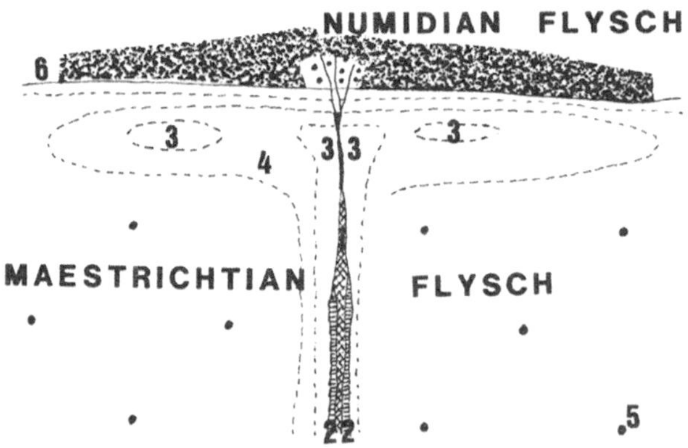

Fig. 3. Diagrammatic scheme showing a possible configuration of the paleoisotherms in the Aïn Barbar geothermal field, with the associated alterations. Note that the model could as well be asymmetrical. 1 = brecciated fault zone; 2 = metasomatic alteration near the ore veins; 3 = epidotic and K-feldspar; 4 = wairakite; 5 = calcite (black dots); 6 = silicification (in Mariguae 1983)

Southern Alps, display a more complex distribution of lead isotope ratios. The results point to the possibility of a remobilization of crustal lead and for a supply of young lead during Tertiary.

Metallogenic, geochemical and isotope ratio data help to correlate the genesis of ore deposits and occurrences in the Western Mediterranean Sea to the main geodynamic events in this area.

Carbonate hosted lead-zinc or fluorite and barite deposits in an epicontinental environment are related both to supergene and tectonic phenomena. It seems however that the initial rifting period and the final closing are the major periods for the supply of metals connected with hydrothermal activity. During the final phase, the boundary of the African Plate suffered lateral displacement and crustal subduction zones played a major part.

The distribution of the elements and therefore the formation of ore provinces are related to major geodynamic events of the initial rifting and of the final stage. Even if within a structural unit, within a particular period, processes which led to the concentration of metals and therefore to the formation of economic deposits were different, the most relevant factors, governing the distribution of provinces, are related to the geodynamic evolution.

References

Badia, D., Bégassat, P., & Fuchs, Y. (1984): Permo-Carboniferous volcanism in France and Western Europe; Metallogenic significance. I. A. E. A. Meeting on uranium deposits in volcanic rocks. El Paso, Texas, 2–6/4/84. in I. A. E. A. Vienna.

Badia, D., Boulègue, J., & Fuchs, Y. (1985): On some chemical characteristics of the hydrothermal phenomenom associated to Permo-Carboniferous volcanism in Western Europe. Eng. III. Strasbourg. 1–4/4/1985.

Bencini, A., Ducchi, V., & Minissale, A. (1980): Distribuzione di alcuni elemti metallici nelle rocce intrusive e carbonatiche associate alle mineralizzazioni a pirite di Gavorrano (Grosseto). Rendiconti Soc. Ital. di Mineralogia e petrologia, *36* (2), 599–609.

Borsi, S., Ferrara, G., & Tongiorgi, E. (1967): Determinazioni con il metodo K/Ar dell eta delle rocce magmatiche della Toscana. Soc. Geol. Italiana, v. 86, 403–410.

Brevant, O., Dupré, B., & Allègre, J. C. (1982): Metallogenic provinces and the remobilization process studied by lead isotopes: lead-zinc ore deposits from the Southern Massif Central, France. Econ. Geol., *77, 3,* 565–575.

Burt, D. M., & Sheridan, M. F. (1980): Uranium mineralization in fluorine enriched volcanic rocks. Dep. Geol., Arizona State University, 494 p.

Corsini, F., Cortecci, G., Leone, G., & Tanelli, G. (1980): Sulfur isotope study of the skarn (Cu, Pb, Zn) sulfide deposit of valle del Temperino. Campiglia Maritima, Tuscany, Italy. Econ. Geol., Vol. 75, 84–96.

Cortecci, G., Klemm, D. D., Lattanzi, P., Tanelli, G., & Wagner, J. (1983): A sulfur isotope study on pyrite deposits of Southern Tuscany. Italy. Min. Deposita, *18,* 285–297.

Durand-Delga, M., & Fontboté, J. M. (1980): Le cadre structural de la Méditerranée occidentale. *In:* Colloque C. 5, Géologie des chaînes alpines issues de la Téthys. XXVI. C. G. I., Paris, B. R. G. M. ed., 67–85.

KÖPPEL, V. (1980): Summary of lead isotope data from ore deposits of the Eastern and Southern Alps. *In:* Mineral deposits of the Alps and of Alpine epoch in Europe. Schneider ed., Springer-Verlag, 162–168.

KOVALENKO, V. I., & KOVALENKA, N. I. (1976): Ongonites (topaz bearing quartz keratophyres) subvolcanic analogue of rare metals Li-F granites (in Russian). Nauk. Press, Moscow, 128 p., Trad., angl.

LATTANZI, P., & TANELLI, G. (1982): Alcune considerazioni sulla genesi dei giacimenti a pirite della Marenna, Toscana. Mem. Soc. Geol. Ital., *22,* 1981, 159–164.

LEROY, J. (1978): Métallogénèse des gisements d'uranium de la division de la Crouzille. Mem. Sciences de la Terre, *36,* 275 p.

MARIGNAC, C. (1983): The polymetallic ore veins of Aïn Barbar (Algeria) as a consequence of alpine geothermal activity. Min. Deposits of the Alps and of alpine epoch in Europe. Schneider ed., Springer-Verlag, 298–311.

PETRASCHECK, W. E., ERKAN, E., NEUWIRTH, K. (1974): Permo-Triassic uranium ore in the Austrian Alps. Paleographic control as a guide for prospecting. *In:* Formation of Uranium Ore Deposits. I. A. E. A. Vienna, 291–298.

PHILIPPE, S. (1984): Géochronologie U-Pb des minéralisations uranifères des gîtes de Val Vedello et de Novazza et de leur encaissant ignimbritique (Alpes Bergamasques). D. E. A.– U. S. T. L. Montpellier, 40 p.

RAJLICH, P., LEGIERSKY, J., & SMEJKEL, V. (1983): Stable isotope study of base metal deposits from the Eastern High Atlas. Morocco. Mineral. Deposita. *18,* 161–171.

SUREAU, J. F., AUBAGUE, M., GIOT, D., LENINDRE, Y. M., LHOMER, A., LELEU, M., BARBIER, J., ORGEVAL, J. J., BODEUR, Y., & LE STRAT, P. (1981): Recherche de guides de prospection pour les gîtes Pb-Zn liés aux strates en environnement carbonaté. Rapport Action Concertée D. G. R. S. T. 76/7/0024. BRGM.

TANELLI, G., & LETTANZI, P. (1983): Pyritic ores of Southern Tuscany. Italy. Spec. Publ. Geol. Soc. South Africa, *7,* 1983, 315–323.

TOUAHRI, B., & FUCHS, Y. (this volume) (1984): Les minéralisations à plomb-zinc cuivre du domaine alpin en Algérie.

WOPFNER, H. (1984): Permian deposits of the Southern Alps as a product of initial alpidic taphrogenesis. Geol. Rundschau, Bd. 37, H 1, 259–277.

ZIELINSKY, R. A. (1982): Tuffaceous sediments as source rocks for uranium. A case study of the White River Formation. Wyoming. Journ. of Geoch-Expl., *18,* 285–306.

Genesis of Copper Deposit of Aïn-Sefra Aera, Western Algeria

By O. KOLLI*)

With 9 Figures

Zusammenfassung

Die oberjurassisch-unterkretazische Sandsteinformation von Ain Sephra im Sahara-Atlas enthält in mehreren Horizonten Cu-Erz-führende Schichten des „Red bed"-Typs. Die Metallherkunft wird aus dem Grundgebirge im Süden vermutet.

Abstract

Copper mineralization within the Upper Jurassic and Lower Cretaceous Sandstone of Aïn-Séfra is controlled by the lithology and paleotopography. The mineralization occurs in channel sands of low energy, flood plains where good permeabilities permitted high intergranular flow of saline solutions acting as metal carriers.

Malachite, azurite, delafossite and chrysocolla fill the interstices and fractures in the clastic host rocks, forming halos around the mineralized fossil plant material. Primary sulfide minerals as chalcocite, chalcopyrite and pyrite were probably deposited by groundwater solutions moving through paleochannelsystems after deposition of the host rock. Also, the petrographical datas show that copper sulfide grains occur on the outside of quartz overgrowths.

Mineralization occurs when a favorable reducing environnement, which provided principally by the decomposing plant fragments, is present.

It appears clearly that the mineral emplacement is related to an epigenetic phase.

---

*) Département de Géologie, Institut des Sciences de la Terre, U. S. T. H. B., B. P. 139 Dar el Beida – Alger, Algerie.

The ultimate source of copper may have been older deposits in Precambrian rocks or copper leaching from weathered Precambrian rocks (of the P. III) in the Anti-Atlas and Ougarta mountains in South-West of Aïn-Séfra.

Résumé

Les minéralisations cuprifères de la région d'Aïn-Séfra sont encaissées dans un ensemble détritique bigarré d'âge mésozoïque. Elles sont caractérisées par:

– une localisation particulière à trois niveaux stratigraphiques différents (Portlandien, Néocomien, Aptien).

– un contrôle lithologique et paléotopographique stricte. La minéralisation est contenue dans des faciès blancs et verts à l'intérieur d'une série à dominante rouge. Les faciès sont discontinus et occupent des chenaux caractérisés par leur richesse en débris végétaux et en matière organique.

Macroscopiquement, la minéralisation est, disséminée ou fissurale ou encore concordante avec le litage de la roche ou épigénitique.

L'association minérale est simple et comprend surtout de la chalcocite, de la malachite, de l'azurite, du chrysocolle, de la covellite et de la cuprite. Ces minéraux forment le ciment des réléments détritiques ou bien ils sont associés au ciment carbonaté de la roche hôte. La cristallisation de ces minéraux montre un caratère typiquement épigénétique et s'expliquerait par la circulation d'eaux souterraines qui semblent avoir joué un rôle essentiel dans la mobilisation et le dépôt du cuivre. Ce processus a abouti à la mise en place d'une minéralisation de type roll ou front mobile.

Introduction

Dans l'état actuel de nos connaissances sur les conditions de mise en place des minéralisations en liaison avec des sédiments détritiques grossiers communément désignés sous le terme de «Red-beds», les recherches ont montré que dans la majorité des cas la mise en place des minéralisations est syngénétique. Les études que nous avons mené sur des minéralisations à Cu dans des sédiments clastiques dans la région d'Aïn-Séfra, ont permis de mettre en évidence le caractère franchement épigénétique de la minéralisation.

Les minéralisations d'Aïn-Séfra sont essentiellement cuprifères et d'âge mésozoïque. Elles sont caractérisées par leur grande extension géographique, leur répartition stratigraphique et leurs contrôles paléomorphologiques (chenaux fluviatiles) et lithologiques.

Situation Géographique

La région d'Aïn-Séfra est située dans la partie centrale des Monts des Ksour. Les Monts des Ksour constituent la partie la plus occidentale de l'Atlas Saharien

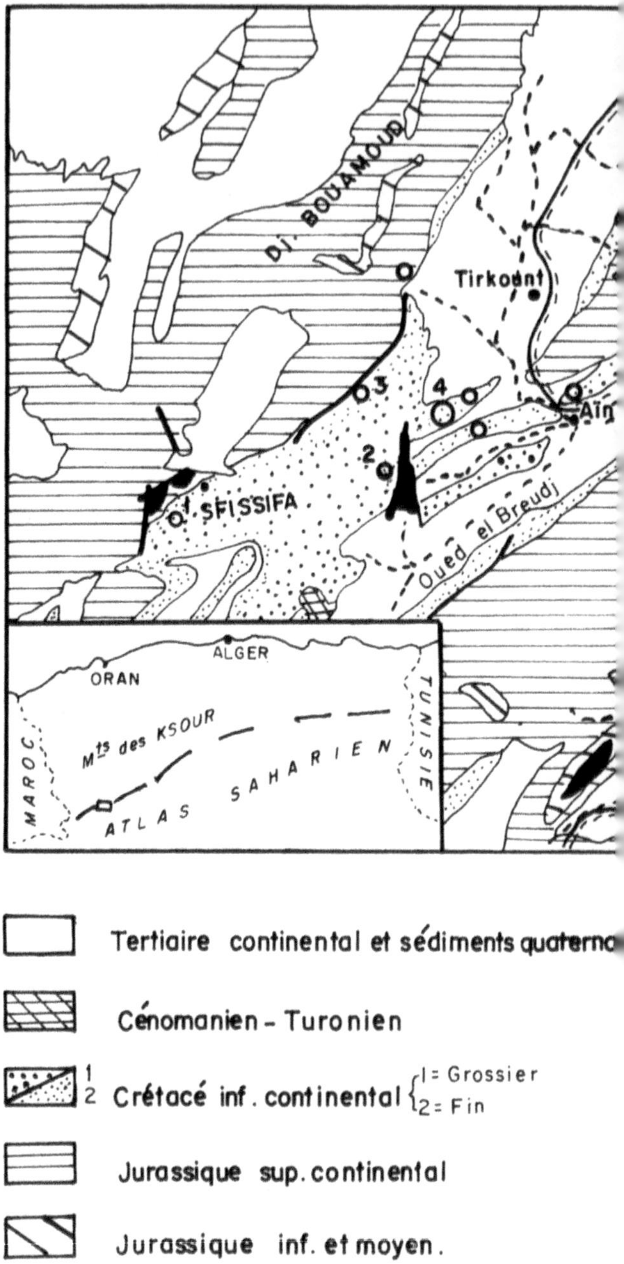

Fig. 1. Localisation du secteur d'étude et distributio

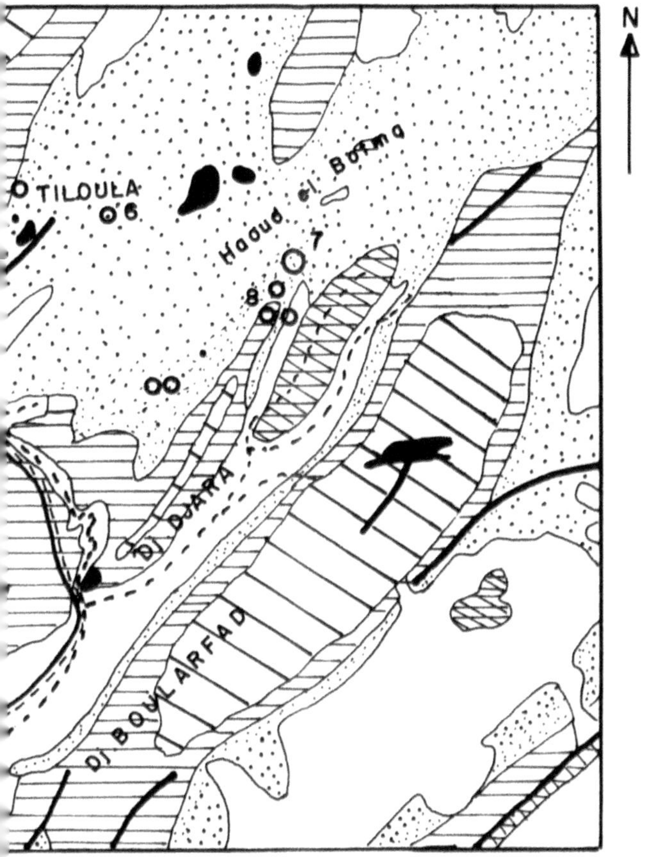

ias, roches vertes

ailles, contacts anormaux

fissifa

L-Hendjir

arat el. Kheil

arat Debba

iloula

dt Tiourtelt

ehirize

jara

Gisement

Indice

Oued

Route

tions dans la partie centrale des Monts des Ksour.

qui trouve son prolongement dans le Haut Atlas au Maroc. Cette partie de l'Atlas Saharien est bordée au Nord par les Hauts Plateaux et au Sud par la plate forme saharienne. Elle est caractérisée par des chaînons allongés et orientés Sud-Ouest–Nord-Est. Ces chaînons sont séparés par de vastes plaines synclinales. Le secteur étudié (Fig. 1) est constitué de deux dépressions principales de plusieurs Kms de long. La première est celle de Tirkount – Ben Hendjir; la seconde est formée par la plaine de Tiout qui se relaye vers l'Est par le Haoud-El-Botma et vers le Nord par la plaine de Tiourtelt. Ces vastes dépressions sont limitées au Sud par les chaînons de Mir-El-Djebel, Dj. Mekter et Dj. Djara et au Nord-Ouest par le Dj. El-Haïrch et le Dj. Morhad. Elles sont séparées entre elles par l'anticlinal du Dj. Aïssa (45 Kms de long).

Cadre Géologique Régional

Dans tout l'Atlas Saharien occidental se développe un ensemble détritique bigarré, rapporté au Jurassique Supérieur Crétacé inférieur. Cet ensemble déborde même sur les régions voisines. Cet ensemble présente des faciès uniformes, azoïques et très monotones aux affleurements. Il offre une étude très complexe du fait que les couches qui le composent et qui s'étendent du Dogger au Cénomanien sont constituées de conglomérats, de grès et d'argiles pélitiques. Les repères chronostratigraphiques sont rares et les seuls critères qui permettent des découpages assez constants sur toute l'étendue des Monts des Ksour sont pétrographiques et morphologiques. Aussi, il est devenu classique de diviser à la suite de J. P. Bassoulet (1973), cet ensemble épais de plus de 2500 m en trois grandes unités lithostratigraphiques (Fig. 2). Cet ensemble détritique se place à la suite normale du contexte d'évolution du domaine atlasique que l'on observe à partir du Bathonien et au cours duquel la regression de la mer venant de l'Est et du Nord (G. Choubert et A. Faure-Muret, 1962) va s'accentuer jusqu'à l'Albien. Durant toute cette période les Monts des Ksour sont caractérisés par une sédimentation fluvio-déltaïque. Celle-ci passe latéralement aux faciès marins de la zone mobile plus au Nord et à ceux également marins de l'Atlas Saharien central.

Étude des Minéralisations

Il existe environ une centaine de points minéralisés dans les sédiments mésozoïques de l'Atlas Saharien occidental. La minéralisation est essentiellement cuprifère et la chalcocite en est le minéral principal. Les zones minéralisées apparaissent à 3 niveaux stratigraphiques bien précis à l'intérieur de l'ensemble détritique bigarré. Elles apparaissent au Portlandien, au Néocomien et à l'Aptien (Fig. 2). Il existe un contrôle lithologique stricte de la minéralisation. De ce point de vue, la minéralisation se localise exclusivement dans des faciès détritiques blancs ou verts dans une série à dominante rouge. En dehors de ces niveaux aucune minéralisation n'a été décelée. Ces trois niveaux correspondent à des grès de remplissage de chenaux,

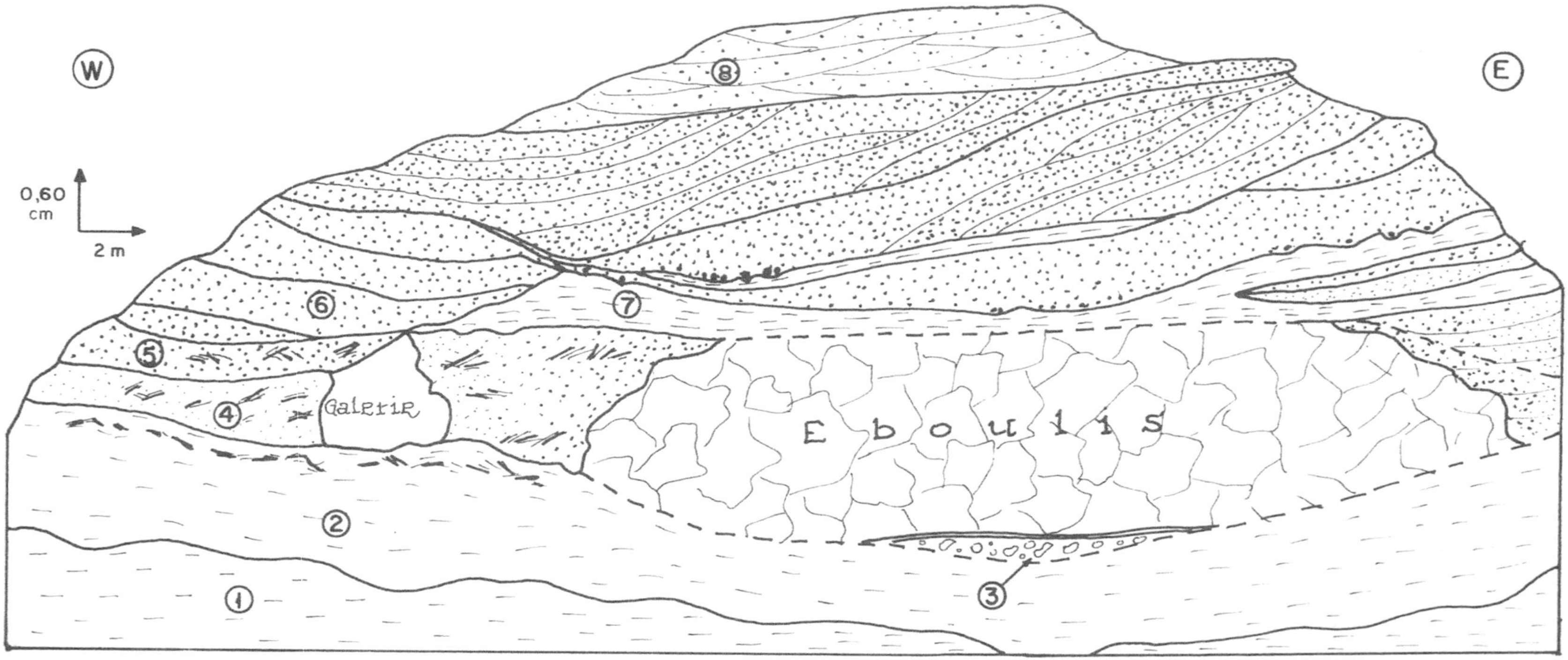

Fig. 2. Exemple d'un empilement de corps lenticulaires dans le système chenalisant minéralisé de Garat-Debba (d'après photo).

① Argiles pélitiques rouges

② Argiles pélitiques vertes á débris végétaux minéralisés

③ Conglomérat à galet d'argile pélitique verte

④ Grés argileux minéralisé à débris végétaux

⑤ Grés blancs quartzeux, á ciment carbonaté, galets mous, et troncs d'arbre mineralisés

⑥ Grés blancs quartzeux á galets mous, ciment carbonaté, non minéralisé

⑦ Argiles pélitiques vertes non minéralisées

⑧ Grés rouges à stratifications obliques, ciment argilo-ferrugineux, non minéralisés caractérisés par rapport aux autres grès rouges par leur bonne porosité et par la présence de débris végétaux et de matière organique. D'allure stratiforme à l'échelle régionale, elles présentent une aspect discontinue à petite échelle et de ce fait elles se rattachent au type pénéconcordant ou type red-beds. Dans ces trois niveaux, la minéralisation est encaissée dans des complexes chenalisants (Fig. 3) et est associées à des argiles pélitiques vertes, à des grès blancs à taches noirâtres de matière organique et surtout à des grès blancs à nombreux débris végétaux. Les complexes chenalisants présentent dans leurs parties affleurantes une extension latérale variable de 15 à 100 m et une puissance de 1 à 6 m. Ils correspondent à des paléochenaux et sont constitués par l'accumulation de lentilles unitaires gréseuses associées à des petits niveaux d'argile pélitique verte ou de siltstone vert. Les lentilles montrent souvent de nombreux débris végétaux épigénisés en malachite, azurite, chalcocite et oxydes de fer. L'imbrication des lentilles de grès blanc donne des structures entrecroisées où les sédiments fins sont ravinés par le matériel plus grossier.

Dans ces lentilles la minéralisation apparaît soit disséminée sous forme nuageuse soit suivant les plans de litage (Fig. 4). Dans certains cas, elle remplit des

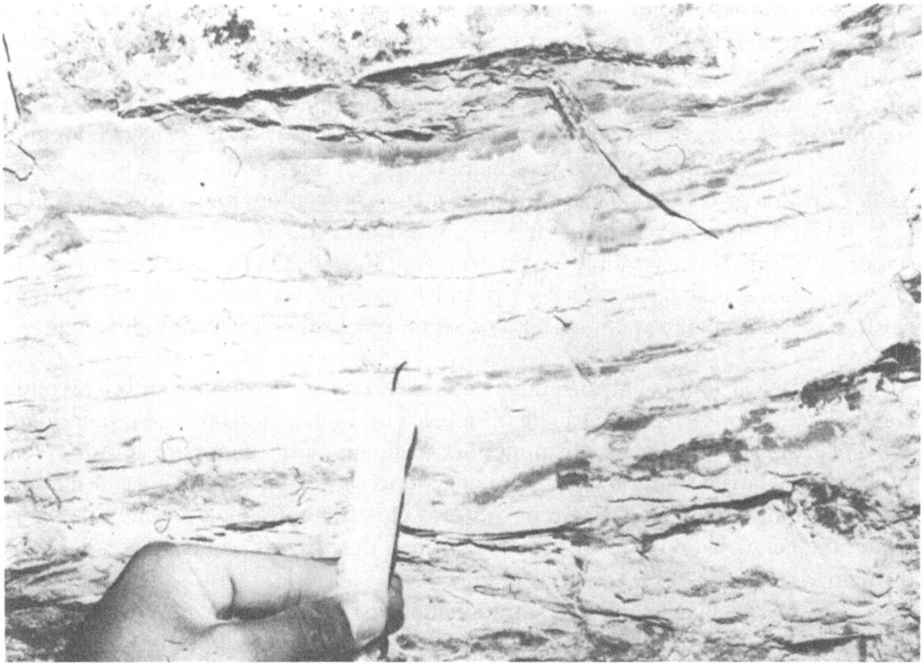

Fig. 3. Photographie montrant, de haut en bas, les relations de la minéralisation avec les roches encaissante:
- des grès blancs, à base érosive, dans lesquels la minéralisation est disséminée sous forme nuageuse.
- des argiles illitiques straftifiées où la minéralisation se dispose dans les plans de litage et dans des petites fractures.
- des argiles gréseuses riches en débris végétaux minéralisés. Provenance: Gerat-Debba.

petites fractures dans les grès et les argiles illitiques et elle épigénise partiellement ou totalement les débris végétaux (Fig. 5).

La minéralisation est constituée pour l'essentiel de malachite, d'azurite, de chalcocite et de covellite auxquels sont parfois associés de la chalcopyrite, de la pyrite, de la délafossite, du chrysocolle, de la cuprite, de la goethite et probablement de la ténorite (SONAREM, 1973). Malachite, azurite et chalcocite apparaissent se concentrer dans les débris végétaux carbonisés. Ces minéraux apparaissent aussi dans le ciment carbonaté des grès hôtes qu'ils remplacent. Dans d'autres cas, ils cimentent les grains détritiques de la roche et apparaissent le long de petites fractures, donnant des halos irréguliers autour des sulfures primaires (Fig. 6).

Chalcopyrite, pyrite et parfois la chalcocite apparaissent disséminés dans le ciment carbonaté des grès. Dans certains cas, la chalcocite remplace la chalcopyrite. La covellite apparaît toujours sous forme de remplacement des sulfures primaires.

Pétrographie des Roches Hôtes de la Minéralisation

Les sédiments qui portent la minéralisation comprennent aussi bien des microconglomérats que des grès à grain moyen, des argiles illitiques et des siltstones. La couleur de ces sédiments et toujours blanche ou verte. Une analyse modale effectuée sur les grès a permis de les considérer comme étant des orthoquarzites comparables à ceux définis par PETTIT-JOHN, F. J., et al. (1972). Sur le plan microscopique, les grès montrent des grains détritiques où les grains de quartz prédominant. Certains grains de quartz sont déformés et présentent une extinction roulante et nombre d'entre eux sont soulignés par un accroissement en silice secondaire important (Fig. 7). Les grains, en général, sont limpides et quelques uns présentent de nombreuses inclusions diverses parmi lesquelles les inclusions aciculaires de rutile sont facilement reconnaissables, d'autres grains montrent des figures de Boehm.

Ces grains sont constitués pour l'essentiel de quartz monocristallin arrondis à sous-arrondis dont la taille varie de 0,1 à 0,4 mm. Le reste du cortège minéralogique est constitué de feldspaths, de minéraux argileux et de minéraux lourds.

Les feldspaths sont aussi arrondis et représentés par le l'orthose et du microcline. Ces minéraux apparaissent peu altérés et leur bonne conservation rapelle ceux communément observés*) dans les formations détritiques cambriennes des Monts d'Ougarta.

Les plagioclases sont occasionnellement présents et sont représentés par de l'albite qui dans certains cas apparaît diagénétique. Ceci attesterait alors de la forte salinité du milieu de formation. Accessoirement, on note la présence de paillettes de micas (biotite et muscovite) altérés, de tourmaline, de rutile, d'épidote, d'hématite, de magnétite ainsi que de zircon. Tous ces minéraux apparaissent fortement usés et s'observent en grains arrondis à sub-arrondis et témoigneraient du caractère polycyclique des grès minéralisés de la région d'Aïn-Séfra.

---

* Communication orale: F. ARBEY.

Fig. 4. Photographie montrant une partie d'un tronc d'arbre complètement épigénisé par la malachite, l'azurite et la chalcocite. Provenance: Garat-Debba.

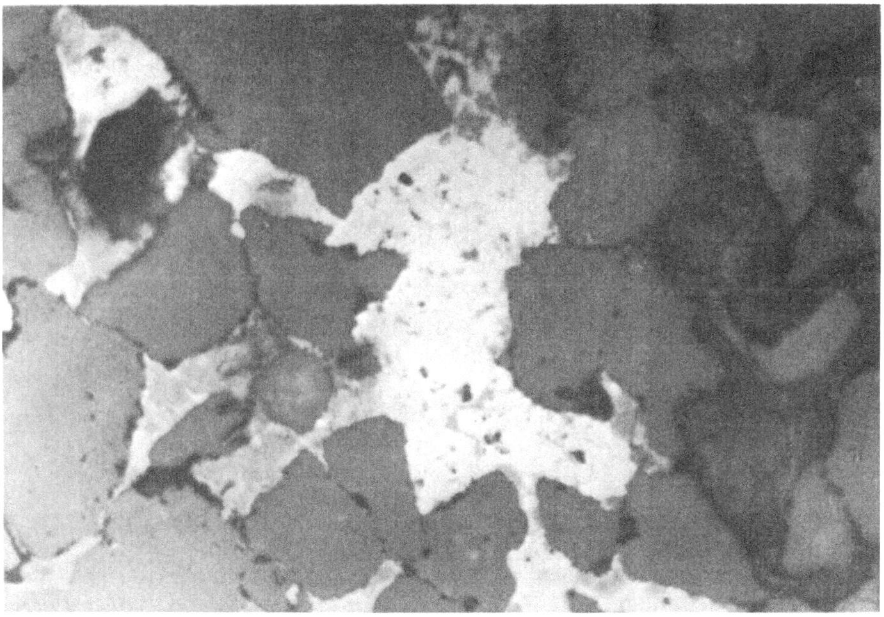

Fig. 5. Microphotographie montrant de la chalcocite (en blanc), avec sur les bords de la covellite (gris blanc), qui cimente les grains détritiques. Section polie, lumière réfléchie: × 10. Provenance: Tiloula.

Il et aussi intéressant de signaler la fréquence relative des grains de zircon qui est à mettre en parallèle avec les formations volcaniques du Pr. III affleurants au S-W de la région étudiée et qui sont riches en zircon ainsi que les sédiments cambriens transgressifs sur ces roches volcaniques.

D'un autre côté l'étude de ces grès a permis de mettre en évidence que leur structure était à l'origine assez poreuse. Ils ont par la suite étaient le siège de phénomènes diagénétiques et épigénétiques importants. Le phénomène le plus spectaculaire est celui du développement des silicifications secondaires qui a conféré à ces grès une structure plus ou moins jointive. En plus de cette cimentation siliceuse, l'étude a montré la présence de plusieurs types de ciments: sulfaté à barytine, carbonaté à calcite et dolomite, cuprifère (sulfurs et carbonates de Cu) et ferrifère. Tous ces types de ciment, et en particulier le type cuprifère, présentent une distribution géométrique en fonction de la porosité résiduelle réalisée par le ciment siliceux. Ils remplissent les interstices granulaires. Les minéraux cuprifères apparaissent toujours en dehors des auréoles de nourrissage en quartz et parfois se surimposent à la structure quartzeuse (Fig. 7). L'observation de certains débris végétaux fossiles montre une structure cellulaire complétement silicifiée, des cellules non déformées et des minéraux cuprifères qui se surimposent à cette structure cellulaire ou bien ils s'insinuent dans les petites fractures postérieures à la silicification (Fig. 8). Cette dernière observation montre clairement que de la silice était présente au moment du

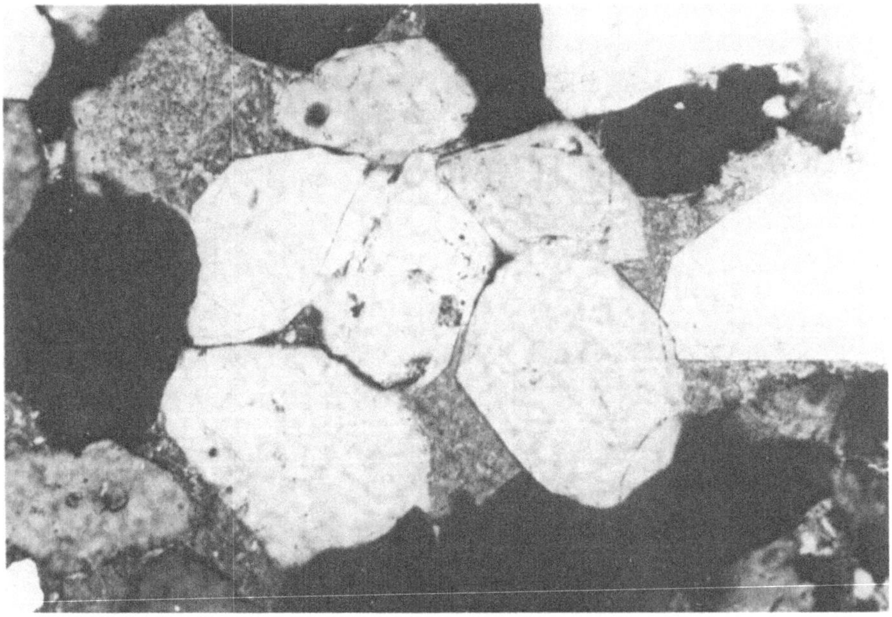

Fig. 6. Microphotographie d'un grès à ciment de carbonates de Cu. La malachite (en gris noir) remplit les vides résiduels, à contours géométriques, laissés par le nourrissage important des grains de quartz dont le noyau détritique apparaît nettement. Lame mince, lumière analysée et polarisée: × 6,3. Provenance: Tiloula.

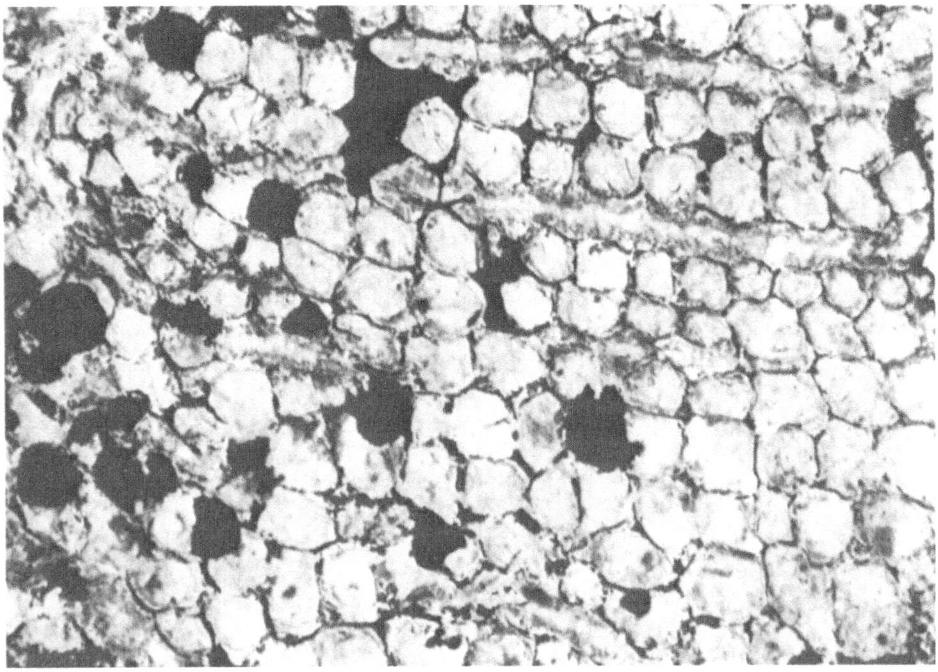

Fig. 7. Microphotographie montrant une structure cellulaire sillicifiée sur laquelle se surimposent les minéraux cuprifères (chalcocite et malachite). Lame mince, lumière naturelle: × 6,3. Provenance: Méhirize.

dépôt des sédiments puisque les débris organiques n'ont pas été détruits par oxydation et leur enfouissement étant probablement rapide. Toutes les observations réalisées montrent à l'évidence le caractère interstitiel des minéraux cuprifères et leur mise en place postérieurement à la silicification de la roche encaissante. La succession paragénétique est difficile à mettre en évidence avec précision cependant les observations accumulées permettent de proposer la succession qui est résumée au tableau 1.

Conclusion

Aspect génétique de la minéralisation et source du Cu.

Les observations accumulées tout au long de l'étude de la minéralisation et de son enveloppe permettent de conclure à une postériorité de la minéralisation (sulfures de Cu, carbonates et silicate de Cu) par rapport à la roche hôte. La minéralisation est discontinue à toutes les échelles d'observation et l'introduction tardive du cuivre par rapport aux sédiments hôtes ne peut faire aucun doute. Il apparaît clairement que la minéralisation est liée à des circulations souterraines d'eau vadose,

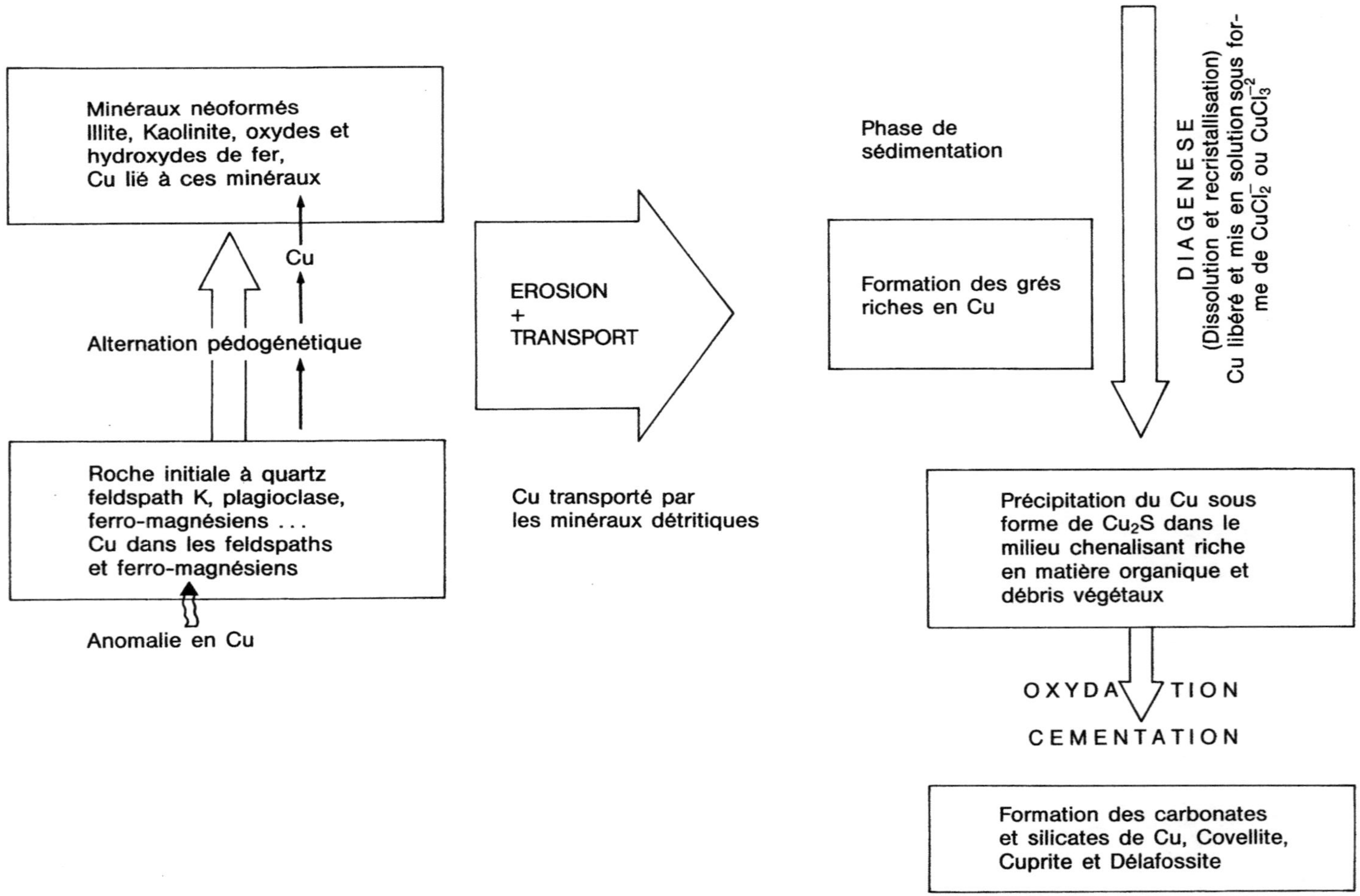

Fig. 8. Schéma résumant l'aspect génétique de la minéralisation et la source du Cu.

probablement d'origine météorique, comme l'atteste le ciment calcitique et dolomitique des grès hôtes. Il est alors vraisemblable que cette minéralisation s'apparente à celle du «type roll» (ou front mobile), type bien connu dans les concentrations uranifères encaissées dans des formations détritiques identiques à celle de la région d'Aïn-Séfra.

Quant à la source du Cu concentré dans les chenaux, il est possible que cet élément ait été lixivié par les nappes oxydantes et empruntés aux autres couches de grès. En effet, à la suite des travaux de WALKER T. R. (1976) et de HOLMES I. et al. (1983), il est largement admis que les silicates ferromagnésiens (pyroxènes, amphiboles, biotites, muscovites . . .) les feldspaths, les argiles (illite, kaolinite . . .) et les oxydes et hydroxydes de fer sont susceptibles d'être complétement détruits lors de la diagénèse ou changer d'état dans le cas des hydroxydes par exemple qui se transforment en oxydes. Par ailleurs, ces mêmes minéraux peuvent renfermer d'importantes quantités de Cu à l'état de trace; une moyenne supérieure à 100 ppm. Lorsqu'on pense au puissant ensemble détritique bigarré d'Aïn-Séfra qui fait 2500 m en moyenne; il est facile d'imaginer l'énorme stock de Cu qui aurait pu être extrait lors de la diagénèse des minéraux énumérés précédemment et dont certains sont absents et d'autres présents à l'état de trace dans les grès d'Aïn-Séfra. Le Cu extrait lors de la diagénèse serait véhiculé par les nappes acides souterraines et se redéposerait dans les grès blancs riches en matière organique et en débris végétaux carbonisés. Ces considérations ont permis d'envisager que l'origine proche du Cu est constituée par la roche encaissante. Quant à l'origine lointaine du métal, nous pensons que c'est l'origine de la roche encaissante qui est celle du cuivre.

A ce propos, l'étude du contexte géologique et paléogéographique régional montre un fait remarquable qui est constitué par la présence au Sud-Ouest du domaine étudié, de roches volcaniques (andésites en particulier) du Panafricain qui représentent une anomalie positive en Cu et qui a fort bien pu fournir par le biais de l'altération pédologique un stock métal important au bassin de sédimentation. A ce sujet l'étude du cortège minéralogique de l'encaissant (KOLLI, O., et ABDEL-WAHAB, S., 1982) a permis d'envisager que pendant le Jurassique supérieur et le Crétacé inférieur, le climat qui régnait sur le continent qui a fournit ces sédiments devait être chaud et humide. D'autre part, CONRAD, C. (1969) signale à 600 Kms plus au Sud (région de Tidikelt) de la zone étudiée des rubéfactions et des cuirassements ferrugineux qui affectent la série paléozoïque pénéplanée avant le dépôt du Continental Intercalaire, c'est à dire à la base du Crétacé. Par ailleurs nous pouvons aussi affirmer que l'ensemble détritique bigarré d'Aïn-Séfra correspond aux produits du démantélement des profils d'altération plus ou moins importants qui se seraient formés au cours de la période biostasique. Il sarait alors logique de penser que pendant l'alteration chimique continentale sous couvert végétal, les minéraux néoformés comme les argiles, les oxydes et hydroxydes de fer ont du retenir le Cu mobilisé. Nous pouvons donc proposer un schéma de transport du Cu vers le bassin de sédimentation (Fig. 9), sous forme adsorbée sur les particules détritiques de nature différente (argiles, oxydes, hydroxydes, silicates ferromagnésiens, feldspaths . . .) après la destruction du couvert végétal dans le sens D'ERHART, H. (1956).

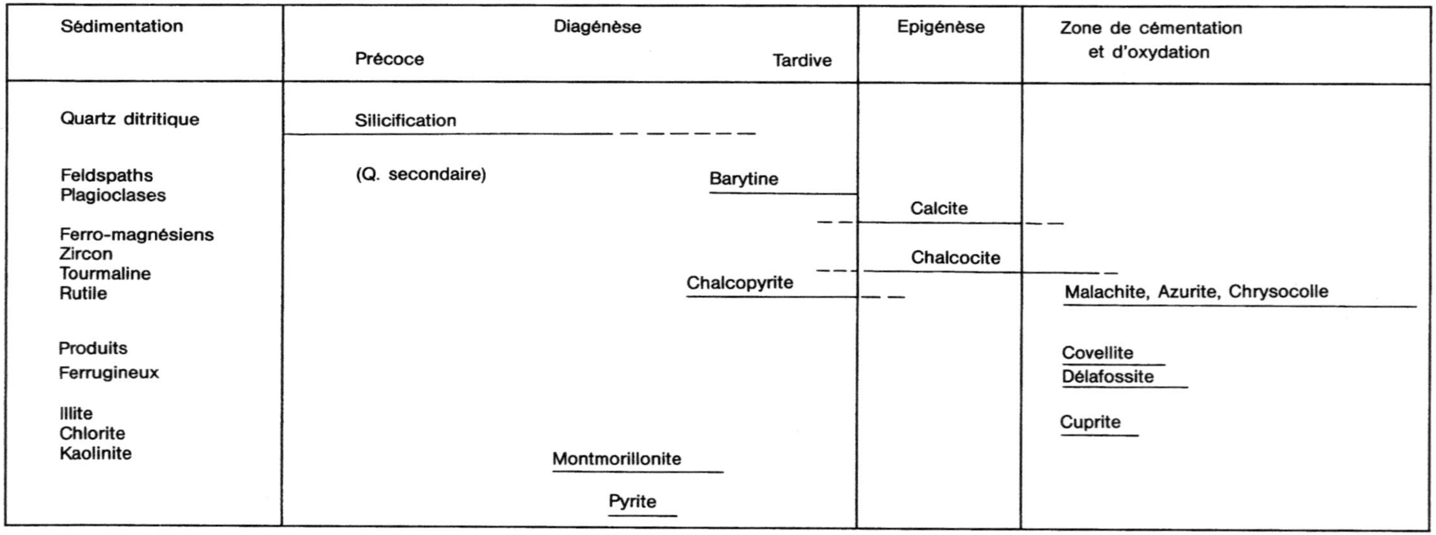

Fig. 9. Schéma récapitulatif résumant les phénomènes syngénétiques, diagénétiques et épigénétique en relation avee l'apparition des divers minéraux.

Remerciements

L'auteur remercie les Professeurs Y. FUCHS et N. ÖZLÜ, respectivement de l'Université Pierre et Marie Curie de Paris VI et de l'U. S. T. H. B. – Alger, pour la lecture critique du manuscrit.

Bibliographie

BASOULLET, J. P. (1973): Contribution à l'étude stratigraphique du Mésozoïque de l'Atlas Saharien occidental, Algérie. Thèse Doct. Fac. Sci. Paris.

CHOUBERT, G., et FAURE-MURET, A. (1962): Evolution du domaine atlasique marocain depuis les temps paléozoïques, M. h. S. Soc. Geol. Fr. (livre mémoire P. Fallot), t 1; 147–527.

CONRAD, C. (1970): L'Evolution continentale post-hercynienne du Sahara algérien – Saoura, Erg – Chech, Tanezrouft, Ahnet, Mouydir. Ed. C. N. R. S., Ser. Géologie N° 10, 527 p.

DELFAUD, J. (1974): La sédimentation déltaïque ancienne: Exemples Nord Sahariens. Bull. Centre Rech. Pau – S. N. P. A., *8, 1,* 241–263.

ERHART, H. (1956): La génèse des sols en tant que phénomène géologique. Collection «Evolution des Sciences», Ed. Masson, Paris N° 8.

HARSMAN, E. N. (1972): Geology and urenium deposits Shirly Bassin area, Wyoming. Geol. Serv. Pap. 745, 82 p.

HOLMES, I., CHAMBERS, A. D., IXER, R. A., TURNER, P., and VAUGHAN, R. J. (1983): Diagenetic processes and the mineralization in the Triassic of Central England. Min. Deposita, Vol. 18, *2,* 365–378.

KOLLI, O., et ABDEL-WAHAB, S. H. (1982): Pétrographie et diagénèse des grès de la région d'Aïn-Séfra (W. Saïda, Algérie) 4ème Sém. Nat. Sci. Terre – Alger.

PETIT-JOHN, F. J., POTTER, P. E., and SIEVER, R. (1972): Sand and Sandstones. Springer-Verlag (Berlin-Heidelberg-New York) 619 p.

ROSE, A. N. (1976): The effect of cuprous chloride complexes in the origin of red-bed copper and related deposits. Econ. Geol. Vol. 71, 1036–1048.

TIXERONT, M. (1973): Lithostratigraphie et minéralisations cuprifères et uranifères syngénétiques et familières des formations détritiques permo-triasiques du couloir d'Argana, Haut Atlas occidental (Maroc). Notes Serv. Géol. Maroc. *33,* 249, 147–177.

WALKER, T. R. (1976): Diagenetic origin of continental red-beds. *In:* FOLKE, H. (ed.), The continental Permien in Central, West and South Europe. D. Reidel Publishing Co; Dordrecht-Holland, 240–282.

WOODWARD, E. A., KAUFMAN, W. H., SCHUMACHER, O. L., and Talbott, L. W. (1974): Statabound copper deposits in Triassic sandstones of Sierra Nacimento, New Mexico. Econ. Geol., *69,* 108–120.

WOPFNER, H., GRIESECKE, S., KOCH, J., and FELS, H. (1983): New aspects on metal deposits of the Groeden sandstone (South Tyrol, Italy). *In:* Mineral Deposits of the Alps, ed. Springer-Verlag, Berlin-Heidelberg-New York, 60–69.

Les Minéralisations de Plomb, Zinc, Cuivre du Domaine Alpin en Algérie

Par B. Touahri*), Y. Fuchs*)

With 9 Figures

Zusammenfassung

Die hauptsächlichen Kennzeichen der Pb-Zn-Cu-Lagerstätten Algeriens werden dargestellt. Ihre räumliche Verteilung ist durch die tektonische Entwicklung in der Zeitspanne zwischen Trias und Miozän bestimmt. Die Pb-Zn-Vorkommen im Tell-Atlas treten in Carbonat-Gesteinen auf und gehören zum Plattform-Typ, während die tertiäre polymetallische Gruppe mit vorherrschendem Kupfer an einen Kalk-Alkali-Magmatismus gebunden ist.

Résumé

On présente un résumé des caratères des minéralisations Pb-Zn-Cu du domaine alpin en Algérie: morphologie, minéralogie, relations avec les roches encaissantes et avec le magmatisme.

La distribution des minéralisations est liée à l'évolution structurale et paléogéographique depuis le Trias jusqu'au Miocéne.

On y distingue:

– Les minéralisations Zn-Pb associées aux séries sédimentaires qui constituent les différentes unités structurales de la chaîne, sont pour un grand nombre d'entre elles dans des environnements carbonatés de plate-forme mobile antérieurs à la phase orogénique paroxysmale.

– Les minéralisations polymétalliques liées au magmatisme calco-alcalin tardi-paléogène, caratérisées par l'importance du cuivre. Ces minéralisations, stratiformes ou filoniennes, sont associées à un magmatisme calco-alcalin effusif et hypovolcanique intrusif.

---

*) Laboratoire de Géochimie et Métallogénie, U.P.M.C., Tour 16, E. 5, 4, Pl. Jussieu, 75230, Paris, Cédex 05.

Abstract

The main characteristic features of the Pb-Zn-Cu ore deposits of the alpidic zone in Algeria are described: morphology, mineralogy, relation to the neighbourhood rocks and to magmatism.

The repartition of ore occurrences is related to the structural and paleographical evolution from the Triassic to the Miocene.

One can observe:

– A lead, zinc group associated with the sedimentary formations which are forming the different structural units of the Tell Mountains ridge. The most of these lead-zinc occurrences are hosted in carbonatic rocks on a mobile plat-form during a preorogénic phase.

– A polymetallic group is related to the calc-alkaline magmatic activity (partly effusif and partly not) during late Paleogene stage and to the closely related hydrothermal activity. These deposits are both of stratiform and vein type. They are characterized by the relatively high amount of copper.

Introduction

On tente une «régionalisation» stratigraphique et structurale des minéralisations de plomb-zinc-cuivre de la chaîne tellienne en Algérie. Pour cela on tiendra compte des progrés récents en matière de géologie qui ont permis á W. Wildi (1983) de préciser le schéma stratigraphique et structural de la chaîne Tello-Rifaine.

La position de certaines minéralisations devra, cependant, être précisée par la suite.

En métallogénie, les références sont plutôt rares, par contre le dossier exploration minière est riche, grâce à l'intérêt que l'Etat lui accorde.

L'important travail d'inventaire effectué par V. Vydrine (1972), diverses études régionales et de gîtes réalisées depuis 1966, l'ouvrage de J. Glaçon (1967), la synthèse de A. Popov (1968), et les quelques thèses de métallogénie présentées ces dernières années, constituent l'ensemble des données disponibles.

1. Cadre structural du nord de l'Algérie

La structure actuelle du Nord de l'Algérie est le résultat d'une évolution géologique complexe du Trias jusqu'au Miocène. Deux schèmas structuraux ont été proposés, le premier par M. Kieken (1962), le second par M. Durand Delga (1969). Celui-ci vient d'être précisé par W. Wildi (1983), en prenant en compte les résultats des travaux les plus récents. Il distingue (Fig. 1):

1.1. La Chaîne Tellienne

Elle est limitée au Sud par le front des nappes telliennes et elle comprend du Nord au Sud (Fig. 1):

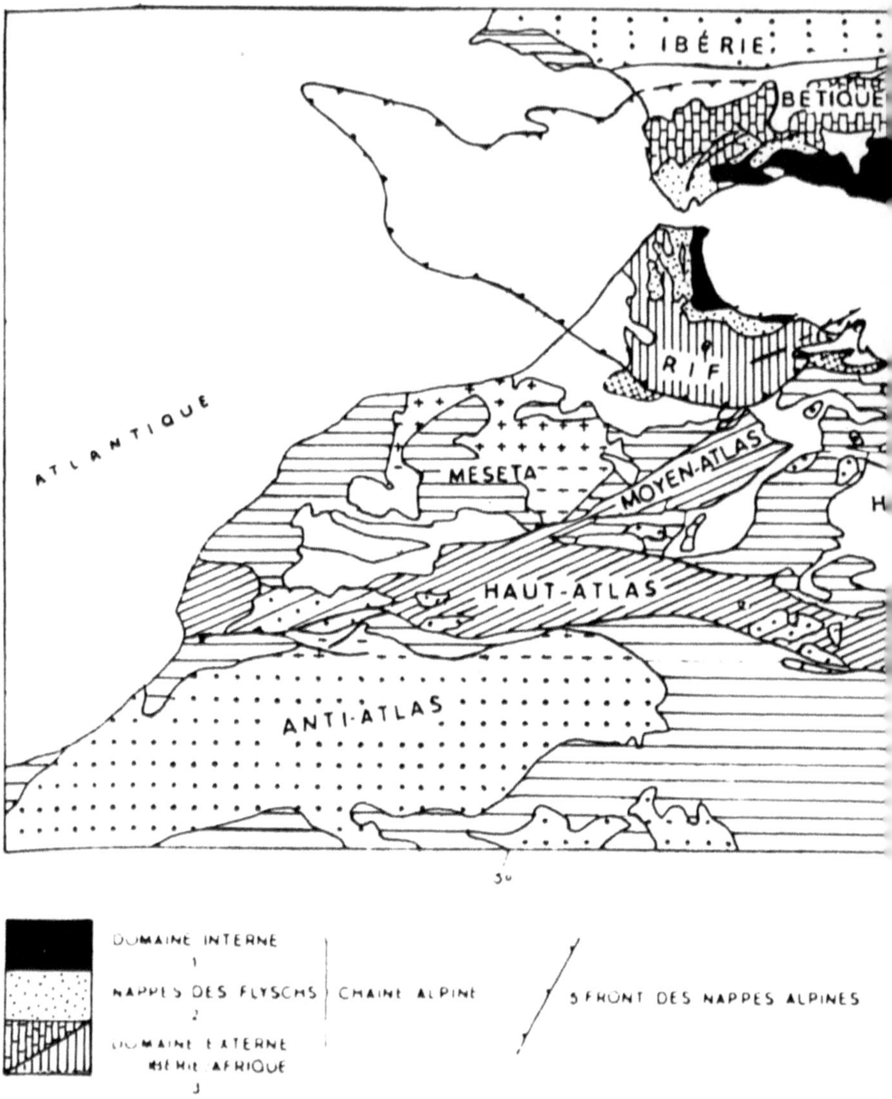

Fig. 1. Cadre structural d

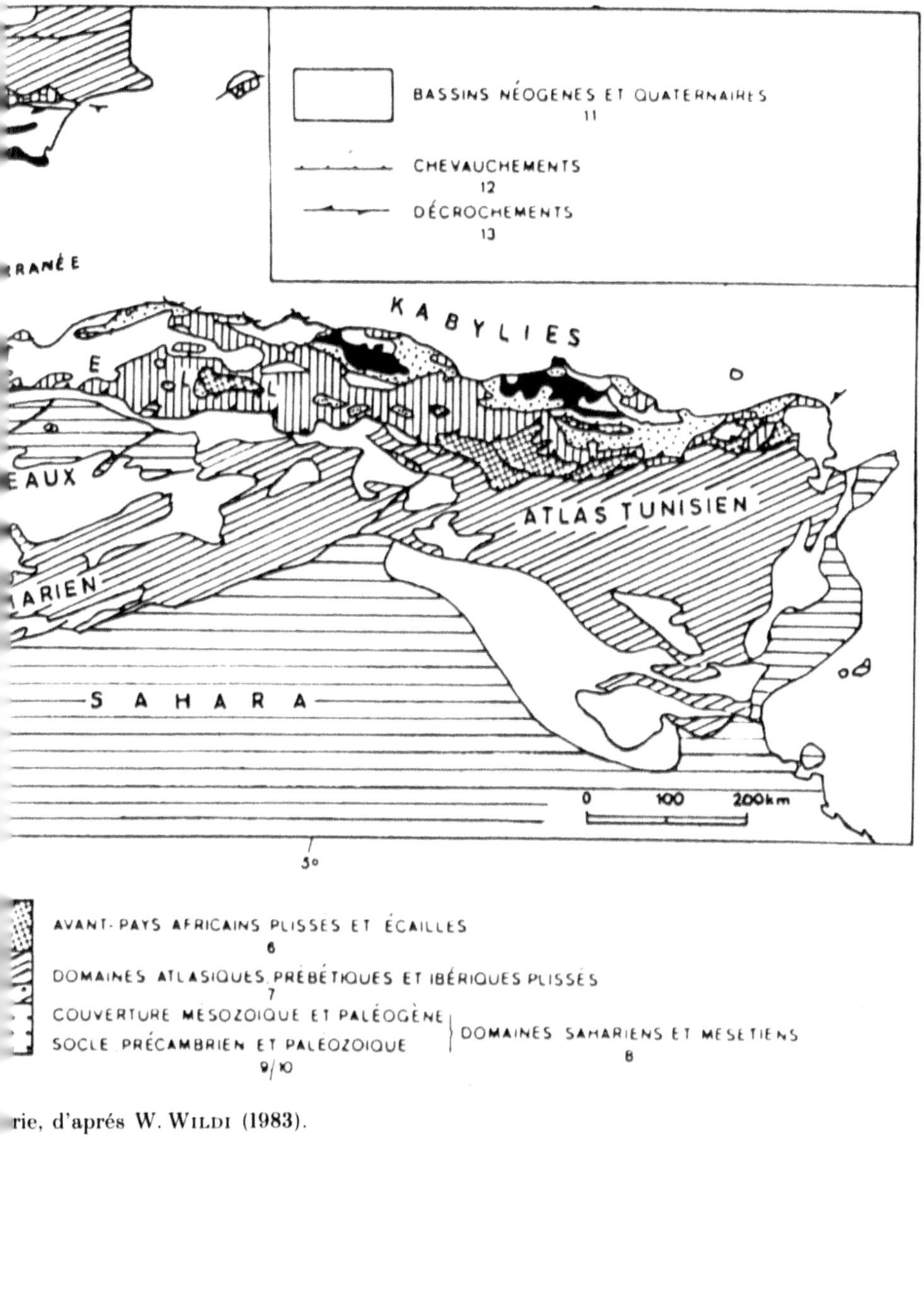

rie, d'aprés W. Wildi (1983).

a) Le domaine interne: il est formé des éléments issus de la dilacération de la plaque d'Alboran. Il affleure en Grande et Petite Kabylie et constitue les massifs d'Alger, du Chénoua, de Cherchell, et de Ténés. Il comprend le socle cristallophyllien kabyle, d'âge Précambrien, et sa couverture sédimentaire, la Dorsale kabyle ou Chaîne calcaire, formée de termes paléozoïques et trias à éocènes.

b) Le domaine des Flyschs: il est lié à la bordure sud et éventuellement ouest de la plaque d'Alboran. Il comprend des sédiments marins profonds de nature flyschoïde d'âge jurassique moyen à oligo-miocène.

c) Le domaine externe tellien: il représente la marge téthysienne du continent africain. Les phases tectoniques tertiaires l'ont découpé en trois types d'unités qui sont du Nord au Sud: les unités ultra-telliennes, les unités telliennes s. s., les unités péni-telliennes, et les unités méridionales à Nummulithes.

d) Le sillon externe sud-tellien: il est constitué par les bassins sédimentaires qui ont reçu au Miocène les nappes de glissement et les klippes sédimentaires d'origine septentrionale.

1.2. Les Avant-Pays africains

Ils bordent la chaîne tellienne au Sud et ils correspondent à la marge continentale de l'Afrique, on distingue:

a) Les Hauts Plateaux, avant-pays stable, tabulaire, et de comportement tectonique rigide.

b) Les Atlas Tunisien et Saharien, les Monts du Hodna, du Belezma, et des Aurés sont des avant-pays autochtones et à series mésozoïques épaisses.

c) Les Sellaouas, à l'Est en avant de l'Atlas Tunisien, et les Traras et le Fillaoucène, à l'Ouest en bordure nord des Hauts Plateaux, sont des avant-pays parautochtones plissés et écaillés vers le Sud.

d) L'ensemble allochtone Sud-Sétifien et l'unité néritique Constantinoise, sont de grandes unités rigides parautochtones, qui chevauchent les avant-pays de l'Algérie orientale.

2. Les minéralisations Pb-Zn-Cu dans la Chaîne Tellienne

C'est dans la chaîne tellienne, à l'Est du méridien de Blida, et dans les avant-pays de l'Algérie nord-orientale, qu'on trouve la plus forte densité d'indices affleurants avec environ 68% de l'accumulation métal, connue en Pb+Zn+Cu de tout le Nord du pays, qui s'élève à 5.3 millions de tonnes environ.

Sur une quarantaine de gîtes de plus de 10.000 t, douze renferment plus de 100.000 t chacun, deux seulement dépassent 500.000 t (les gisements de El Abed et de Oued El Kébir).

Les avant-pays renferment 56% de l'accumulation métal connue de tout le pays, les complexes post-nappes 30%, les unités telliennes 13%.

Dans les avant-pays, où dominent les minéralisations du type sous-inconformité en environnement carbonaté épicontinental, ce sont le Crétacé inférieur à l'Est et le Dogger à l'Ouest, qui portent la plus forte concentration (43%).

2.1. Le socle cristallophyllien de Grande et Petite Kabylie

Les minéralisations se présentent en amas veinulés et d'imprégnations dans des calcaires métamorphiques, ou en filons et filonnets dans des gneiss et des micaschistes.

Les premières sont souvent accompagnées par de la barytine et ont les caractères de gîtes concordants. Leur minéralogie est simple: blende (généralement dominante), galène, pyrite et accessoirement chalcopyrite et stibine. On les interprète comme des minéralisations stratiformes postérieurement remobilisées par le métamorphisme et le magmatisme.

Les secondes, à gangue de quartz, sont:

– en Grande Kabylie, à chalcopyrite, blende, pyrite, stibine et fluorine: l'étain peut atteindre 800 ppm et l'or, 0.3 ppm.

– en Petite Kabylie, à blende, galène, cuivres gris; l'or titre jusqu'à 3–4 ppm dans les minerais, et l'argent dans les concentrés atteint 110 ppm.

2.2. La Dorsale kabyle ou Chaîne calcaire

Les minéralisations sont mal connues; on signalera pour mémoire, le district du Djurdjura, où les indices de zinc (dominant) et de plomb, sont associés à des structures où sont impliqués, des calcaires dolomitiques et des dolomies de plateforme du Lias des séries internes et médianes de dorsale. Ces minéralisations peuvent être remaniées en lentilles stratiformes dans l'Eocène carbonaté épicontinental.

2.3. Le domaine des Flyschs

Les minéralisations se répartissent selon une bande qui longe la bordure méridionale des massifs kabyles.

En Petite Kabylie, elles sont associées aux niveaux carbonatés du Lias et du Crétacé, et elles présentent différents aspects:

– amas veinulés brêchiques ou d'imprégnation, à blende dominante, galène argentifère et cuivres gris. La localisation des minéralisations est contrôlées soit par des zones silicifiées, hard-ground, à la partie supérieure du Lias sous un recouvrement de marnes du Crétacé inférieur, soit par des vides karstiques à l'intérieur des dolomies intensément fracturées.

– amas filoniens dans le Crétacé et le long des failles qui le mettent en contact avec le Lias dolomitique, ou l'Eocène. Outre la blende et la galène apparaissent la chalcopyrite et les cuivres gris argentifères.

Ce dernier type présentent des analogies quant à sa composition, et sa localisation structurale avec les minéralisations dans les unités haut-telliennes de la partie septentrionale des Babors (voir paragraphe suivant).

2.4. Le domaine externe tellien

On constate que la plupart des minéralisations est associée aux unités haut-telliennes.

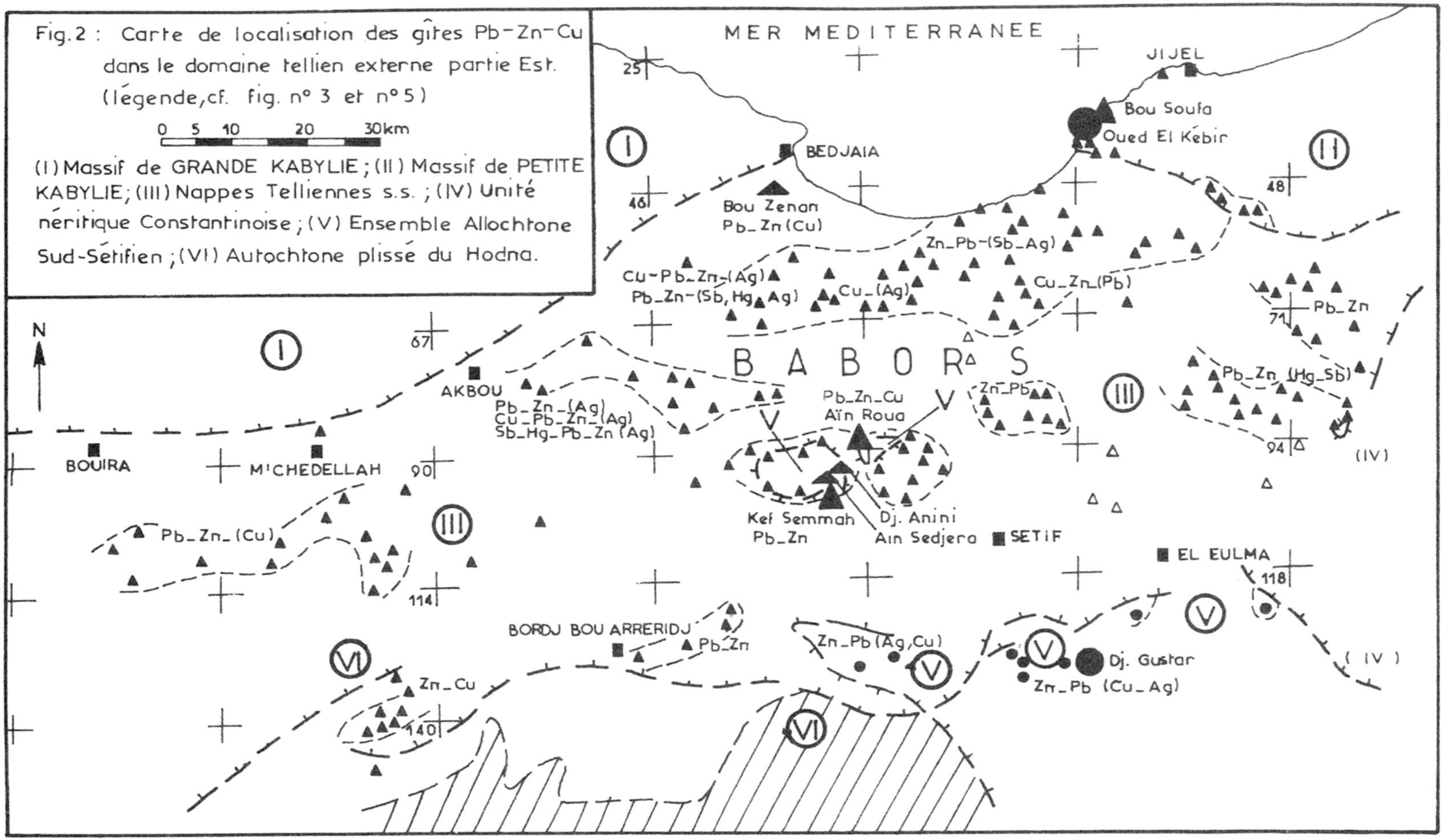

Fig. 2 : Carte de localisation des gîtes Pb-Zn-Cu dans le domaine tellien externe partie Est. (légende, cf. fig. n° 3 et n° 5)

(I) Massif de GRANDE KABYLIE ; (II) Massif de PETITE KABYLIE ; (III) Nappes Telliennes s.s. ; (IV) Unité néritique Constantinoise ; (V) Ensemble Allochtone Sud-Sétifien ; (VI) Autochtone plissé du Hodna.

Celles associées aux unités mi et bas-telliennes sont encaissées dans des formations carbonatées d'âge jurassique à èocène. Elles peuvent localement être remaniées dans le Miocène.

a) Les Babors: Les minéralisations sont soit liées à des brêches tectoniques constituées de dolomies du Lias et qui longent des accidents Est-Ouest qui mettent en contact du Jurassique inférieur avec des marnes du Crétacé inférieur, soit en lentilles et amas de faibles extensions encaissés dans les dolomies du Lias (Fig. 2; GLAÇON, 1967).

Les premières sont à blende et pyrite avec apparition, en profondeur, de cuivres gris, de chalcopyrite, et d'arsénopyrite; les secondes sont à cuivres gris argentifères, pyrite, blende, galène avec parfois du cinabre.

Ce dernier type se distingue aussi par sa gangue composée de barytine, de sidérite et de quartz.

b) L'Atlas de Tablat: l'allure générale des gisement d'un point de vue morphologie est la même. Ce sont des filons de brêches tectoniques composés d'éléments de l'encaissant qui est représenté par des marnes et calcschistes du Crétacé. Une particularité de ce district est que l'on y remarque une zonalité Est-Ouest: les minéralisations sont cuprifères à l'Ouest, et zinci-plombifères et argentifères à l'Est, et où la gangue est riche en barytine (Fig. 3).

c) L'Ouarsenis, ou mine de Bou Caïd: les corps minéralisés sont souvent discordants par rapport à leur encaissant représenté par des calcaires dolomitiques et des dolomies du Lias. Celui ci est renversé et repose en contact anormal sur une série flyschoïde de type tellien et d'âge crétacé.

La localisation des corps minéralisés (en run, amas et colonnes) est contrôlée par des failles et des fractures de direction N-S, N-E, et N-W, marquées par des zones de brêchification et des cavités karstiques.

Celles-çi sont souvent de veritables «grottes aveugles» remplies de minerais oxydés, et enrobée de minerais sulfurés brêchiques dans l'encaissant carbonaté.

Selon les quartiers de la mine les minerais sont différents: barytine avec trés peu de sulfures, lentilles pénéconcordantes à barytine et sulfures, minéralisations brêchiques à gangue de carbonates avec peu de barytine et à sphalérite dominante.

d) Le gîte de de Koudiat Stah: la minéralisation, encaissée dans des calcaires éocènes appartenant aux unités bas-telliennes, est contrôlée par des joints marneux (à bitume et dissémination de cinabre) et par la fracturation. Aux niveaux minéralisés succèdent des marnes et argiles de même âge et qui correspondent à un épisode transgressif sur des dépôts de plate-forme tidale. Il s'agit là d'une minéralisation du type sous-inconformité en environnement épicontinental.

2.5. Les minéralisations associées aux phénomènes magmatiques tardi-paléogènes

a) Généralités:

Les principales expressions de ce magmatisme sont:

– des complexes volcano-sédimentaires, régions de Cap de Fer-Edough, de Jijel-Bejaïa, d'Oran et de Ténés.

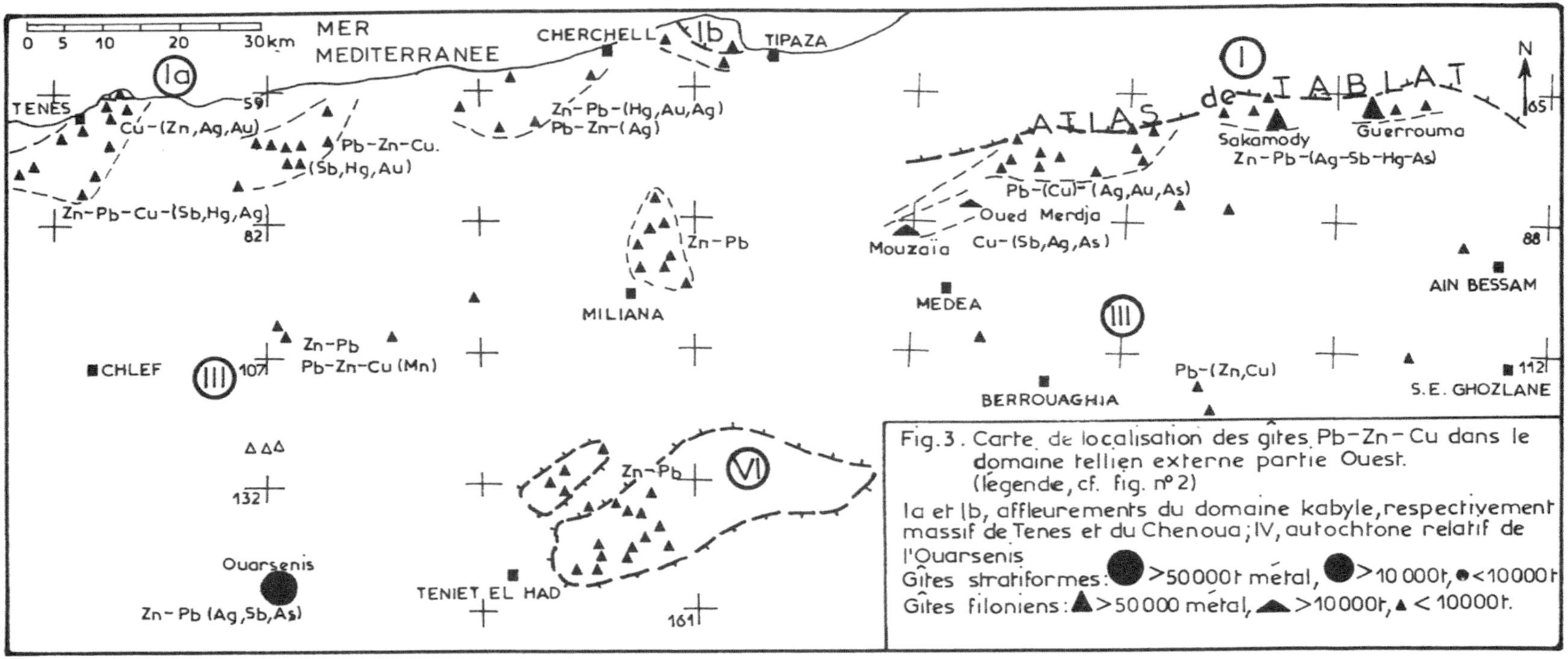

Fig. 3. Carte de localisation des gîtes Pb-Zn-Cu dans le domaine tellien externe partie Ouest. (légende, cf. fig. n° 2)

Ia et Ib, affleurements du domaine kabyle, respectivement massif de Tenes et du Chenoua; IV, autochtone relatif de l'Ouarsenis

Gîtes stratiformes: ● >50000 t métal, ● >10 000 t, • <10000 t

Gîtes filoniens: ▲ >50000 métal, ▲ >10000 t, ▲ < 10000 t.

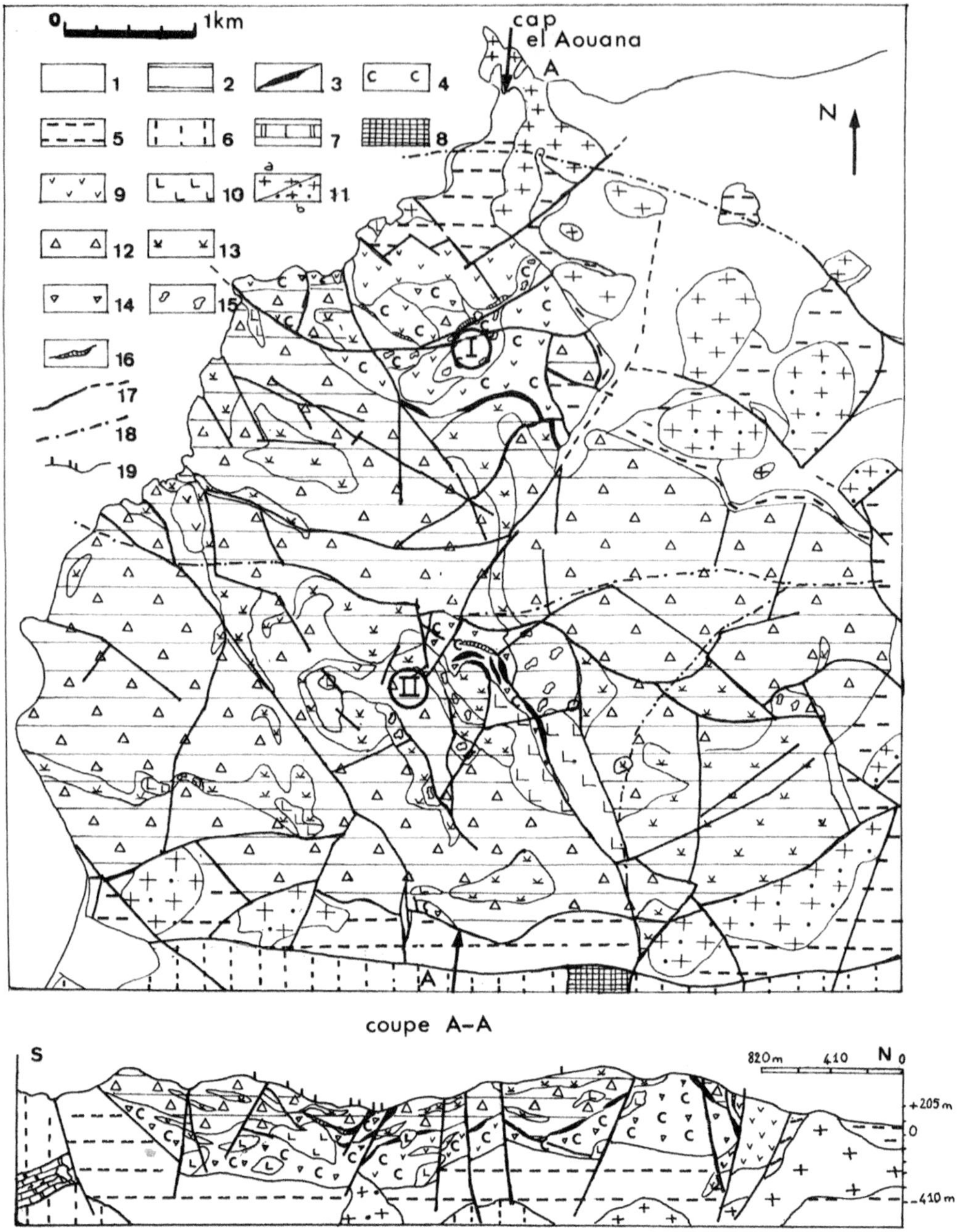

Fig. 4. Carte géologique simplifiée du Massif d'El Aouana, et coupe géologique (A-A) Nord-Sud; d'aprés G. MATSAK, 1978. I = gîte de Bou Soufa; II = gîte de Oued El Kébir; 1 = alluvions, sables, éboulis du quaternaire; 2 = formation de Port Maria; 3 = niveau volcano-sédimentaire; 4 = formation de Bou Soufa; 5 = grés, marnes, calcaires, de l'Oligocène et du Miocène; 6 = grés, marnes et calcaires du Cénomanien; 7 = calcaires domérien à oxfordien; 8 = conglomérats, grés, argiles gypsifères, dolomies, du Trias; 9 = intrusions andésitiques; 10 = intrusions dacitiques; 11 = a) intrusions périphériques grenues, b) microgrenues, diorites et microdiorites; 12 = brêches et tufs andésitiques; 13 = coulées andésitiques; 14 = brêches et tufs andésitiques et dacitiques; 15 = brêches d'explosion; 16 = niveaux mineralisés; 17 = failles et leurs prolongements supposés; 18 = failles interprétées d'aprés la géophysique; 19 = forages.

– des filons de rhyolites et des épanchements du type ignimbritique en Petite Kabylie.

– des sills et dykes de diorites, de microgranites, et de rhyolites dans la région du Cap de Fer-Edough.

– des intrusions de granites calco-alcalins au Cap Bougaroun, en Petite Kabylie (massifs de Boudoukha et d'El Milia), dans les régions de Thénia et Dellys en Grande Kabylie, d'Oued Amizour, de Jijel et d'Oran.

Ces phénomènes magmatiques seraient contemporains de la transgression de la fin du Miocène inférieur et du début du Miocène moyen (Burdigalien).

Ils succèdent à la fermeture de la Téthys occidentale (au Lutétien supérieur-Priabonien) qui conduit à la collision entre la partie orientale d'Alboran et la marge africaine tellienne.

Les minéralisations Pb-Zn-Cu associées à ces processus sont trés fréquentes tout le long du littoral maghrébin. On donnera un bref aperçu sur les gîtes les plus importants connus en Algérie.

b) Les gîtes du Massif d'El Aouana:

– cadre général (MATSAK, 1978; GLAÇON, 1967): le complexe volcano-sedimentaire (Fig. 4) comprend deux assises, l'une inférieure, celle de Bou Soufa comprend des andésites, des tufs andésitiques, des niveaux de conglomérats et de brêches pyroclastiques, des niveaux argileux à faunes du Miocène inférieur. Elle repose sur du Miocène argilo-gréseux et carbonaté. L'autre, supérieure, celle de Port Maria comprend des basaltes andésitiques recoupés par des dômes hypovolcaniques de même composition; les tufs et brêches sont plus rares que dans la première formation.

Entre les deux assises, il y aurait une faible discordance angulaire soulignée par un niveau sédimentaire argileux surmonté par une brêche explosive du type ignimbritique.

Tout le complexe est recoupé par des dômes de dacite et de rhyodacite et leurs brêches associées, et il est ceinturé par des intrusions grenues et microgrenues («ring wall») qui percent les argiles et grés de l'Oligocène qu'elles métamorphisent légèrement.

Le gîte de Bou Soufa: situé au Nord du massif, il est encaissé dans la formation de Bou Soufa. En dehors des zones minéralisées, les roches andésitiques présentent un fond géochimique à 2000 ppm en cuivre.

La minéralisation se serait mise en place en trois phases:

1ére phase: pyrite et quartz en imprégnation diffuse, amas et veinules.

2ème phase: réouverture de la fracturation par la mise en place de intrusions, apparition des minéraux cuprifères (énargite, tétraédrite, chalcopyrite, bornite, chacosine I) et de barytine, de quartz ainsi qu'une deuxième génération de pyrite. La minéralisation prend l'aspect de stockwerks, de filonnets et de veinules. Cette phase donne naissance aux zones riches (2 à 8% en Cu) qui se localisent le long des contacts avec les intrusions et dans des fractures (Fig. 5).

3ème phase: elle est induite par une longue période d'enrichissement supergène. Une deuxième génération de chalcosine se développe et domine sur les sulfures primaires de la deuxième phase. Ces processus supergènes sont trés impor-

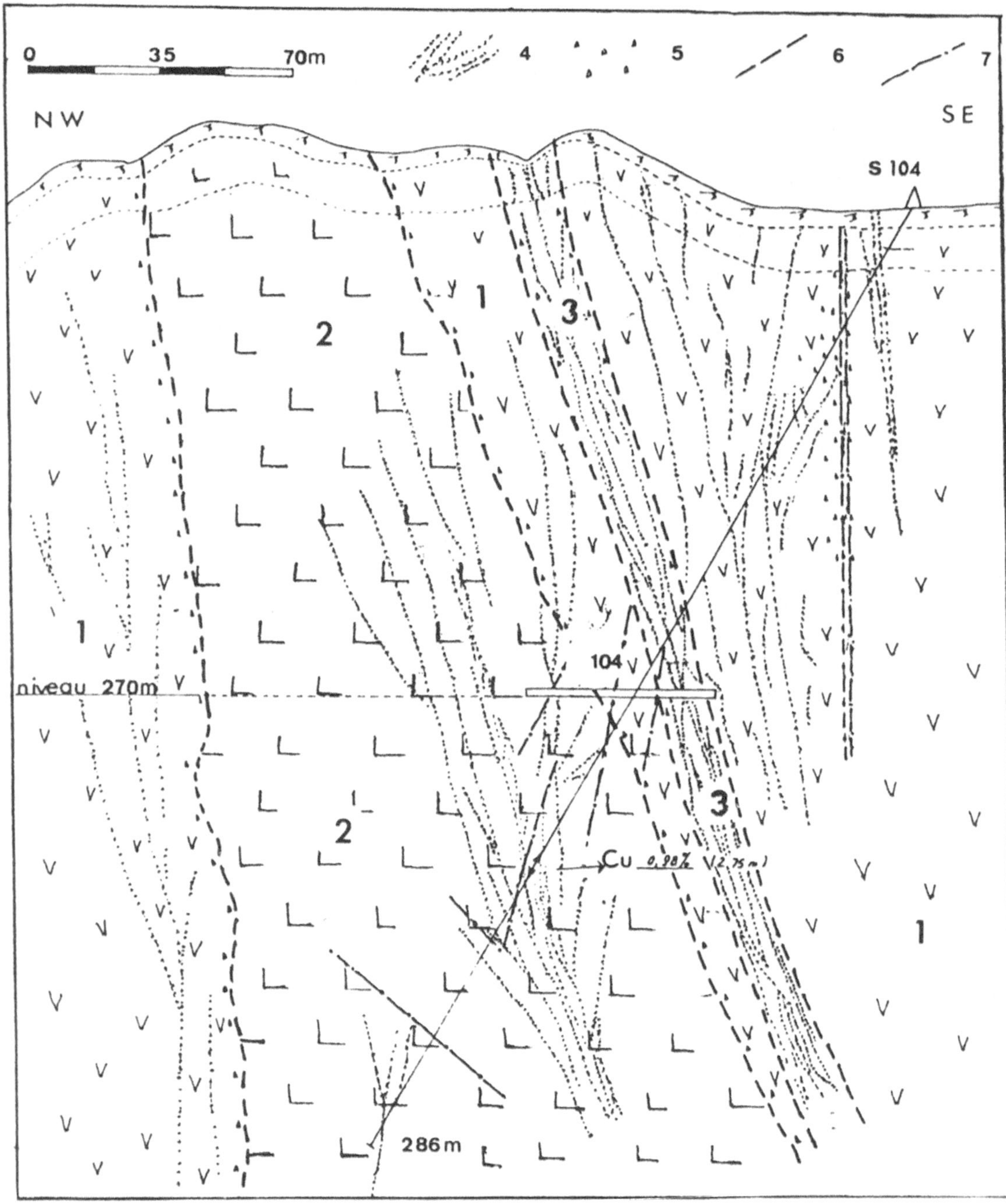

Fig. 5. Le gisement de cuivre de Bou Soufa, allure générale de la minéralisation (D'aprés rapport interne SO. NA. RE. M., inédit, 1971). 1 = formation andésitique de Bou Soufa, trés altérée; 2 = intrusion andésitique, fortement altérée; 3 = zone minéralisée, riche, silicifiée, et barytisée; 4 = brêchification le long des failles des fractures et des contacts entre l'intrusion et la formation de Bou Soufa; 5 = failles; 6 = fractures.

tants à la partie supérieure du gîte où l'or en particulier titre en moyenne 2 g/t alors que dans les niveaux profonds il est toujours à moins de 1 g/t.

Le gîte de Oued El Kébir: il se trouve à la partie centrale du massif au Sud du gîte précédent, et il présente un caractère franchement stratiforme, tout au moins pour la minéralisation dite du niveau principal (Fig. 6), qui constitue l'essentiel des réserves calculées.

Celle çi se localise à la partie supérieure de la formation de Bou Soufa sous un niveau volcano-sédimentaire argileux qui précède un dépôt de brêches ignimbritiques. Le tout étant surmonté par la formation de Port Maria.

Des forages géologiques ont récemment recoupé une minéralisation cuivreuse en imprégnations et veinules (du type griffon), analogue à celle du gîte de Bou Soufa et encaissée dans les mêmes types de roches.

Le minéralisation principale correspond à la superposition d'au moins deux phases (Lang-Villemaire, comm. or.):

– la première syn et diagénétique montre une paragenèse à pyrite, marcassite, cuivres gris, galène, schalenblende, chalcopyrite, et à gangue de roches volcaniques altérées avec quartz, barytine, calcite, sidérite, et dolomite.

– la seconde est représentée par une minéralisation filonienne qui recoupe la première phase. Elle montre une association à chalcopyrite, pyrite, marcassite, énargite et galène. Elle est liée à une phase hydrothermale tardive et est contrôlée par une tectonique syn et post-volcanique à caractère distensif.

c) Les altérations hydrothermales:

Dans une première approche on peut distinguer:

– celles qui affectent l'ensemble du complexe volcano-sédimentaire: ce sont essentiellement une kaolinisation et une silicification.

Cette dernière, plus rare dans la formation de Port Maria, est trés développée dans celle de Bou Soufa. A Oued El Kébir, des quartzites hydrothermaux riches en alunite à leurs parties supérieures, affleurent et ont été recoupés par forage à la partie orientale du gîte.

– celles circonscrites aux limites des gisements: il faut dire qu'à Oued El Kébir, les altérations liées aux processus de minéralisation dans le niveau stratiforme principal, se limitent à une faible silification. Elles sont plus développées en association avec les minéralisations pyriteuses et cuprifères du type griffon du gîte de Bou Soufa et sous le niveau principal de Oued El Kébir. Les roches, surtout au voisinage des fractures, sont intensément silicifiées puis pyritisées et barytisées. La séricitisation est fréquente, et on note aussi la présence de sidérite, d'ankérite, et de magnétite.

En conclusion, les altérations ont débuté à la fin du dépôt de la formation de Bou Soufa et se sont pousuivies postérieurement à celui de la formation de Port Maria. La superposition des différentes phases de minéralisation correspond aux différentes étapes d'évolution dans la nature des fluides hydrothermaux contrôlées par plusieurs phases de réouverture de la fracturation.

d) Les gîtes filoniens polymétalliques de Petite Kabylie:

Des manifestations intrusives, granodiorite, granite, monzonite, d'âge essentiellement Langhien (Bouillin, 1979), se placent à la partie centrale du massif et

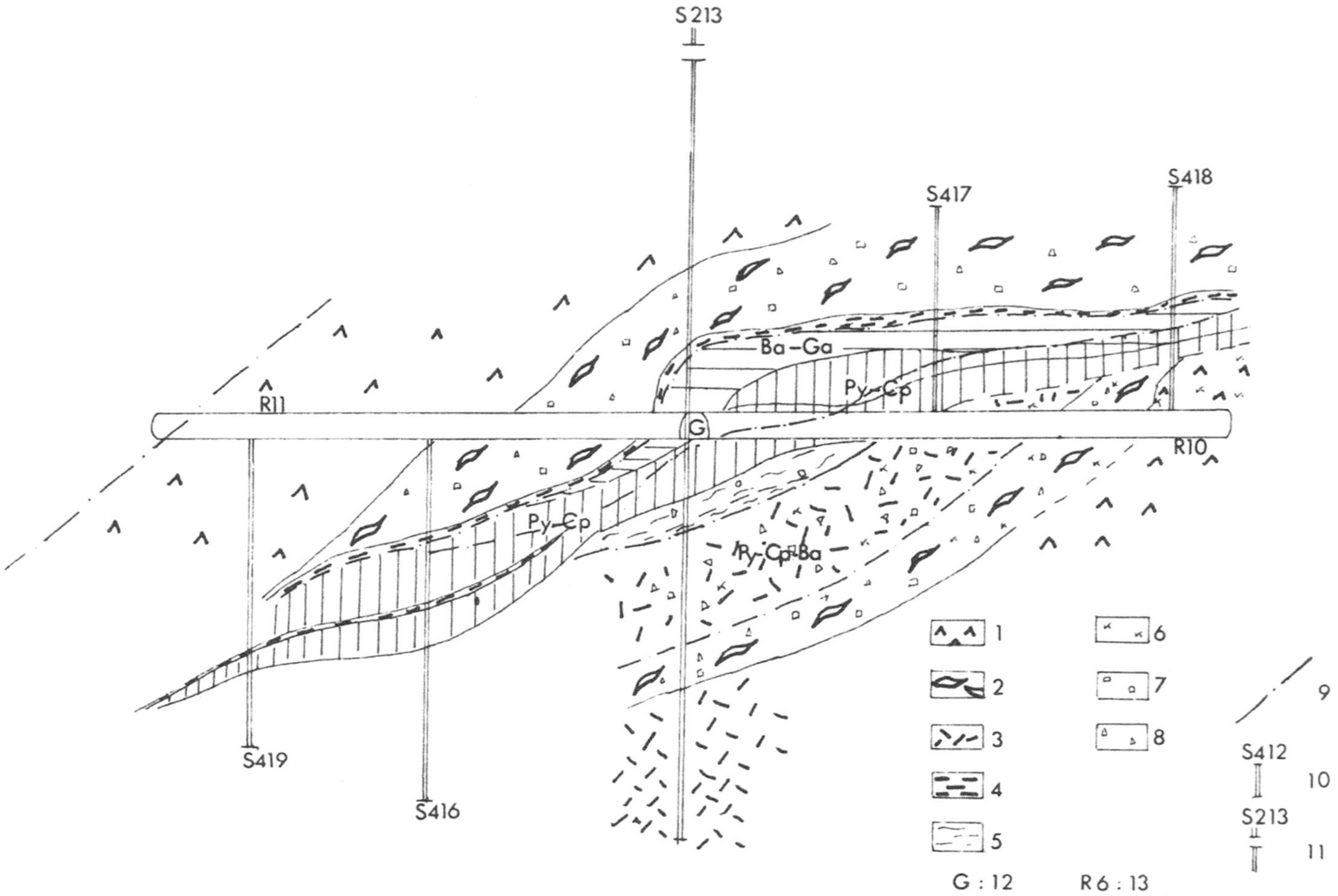

Fig. 6. Le gisement de Pb-Zn-Cu de Oued El Kébir, morphologie des amas minéralisés, et zonalité verticale de la minéralisation. Ba = barytine; Ga = galène; Cp = chalcopyrite; Py = pyrite. 1 = tufs andésitiques; 2 = laves andésitiques; 3 = niveaux argileux, volcano-sédimentaires; 5 = brêches volcano-sédimentaires; 6 = kaolinisation; 7 = pyrite; 8 = brêchification; 9 = failles; 10 = sondages souterrains et 11 = depuis la surface; 12 = galerie principale; 13 = recoupes.

Tableau 1. Caractéres généraux des minéralisations associées au magmatisme tardi-paléogène

	District de THÉNIA	District de BOUDOUKHA	District de SIDIKAMBER	District de OUEDOUDINA
TYPE (les chiffres indiquent les teneurs en %)	Pb (↑ 0.4)-Zn; Cu (↑> 1)	Zn (5.8) Cu (2.3) Pb (1.2)	Zn-Pb-(Cu)	Zn (5–6)-Pb (2–2.5)
MORPHOLOGIE	filons, imprégnations	filons, amas veinulés	filons, amas veinulés	filons
GANGUE	quartz, tourmaline	quartz, barytine, chlorite, carbonates	quartz, calcite, sidérite, barytine	idem Sidi Kamber
ENCAISSANT	granite de Thénia et dykes associés	socle kabyle nappe des flyschs	socle kabyle	socle kabyle
MINÉRALOGIE	Pyrite, chalcopyrite, pyrrhotite, blende	Blende, chalcopyrite galène, pyrite, mispickel, cuivres gris, boulangérite	Pyrite, chalcopyrite, pyrrhotite, blende galène	Blende, galène, cuivres gris, chalcopyrite, pyrrhotite, bournonite
ZONALITE				vers. SW: Zn, Ba ↑; en prof. Pyrite, Zn, Cu, Sb ↑
ELEMENTS TRACES (teneurs en ppm) 1. dans le minerais	Sn ↑ 200, As 1000–10.000, W 300–1000, Au 1–n. 10	As, Sb, Au, Ni, Co, Bi 1900–5000, In 20, Sn ↑ 100, Cd 300–3000,	(1) Conc. Pb 1964/68 Cu 300–500, Mn 300, P 50, Ag 80–120, Au 0,6	
2. dans les concentrés de Plomb		Ag 900–2500	(1) Conc. Zn 1964/68 Cu 2800–3000, Cd 1800–4500, Mn 75	Ag 1500–2000
de Zinc			P 9–85, As 15–20, Ag 35–42, Au ≠1, Ge 20–195, Ga < 30	
de Cuivre		Au 7	In 3–20	
3. dans PbS			615–1000	

(1) = d'aprés M. Oussedik et V. Hérault, 1971. – ↑ indique l'accroissement des teneurs. – Conc. = concentrés. – Prof. = profondeur.

contrôlent la plus forte concentration polymétallique de la région (Fig. 7 et Tableau nº 1).

– le district de Boudoukha: les filons minéralisés recoupent le socle et sa couverture chevauchante de flyschs crétacés à éocènes. Ils recoupent aussi des minéralisations cuprifères veinulées encaissées dans les schistes métamorphiques, et qui sont de même composition que celles filonniennes.

– le district de Sidi Kamber: les filons recoupent les gneiss et les micaschistes du socle qui renferment des minéralisations veinulées concordantes à la foliation des roches. Les teneurs en argent sont plus faibles que pour le district de Boudoukha.

– le gîte de Oued Oudina: analogue aux précédents, il est cependant le plus éloigné des massifs intrusifs. Sa particularité est qu'on y signale une zonalité verticale et horizontale: vers le S-W, la barytine et la zinc augmentent, et en profondeur le plomb diminue, la pyrite et le zinc augmentent en même temps que la minéralisation devient à cuivres gris antimoniés et à bournonite.

e) Le gîte de Aïn Barbar, à l'Est du Massif de l'Edough:

Les filons polymétalliques recoupent une série de flyschs d'âge Maestrichien, er le socle cristallin externe, de type africain, de l'Edough (Vila, 1980; Marignac, 1976). Ces flyschs affleurent en boutonnière parmi la nappe nummidienne, d'âge Oligo-Miocène, qui les surmonte et que les filons, il semble, ne recoupent pas (Fig. 8).

Des dykes de rhyolites et de microgranites recoupent les filons et aux intersections, C. Marignac (1976) signale des enrichissements en étain.

Une zonalité marquée, comme à Kef Oum Théboul (Popov, 1986; Nesterov, 1982), ou certains indices de Petite Kabylie, n'est pas signalée. Dans quelques filons la pyrrhotite et la galène augmentent en profondeur, mais en général, l'importance de la minéralisation diminue en profondeur.

C. Marignac (1983) propose l'hypothèse d'une aire paléogéothermique plusieurs fois réactivée par des phénomènes tectoniques qui ont rythmé l'évolution des fluides hydrothermaux. Il considère une mise en place des minéralisations en trois phases:

1ère phase hydrothermale précoce: chloritisation et brêchification de l'encaissant.

2ème phase: à sphalérite, pyrrhotite (subordonnée), puis chalcopyrite.

3ème phase: composée de plusieurs épisodes tardifs (brêchification, remaniement des phases précédentes, formation de quartzites hydrothermaux), chacun marqué par une reprise de l'activité tectonique.

3. Conclusions générales

Les minéralisations Pb-Zn-Cu de la chaîne alpine en Algérie sont hétérochrones et hétérotypiques: l'âge des formations porteuses s'échelonnent du Protérozoïque au Miocène et au Pliocène (gîte Zn-Sb-Cu de Hammam N'Baïls, Toubal, 1984). Les plus importantes en tonnages dans les aires à plus forte concentration d'indices sont associées soit à une activité magmatique et volcanique à tendance

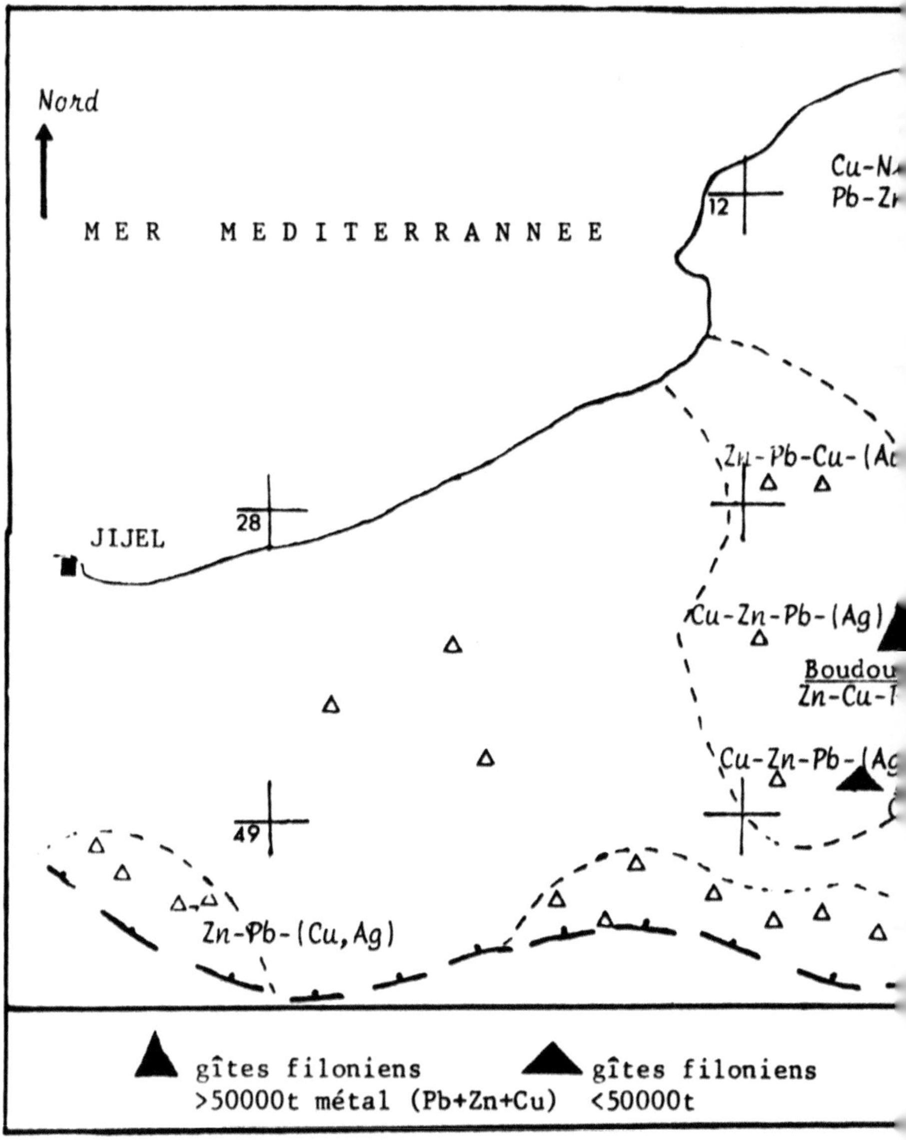

Fig. 7: Carte de
PETITE

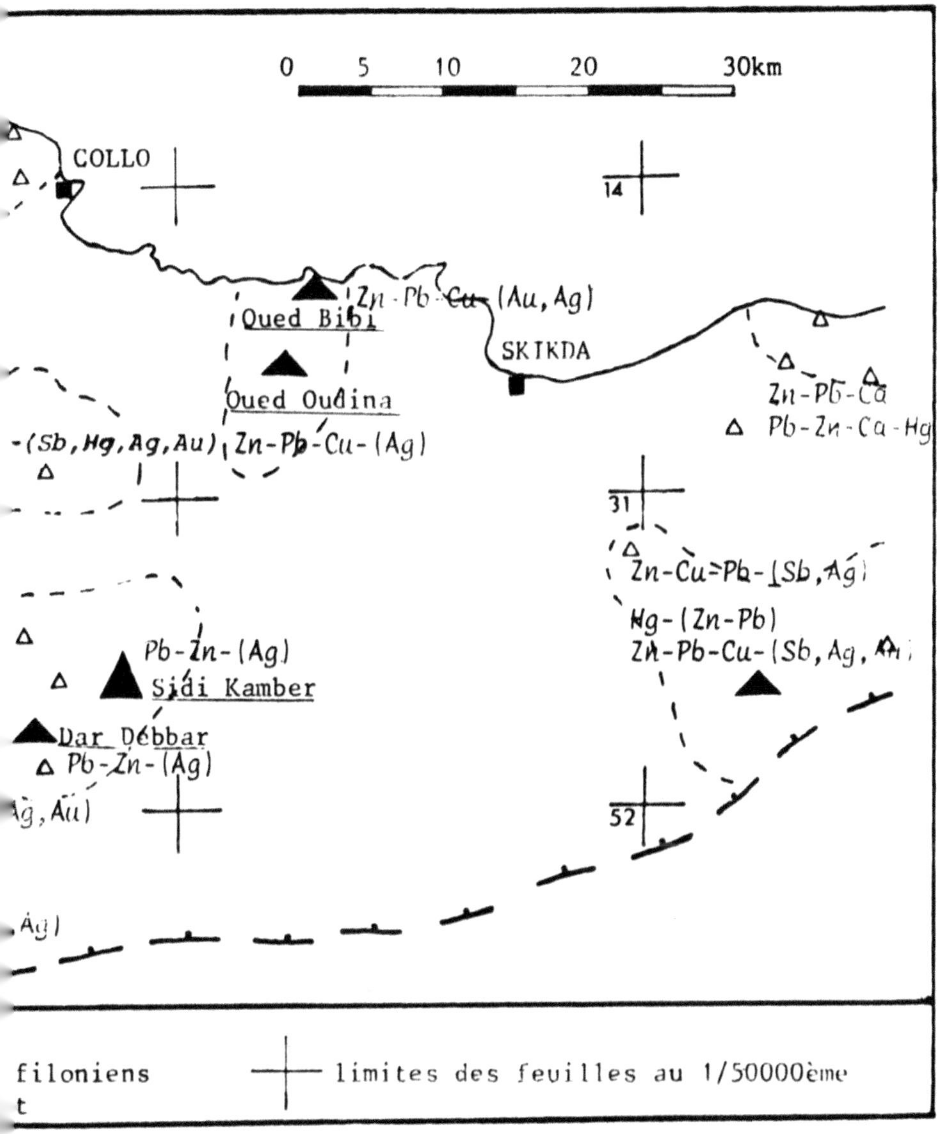

s gîtes Pb-Zn-Cu en
de, cf. Fig. n° 2, 3).

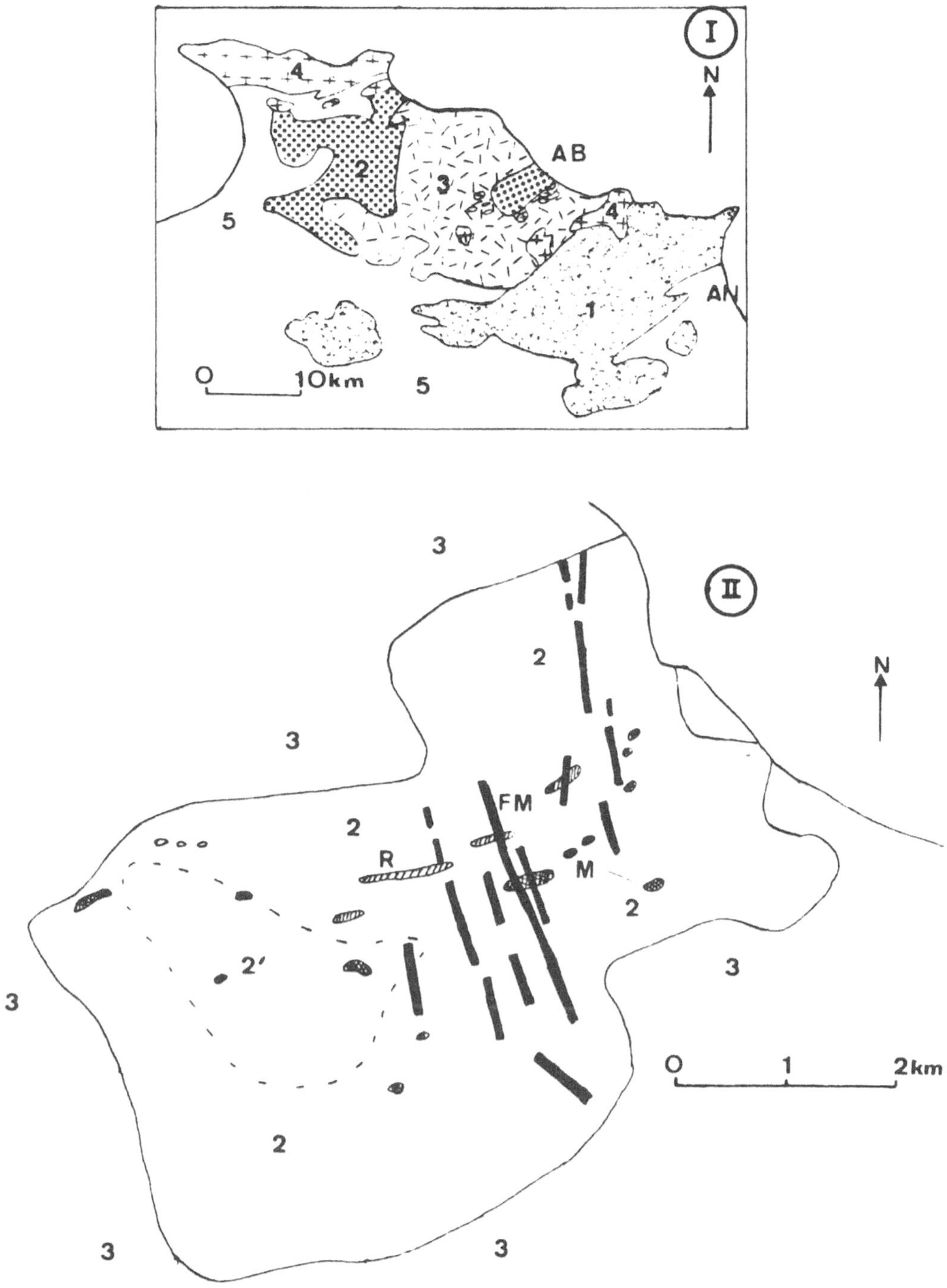

Fig. 8. Le gisement filonien polymétallique de Aīn Barbar. I = cadre régional: 1 = socle cristallin externe de l'Edough; 2 = flysch maestrichien à microbrêches; 3 = nappe numidienne: Oligo-miocéne argilogréseux; 4 = formations magmatiques intrusives, sub-volcaniques et volcaniques tertiaires; 5 = formations sédimentaires post-nappes et Quaternaires; AB = fenêtre tectonique de flysch maestrichien parmi la nappe nummidienne, et encaissant les filons minéralisés, voir détail II. Détail II: 2' = «cornéennes» à niveau de tufs enrichis en carbonates; FM = filons minéralisés; R = dykes de rhyolites; M = dykes de microgranites. D'aprés C. Marignac, 1976.

calco-alcaline, soit à des formations dolomitisées et probablement à des discordances (type sous-inconformité).

a) Les minéralisations associées au magmatisme

Elles peuvent être scindées en deux groupes:

– celles associées au volcanisme effusif de tendance andésitique: elles sont représentées dans le district d'El Aouana et elles sont connues tout le long du littoral maghrébin depuis le Rif, au Maroc et jusqu'en Tunisie.

A Oued El Kébir on définit un amas stratiforme principal à Pb-Zn-Cu, et pyrite, barytine; il repose sur une minéralisation cuprifère (du type griffon) et pyriteuse. Cette dernière constitue le gîte de Bou Soufa et se rencontre à Oued El Kébir.

– celles associées au magmatisme intrusif et subvolcanique tardipaléogene: elles constituent une véritable aire à minéralisation polymétallique caractérisée par l'importance du cuivre. Elles chevauchent des domaines structuraux différents d'Ouest en Est (massifs de Grande et Petite Kabylie, socle externe de l'Edough et sa couverture flyschoïde chevauchante, nappe nummidienne prés de la frontière avec la Tunisie). Si on prend comme référence le schéma génétique de C. Marignac (1976, 1983), proposé pour le gîte de Aïn Barbar, le champ géothermique et donc l'activité hydrothermale a été déterminant dans la formation de ces minéralisations.

b) Les minéralisations associées aux formations carbonatées

Elles sont trés pauvres en cuivre et sont le plus souvent à zinc (dominant) et galène. A l'exception des minéralisations associées aux calcaires métamorphiques du socle kabyle, elles sont encaissées dans des formations dolomitiques de plateforme mobile d'âge lias à éocène.

Celles ci se corrèlent avec les différentes plate-formes qui contrôlent la localisation des minéralisations Pb-Zn du type sous-inconformité des domaines autochtones et parautochtones au Sud du front des nappes telliennes.

Vers l'Est de la chaîne et en bordure du domaine externe tellien et méridionale du domaine interne, le mercure, l'antimoine, l'arsenic et le cuivre (des cuivres gris argentifères), viennent compliquer les paragenèses ce qui reflète l'importance des remobilisations hydrothermales lors de l'orogenèse alpine et le long des principales lignes structurales E-W et N-E.

References

Bouillin, J. P. (1979): La transversale de Collo et d'El Milia (Petite Kabylie): une région clef pour l'interprétation de la tectonique alpine de la chaîne littorale d'Algérie. Mém. Soc. Géol. Fr., N. S., t. LVII, *135,* pp.1–84.

Durand Delga, M. (1969): Mise au point sur la structure du Nord'Est de la Berbérie. Bull. Serv. Géol. Algérie, nouv. série, *39,* pp. 89–131.

Glaçon, J. (1967): Recherches sur la géologie et les gîtes métallifères du Tell Sétifien (Algérie). Bull. Serv. Géol. Algérie, nouv. série, *32,* 2 tomes.

KIEKEN, M. (1962): Les traits essentiels de la géologie algérienne. Livre mém., P. Fallot, Mém. h-série, Soc. Géol. Fr., vol. I, pp. 545–614.

LANG-VILLEMAIRE, C. (1984): Contribution à l'étude pétrographique du massif volcanique d'El Aouana et des minéralisations associées. Rapport de mission, Inivers. Alger, Bab Ezzouar, inédit.

MATSAK, G., et al. (1978): Rapport des travaux de prospection et d'évaluation sur le gisement cupro-polymétallique de Oued El Kébir. Rapport E. RE. M., Algérie, inédit.

MARIGNAC, C. (1976): Mise en évidence des successions paragénétiques dans les principaux filons minéralisés du district filonien polymétallique d'Aïn Barbar (Wilaya d'Annaba, Algérie). Sc., de la terre, Nancy, t. XX, 1975–1976, pp. 333–401.

MARIGNAC, C. (1983): The polymetallic ore Veins of Aïn Barbar (Algeria) as a consequence of alpine geothermal activity. Miner. Deposits of the Alps and of alpine epoch in Europe (ed. by H. J. SCHNEIDER).

NESTEROV, V. (1981): Rapport des travaux de prospection et d'evaluation sur le gisement cupro-polymetallique de Kef Oum Théboul. Rapport E. RE. M., Algérie, inédit.

OUSSEDIK, M., HÉRAULT, V. (1971): Etudes des blendes et galènes Algériennes. Bull. Serv. Géol. Algérie, nouv. série, *41,* pp. 71–93.

POPOV, A. (1968): Les types morphologiques et la répartition des gisements de zinc et de plomb en Algérie. Gisements de plomb zinc en Afrique. Ann. Mines et Géol. Tunisie. n° 23. pp. 103–203.

POPOV, A. (1971): Rapport des travaux géologiques sur le gisement de Bou Soufa. Région de Jijel, Rapport E. RE. M., Algérie, inédit.

TOUBAL, A. (1984): Contribution à l'étude des minéralisations antimonifères du Nord-Est algérien: une province métallogénique hétérochrone. Thèse Doct. 3ème cycle, Univers. Paris VI.

VILA, J. M. (1980): La chaîne Alpine d'Algérie Orientale et des confins algéro-tunisiens. Thèse Doct. és-sciences, Univers. Paris VI.

VYDRINE, V. (1972): Géologie et répartition des gîtes de métaux non-ferreux au Nord de l'Algérie. 2 tomes, fichier des indices et des gîtes de métaux non-ferreux, carte provisoire au 1 : 500.000. Rapport E. RE. M., Algérie, inédit.

WILDI, W. (1983): La chaîne tello-rifaine (Algérie, Maroc, Tunisie): structure, stratigraphie et évolution du Trias au Miocène. Géol. Dyn. et Géograph. Phys., *3,* 24, pp. 201–297.

Les gisements et indices metalliferes de l'Edough (Algérie)

Par DJAMAL-EDDINE AISSA*)

With 8 Figures

Zusammenfassung

Das Massiv von Edough in NE-Algerien stellt den östlichsten Kristallinausbiß im algerischen Küstenbereich dar. Das Massiv hat eine polyzyklische metamorphe Geschichte von afrikanischem Typ; es wird überlagert von allochthonem Flysch. Im Massiv tritt ein kalkalkaliner Magmatismus miozänen Alters auf, an den Erzgänge von Antimon, Silber sowie stratiforme Fe, W, As, Cu, Sn-Vorkommen gebunden sind. Eine gewisse Permanenz einiger Erzelemente ist auffällig. Bemerkenswert ist das Auftreten von Scheelit in einem Pyroxenit. Antimon ist als reichliches Spurenelement weit verbreitet.

Summary

Edough (NE Algeria) is the most oriental cristalline outcrop of the Algerian littoral. This massiv has a polycyclic metamorphic basement of an African affinity, covered by allochthonous flysch. Apart from the sedimentary formations we note the development of a calc-alcaline magmatism of Miocene age (Langhien) in relation with the complex geodynamic evolution of the Western Mediterranean.

This massiv is characterized by the presence of various ore deposits. In strong association with the metamorphic rocks of the basement we find stratabound iron deposits in addition to complex deposits of W (scheelite), As, Fe, Cu, Sn, ... Furthermore ore veins in clear relation with the effects of the Miocene magmatism located in fractures directed N 150 to N 175° E.

Finally deposits of stibium and silver having the form of irregular veins and lenses, are occuring in micaschists with intercalations of marbles.

The ore deposits of the massiv are showing a tectonic and lithological control and a certain local constancy of some elements.

*) Dept. of applied geology – U.S.T.H.B. – B.P. 9 – Dar el Beida – Alger (Algeria).

Résumé

L'Edough (Nord-Est Algérien), constitue le témoin le plus oriental des affleurements cristallins et cristallophylliens du littoral algérien. Ce massif est constitué d'un socle métamorphique polycyclique à affinité africaine, recouvert par des nappes de flyschs allochtones.

Postérieurement à toutes les formations lithologiques, on note le développement d'un magmatisme calco-alcalin d'âge Miocéne (Langhien), en liaison avec l'évolution géodynamique complexe de la méditerranée occidentale.

Ce massif se caractérise par la présence de gîtes et indices métalliféres variés. Ainsi, en association intime avec les roches métamorphiques du socle, on rencontre des gisements stratiformes de fer, ainsi que des gisements et indices complexes á W (scheelite), As, Fe, Cu, Sn.

Par ailleurs, des filons polymétalliques (à Cu, Pb, Zn, Bi), en rapport très net avec les effets du magmatisme miocéne, se sont mis en place dans des fractures d'orientation N 150 à 175° E. Enfin, des minéralisations antimoniféres, argentiféres (et auriféres?), sous forme de filons en chapelets et lentilles gîtent principalement dans une série de micaschistes à intercalations de marbres. En outre, dans ce massif, on a mis en évidence la présence de contrôles tectoniques et lithologiques des minéralisations, ainsi que la présence d'une permanence géochimique en certains éléments.

1. **Cadre geologique** (Fig. 1, 2, 3)

Le massif de l'Edough est constitué d'un socle métamorphique recouvert par des nappes de flyschs allochtones. Tardivement, on note le développement d'un magmatisme d'affinité calco-alcaline d'âge Miocéne.

1.1. Le socle metamorphique

Dans le socle métamorphique, on distingue divers formations dont il est difficile pour l'instant de préciser les termes de passage. Ainsi on distingue de bas en haut, trois séries (Fig. 2):

a) La série des gneiss migmatitiques constituant l'ossature du massif. Des leucogranites recoupent localement ces roches.

b) La série des micaschistes à intercalations de marbres et des gneiss œillés et rubannés.

On note également des intercalations de niveaux de pyroxénites et d'amphibolites dans cette deuxiéme série.

c) La série des schistes satinés à intercalations de quartzites et marbres.

L'évolution de la série des gneiss migmatitiques demeure inconnue. Par contre, les deux derniéres séries métamorphiques (b et c), ont subi d'importantes phases de plissement et de métamorphisme.

Les phases de plissement les plus prédominantes sont celle ayant donné des plis isoclinaux d'axes orientés au N 140° E; et celle ayant plissé l'ensemble pour lui

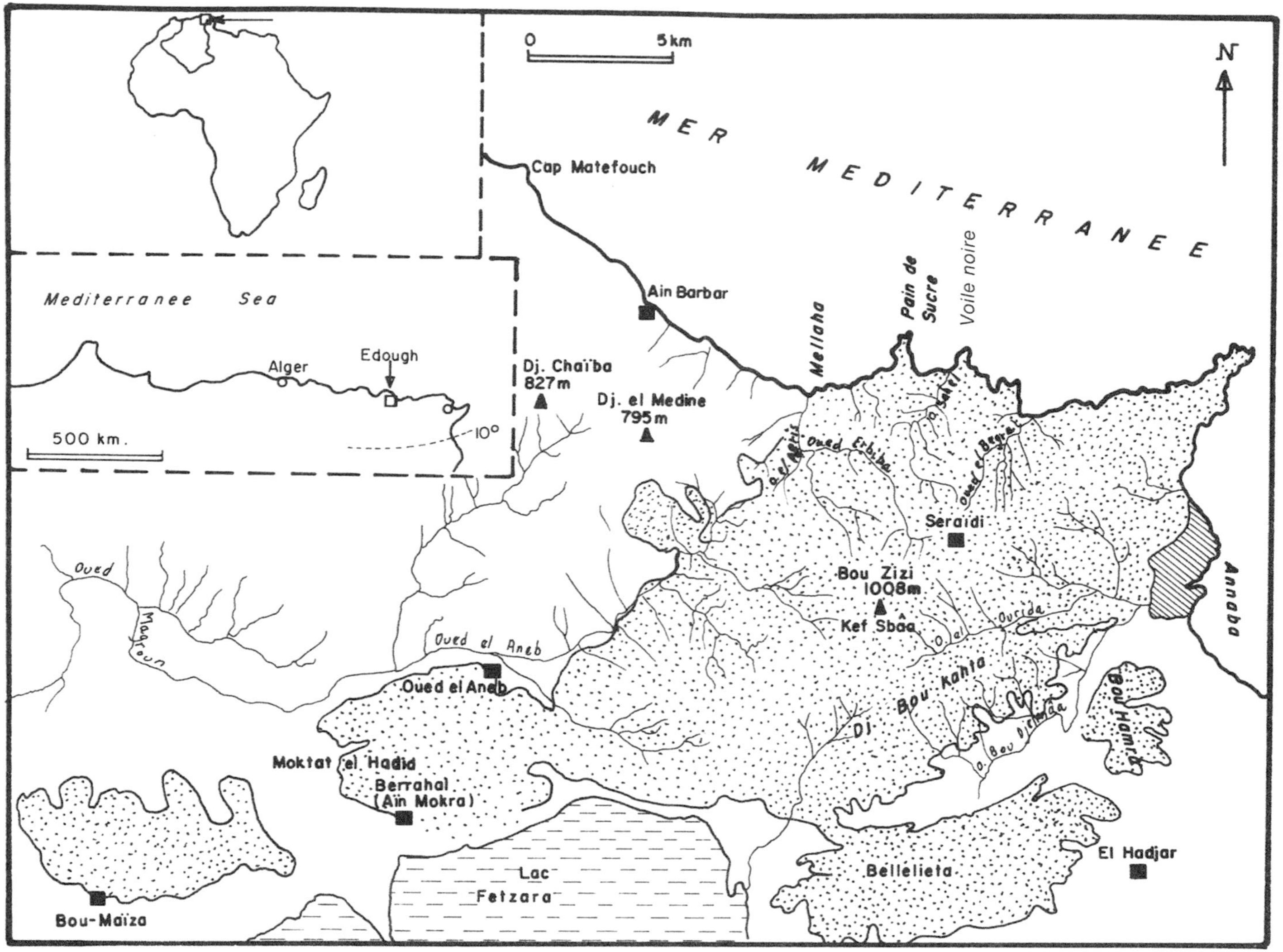

Fig. 1. Situation géographique du massif de L'Edough

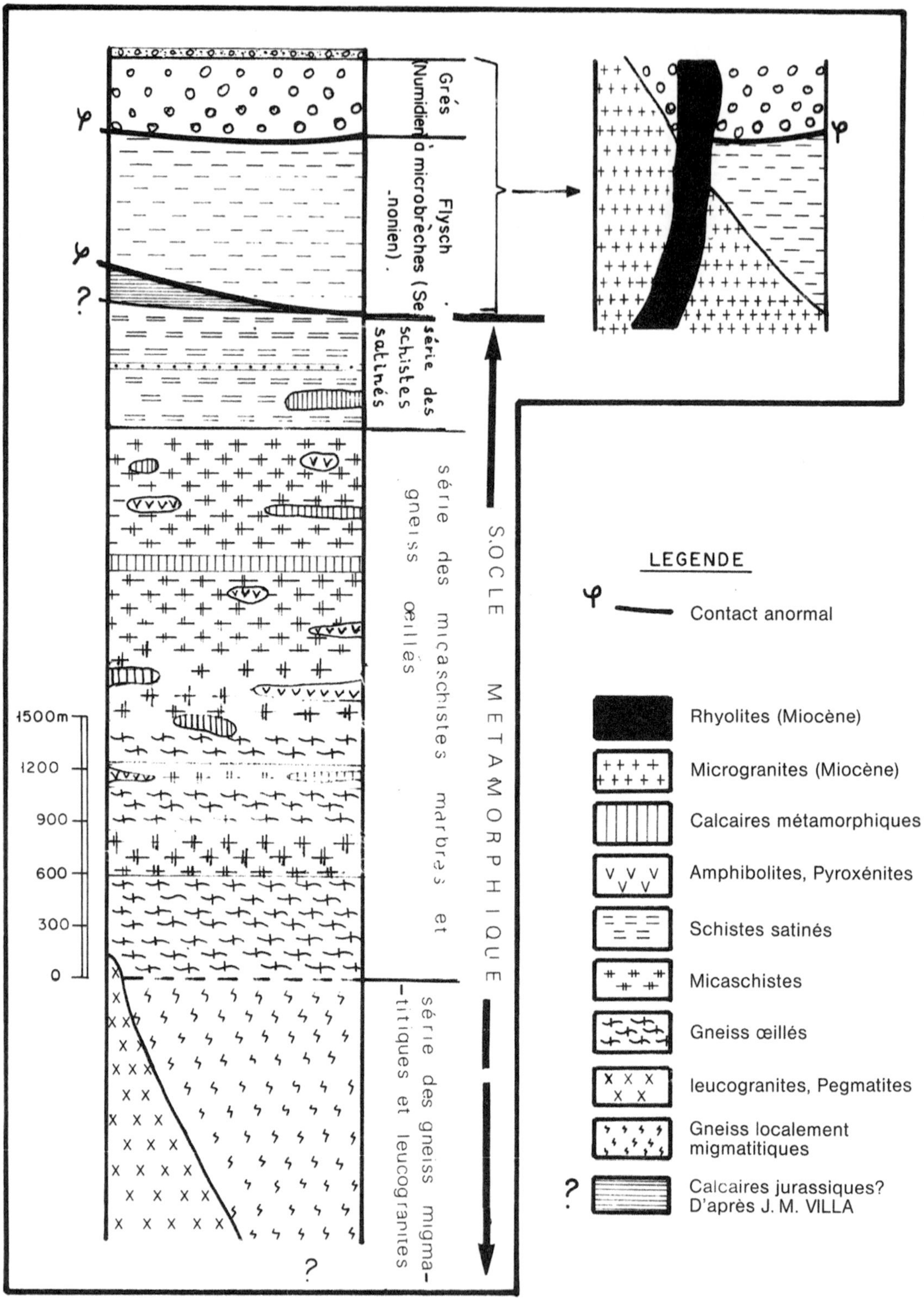

Fig. 2. Colonne lithologique schématique du massif de L'Edough compilé d'aprés les travaux et sondages de l'E.RE.M. (1973–76), Y. Mahdjoub (com. orales) J. Hilly (1962), J. M. Villa (1970) et travaux de l'auteur

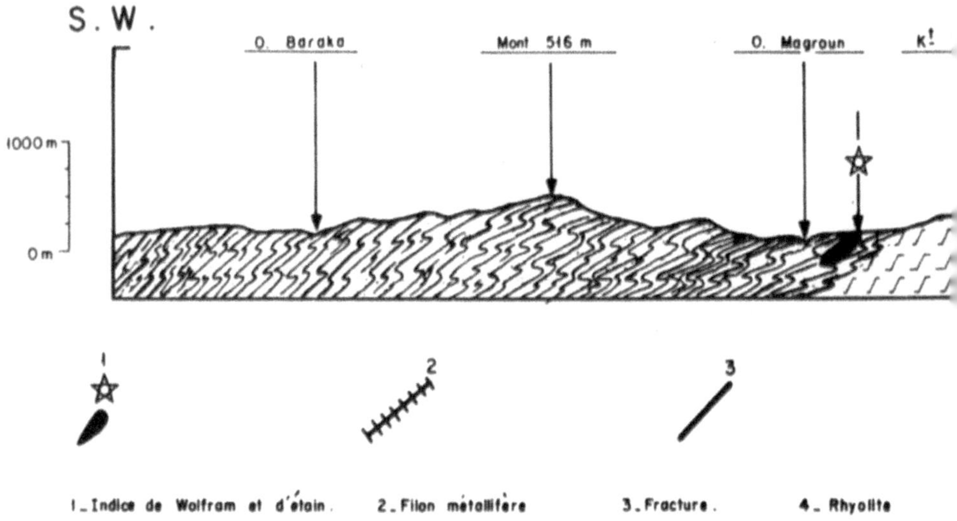

Fig. 3. Coupe géologique schémati

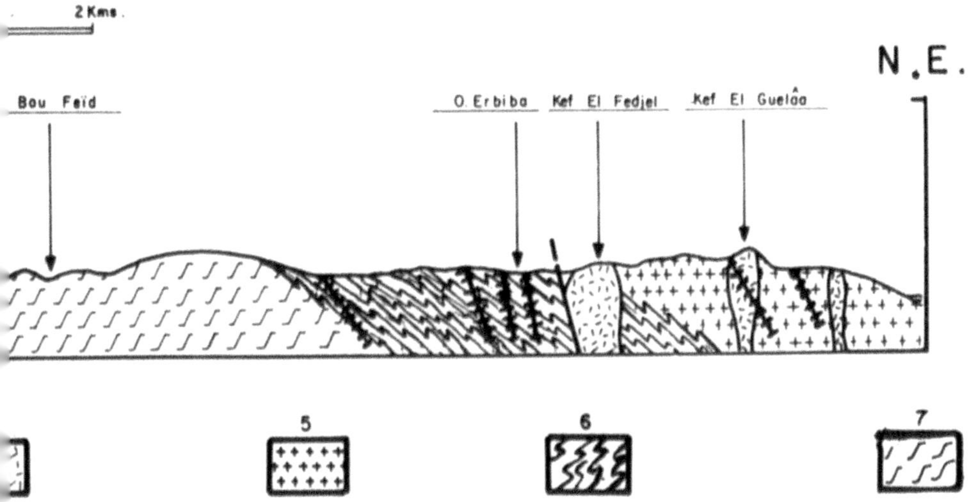

diorite.

6 _ Série de gneiss (Oeillés et Rubannés) et de micaschistes.

7_ Gneiss migmatitiques

massif de L'Edough (C. f. Fig. 1)

donner la structure brachyanticlinale d'orientation approximative Nord-Est–Sud-Ouest, qu'on observe actuellement. Ces deux séries ont également subi un polymétamorphisme mésozonal à catazonal (Température: 500° à 700° C; Pression: 1 à 7 Kb).

En conséquence, beaucoup de travaux restent à reáliser sur la géologie du massif de l'Edough. Il importe en premier lieu de procéder à la datation des roches du socle, car lâge de ce massif reste encore inconnu. De plus, on ne sait pas non plus s'il est allochtone ou en place. En outre, la position structurale de ce massif demeure encore imparfaitement comprise.

1.2. Les nappes de flyschs allochtones

On distingue des flyschs à microbréches rattachés aux séries massylliennes du sénionien détritique; et des flyschs nummidiens (grosses barres de grés avec fines intercalations de pélites). Ces derniers reposent sur les flyschs à microbréches par l'intermédiaire d'une zone bréchique intensément chloritisée.

1.3. Le magmatisme miocene

En liaison directe avec l'évolution géodynamique complexe de la méditerranée occidentale, un important magmatisme d'affinité calco-alcaline s'est développé dans le massif de l'Edough. Les roches magmatiques rencontrées sont des microgranites formant des petits massifs à tendance laccolitique, et des rhyolites formant des dykes et des massifs.

Ces roches ont subi des transformations hydrothermales plus ou moins importantes (séricitisation, adularisation, épidotisation, tourmalinisation . . .).

1.4. La tectonique cassante

L'immense et dense couvert végétal, et l'absence de niveaux repéres ont rendu ardue l'étude de la tectonique cassante. Néanmoins, on a pu mettre en évidence quatre familles d'accidents tectoniques importants:

- failles de direction Est-Ouest,
- failles décrochantes dextres de direction N.W.-S.E.,
- failles décrochantes senetres de direction N.E.-S.W.,
- faille décrochante dextre de direction Nord-Sud.

En outre, on rencontre également des fractures secondaires; dont la plupart sont des fractures d'extension; orientées N 150 à N 175° Est, avec un pendage variant de 50° à 90°. Ces derniéres fractures sont les hôtes de minéralisations métalliféres plus ou moins importantes, et constituent un remarquable réseau en relais de filons sensiblement paralléles (Fig. 1).

2. Les gîtes et indices métalliféres du massif de l'Edough

Le massif de l'Edough renferme des indices et gisements métalliféres variés.

2.1. Les gîtes et indices du socle metamorphique

2.1.1. Les gisements stratiformes de magnetite (M. Dussert, 1912, et E. NA. RE. M., 1968)

Les indices et gisements de magnétite (une dizaine) sont tous situés au pied des versants Sud et Sud-Ouest du massif de l'Edough. Tous les gisements en partie exploités (écrémés) pendant les deux guerres mondiales, sont depuis restés en inactivité.

Les gisements les plus importants sont liés aux niveaux de marbres intercalés dans les micaschistes à grenat (Marouania, Aīn Mokra, Bou Hamra, Meboudja, Beleliêta). Les corps minéralisés se présentent généralement sous forme de couches lenticulaires plus ou moins continues dans les marbres, ou dans les zones de passage marbres-micaschistes. La puissance des couches de magnétite varie de 30 cm à 35 m.

Souvent, la magnétite se substitue irréguliérement aux marbres. On observe parfois des «reliques» de marbres ou micaschistes au sein du minerai. Dans ces gisements de fer, le minéral principal est constitué par de la magnétite (plus ou moins martitisée) avec des formes octaédriques fréquentes. A l'intérieur des cristaux de magnétite, on observe souvent des reliques et des fantômes de carbonates (substitution).

Par ailleurs, de trés belles rosettes d'hématite (Eisenrosen), sont observées localemant dans ces gisements.

En outre, dela pyrite, galéne, chalcopyrite, sont présents en trés petites quantités, et sont vraissemblablement postérieurs à la magnétite.

D'autre part, de la sidérite, a été signalée dans le gisement de Bou Hamra par M. Dussert (1912).

De plus, des teneurs importantes en antimoine ont été signalées par l'E. Rem (1968) dans les minerais de fer de Marouania et Djehaf; et beaucoup de Nadorite à été décelée par batée autour de ces deux gisements.

La gangue des gisements de Fer de l'Edough est constituée principalement par de la calcite, et de faibles quantités de séricite, chlorite, muscovite, quartz, épidote.

Divers hypothéses de genése ont été avancées, cependant la plus vraissemblable serait la suivante: le fer se serait déposé synchrone avec la sédimentation des roches carbonatées (calcaires, marnes) sous forme de sidérite (ou d'hématite), par la suite, le métamorphisme général à développé les importantes concentrations de magnétite.

A Djehaf et Bourbaia, on rencontre aussi des gisements stratiformes de magnétite de moindre importance associés à des amphibolites, mais dont la genése demeure obscure.

2.1.2. Les Indices de manganese

Dans la région de l'Oued Sahel (Nord-Est du massif), M. Dussert (1912) signale la présence de lentilles métriques de pyrolusite et divers carbonates de manganése intercallées dans des amphibolites et des gneiss.

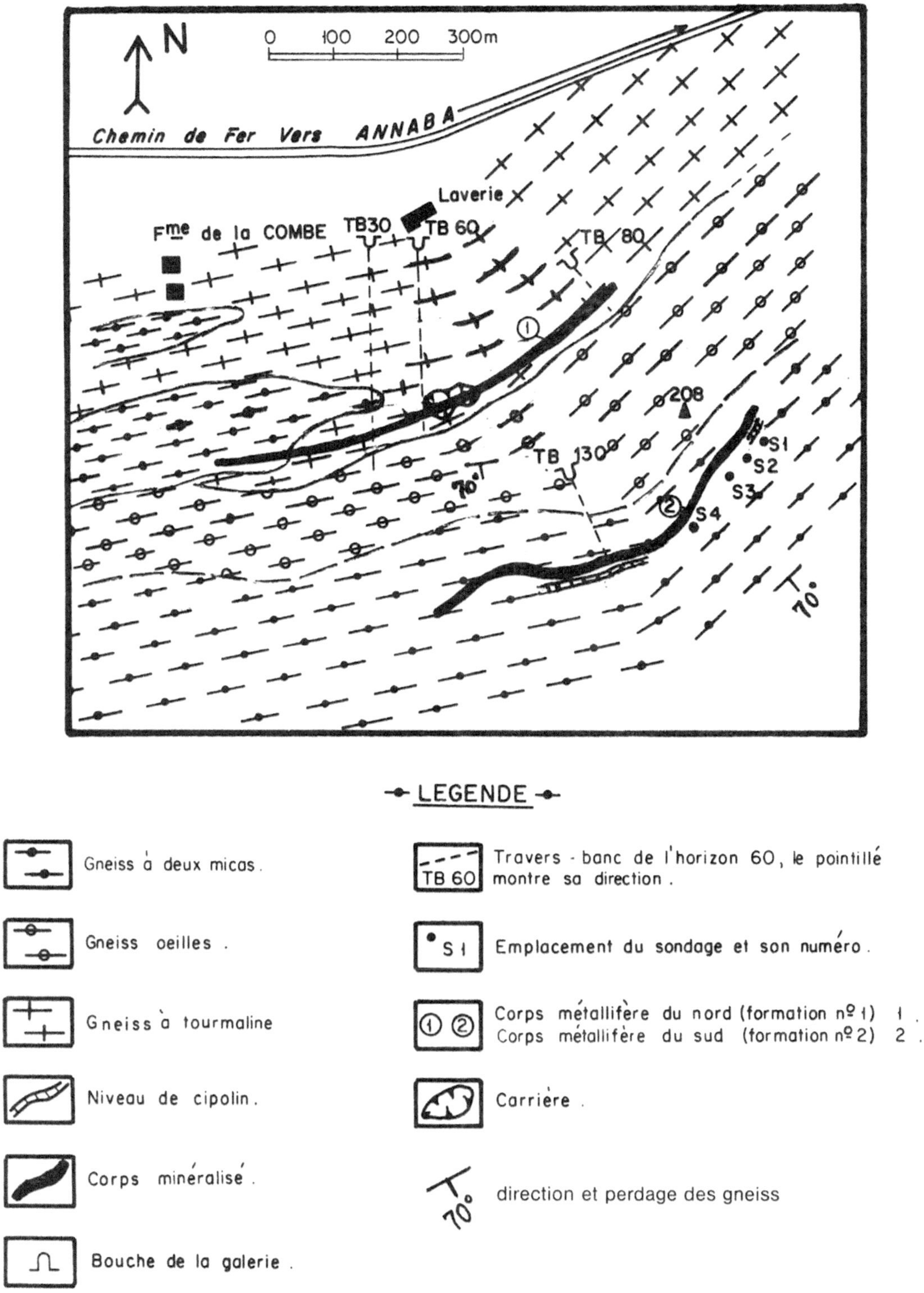

Fig. 4. Schema géologique du gisement arseno-wolframifere de Karezas (d'aprés Sonarem, 1969 – modifié)

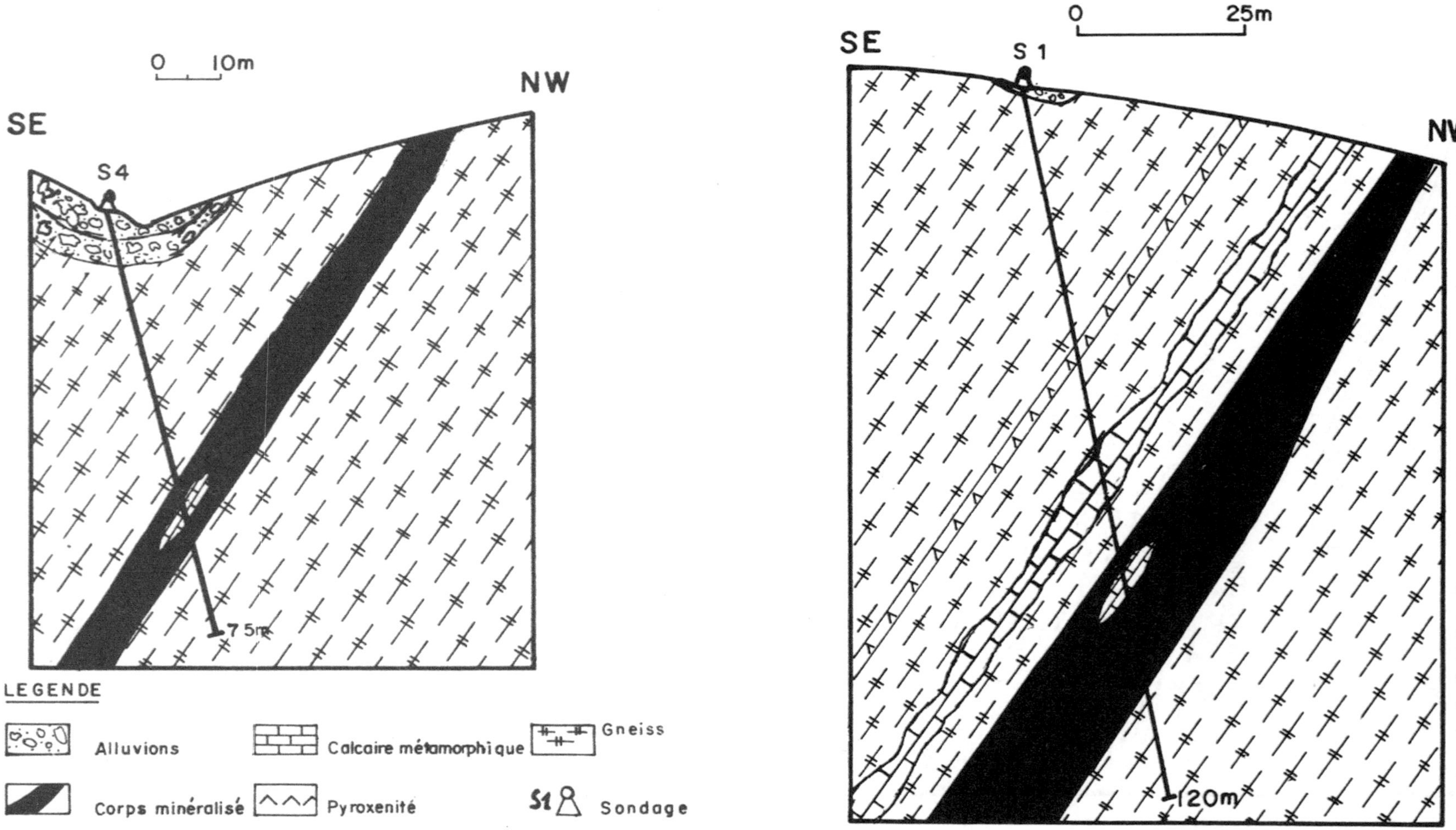

Fig. 5. «allure stratiforme» des corps minéralisés dans le gisement de Karezas (d'aprés les sondages executés par l'E.RE.M. 1968–69)

2.1.3. Les gîtes et indices de wolfram et d'étain

A Le gisement arséno-wolframifére de Karezas

Le gisement arséno-wolframifére de Karezas est situé à 12 km au Sud-Ouest de la ville de Annaba. Il est localisé au pied du versant Sud du massif cristallophyllien de l'Edough, au sein de séries métamorphiques complexes (Fig. 1).

Le gisement est constitué essentiellement par deux corps minéralisés approximativement paralléles et distants de troiscent métres. Ceux-ci sont suivis en direction sur sept cent métres environ, avec des puissances apparentes variant entre un et dix métres (Fig. 4).

Dans l'ensemble, les corps minéralisés adoptent une allure «stratiforme» dans les roches encaissantes (pyroxénites, . . .) et ont une direction variant de Nord-Est–Sud-Ouest à Est-Ouest avec un pendage variant, de 60° à 80° Sud. Les minéraux métalliféres rencontrés dans ce gisement sont par ordre d'importance: scheelite, lollingite, arsénopyrite, pyrrhotite, pyrite, chalcopyrite, cubanite, bismuthinite, Safflorite, haussmanite, magnétite, ilménite, wolframite, érythrite (Fig. 5).

La gangue est constituée par les minéraux suivants: quartz, clinopyroxénes; amphiboles, feldspaths potassiques et plagioclases, fluorite, sphéne, biotite; chlorite, épidotes, calcite, grenats calciques, Lepidolite, allanite, apatite.

Eléments en trace: Au: 4,5 gr/T; U: 7 gr/T (par analyse spectrale).

Deux époques de minéralisation sont mises en évidence dans ce gisement de genése trés complexe (Fig. 6).

B Les indices de wolfram et d'étain du Bouzizi

Les indices de wolfram et d'étain du Bouzizi sont situés dans les zones de Koudiat Slimat et Djemâ el Grara (zone centrale du massif de l'Edough) (Fig. 1 et 3).

La minéralisation se présente en corps stratiformes intercallés dans une série de gneiss à niveaux d'amphibolites et marbres. On dénombre au moins trois corps minéralisés de puissance pluridécimétrique, d'orientation N 90° à N 120° E, et de pendage moyen 20° à 25° Sud. Leur extension horizontale moyenne est d'ordre kilométrique. Les corps minéralisés, trés durs, et trés compacts, présentent des plages et lits millimétriques de scheelite (lampe U.V.), et des disséminations de grains de pyrrhotite. Le microscope révéle également la présence de plages micrométriques de chalcopyrite, et de rares cristaux isolés de bismuthinite et de molybdénite.

En outre, les analyses spectales semiquantitatives des echantillons prélevés dans les corps minéralisés, ont toutes révélé des teneurs assez élevées en étain: de 0.01% à 1% d'étain (on reléve parfois des teneurs exceptionnelles de l'ordre de 4%).

Mis à part un peu d'étain de bois rencontré dans les chapeaux oxydés des corps minéralisés, on n'a pu observer aucun minéral d'étain au microscope. L'étain ne serait-il pas exprimé minéralogiquement? Des études plus poussées le révéleront ultérieurement.

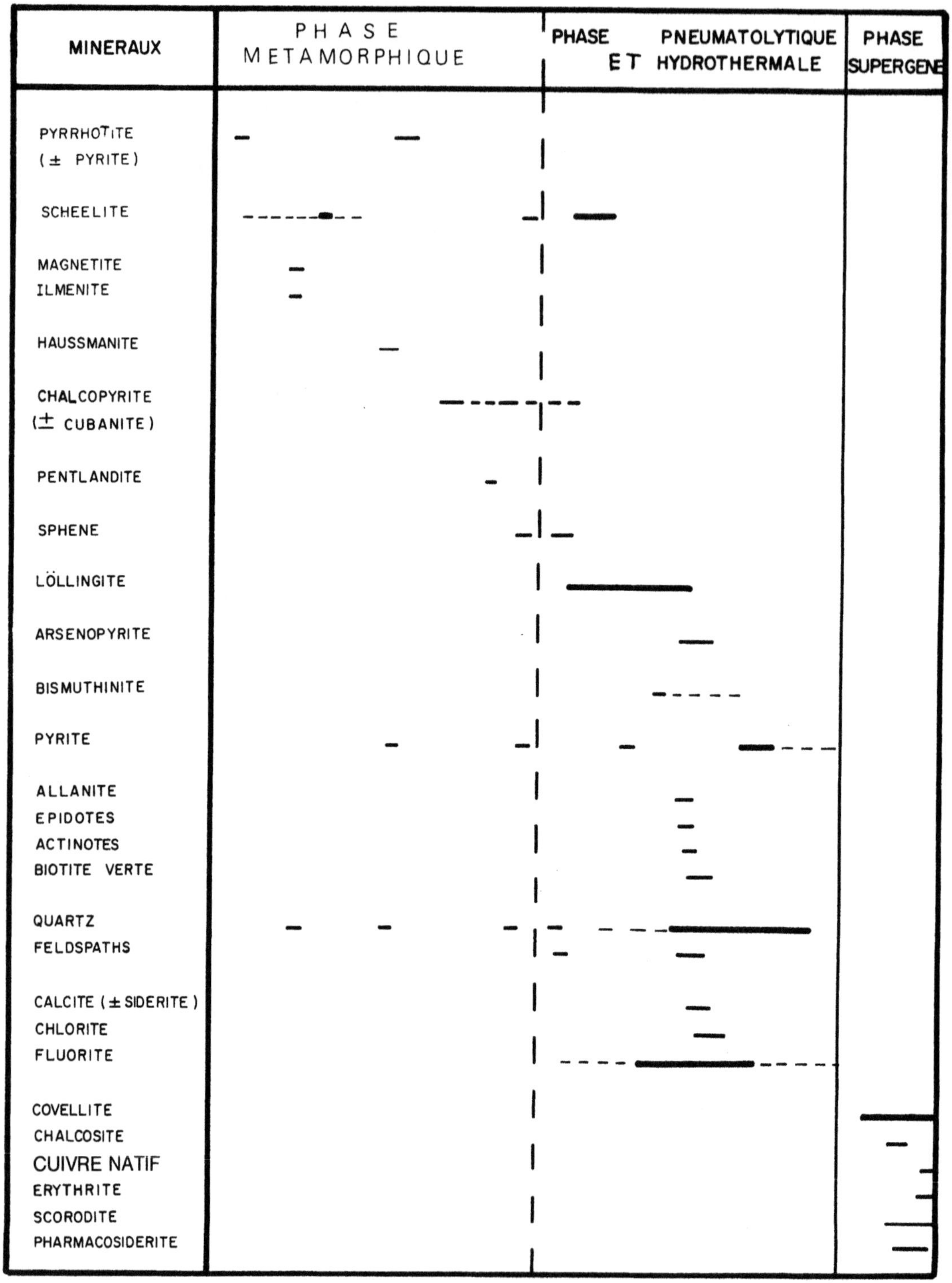

Fig. 6. Tableau des successions minérales dans le gisement de Karezas

Tableau A: caractéres principaux de

FILONS ou ZONE FILONIENNE		Direction (D) et Pendage (α)	Extension horizontale filons et filonnets (L) et puissance (p)
SAF–SAF	2 filons + une vingtaine de filonnets	D: N 145 à N 175° E α: 60° N. E.	L: 100 m à 2500 m p: 10 cm à 120 cm
MELLAHA	2 filons + une vingtaine de filonnets veinules	D: N 130 à N 175° E α: 45 à 85°	L: 10 m à 300 m p: 100 cm à 50 cm
EDDIKRA	1 filon + une dizaine de filonnets veinules	D: N 150° E α: 85° E	L: 30 m à 1300 m p: 10 cm à 200 cm
FERNANA EL KEBIRA	1 filon + dizaine de filonnets	D: N 140 à N 170° E α: 55 à 80° E	L: 30 m à 60 m p: 10 à 50 cm
GROS CHENE (Kt El Ahrach)	2 filons principaux	D: N 150 à N 165° E α: 45 à 75° E	L: 160 m p: 10 à 70 cm
GUELAÄ	1 filon + une dizaine de filonnets	D: N 140 à N 160° E α: 85° W	L: 30 m à 60 m p: 30 cm à 120 cm
BERDA	2 filons	D: N 160 à N 170° E α: 70 à 90° E	L: 60 m p: 50 cm
MOUSSA	1 filon	D: N 140 à N 165° E α: 70° E	L: 1200 m p: 30 cm à 60 cm

étalliques du Nord du massif de l'Edough

Roche encaissante	Association minérale	Eléments en traces (en P. P. M.)	
gneiss	marmatite, chalcopyrite, pyrite, pyrrhotite, galéne, bismuthinite, arsénopyrite, quartz, séricite, chlorite	Ag, Au, Sn, W, Sb	
flysch à microbréches (Massyllien)	marmatite, chalcopyrite, pyrite, arsénopyrite, galéne, bismuthinite, quartz, épidote, calcite, chlorite, adulaire, fluorite	Sn: 100 à 300 Sb: 30 à 300 W: 30 à 300 Ti: 3000	Co: 80 Cd: 40 Au: 1 Ag: 20
micaschistes gneiss microgranites	pyrite, arsénopyrite, sphalerite, chalcopyrite, galéne, quartz, séricite	Sn: 40 à 600 Sb: 60 à 300 W: 100 à 3000 Co: 100 à 300	Ag: 40 à 150 Au: 0,3 à 2 Bi: 100 à 300 Cd: 20
micaschistes gneiss	pyrite, arsénopyrite, sphalérite, galéne; (localement stibine), quartz, séricite	Sn: 40 à 100 Sb: 60 à 10^6 Bi: 200	W: 1000 Au: 0,2 à 1
micaschistes gneiss rhyolytes	pyrite, arsénopyrite, marcassite, quartz, adulaire, fluorite, chlorite, calcédoine	Sb: 50 à 1000 Ag: 15 à 100 Bi: 15 à 100	Ag: 15 à 100 Ti: 3000 Co: 30
rhyolites microgranites micaschistes	pyrite, chalcopyrite, galéne, sphalérite, arsénopyrite quartz, chlorite, séricite	Sb: 300 à 500 Ag: 45 à 80 Sn: 30 à 60	Au: 0,3 à 1 Bi: 80 Ti: 800
rhyolite	pyrite, chalcopyrite, galéne, sphalérite, arsénopyrite, quartz, chlorite, séricite	Sb: 100 à 8500 Sn: 150 à 400	Bi: 150 à 200 Ag: 50 à 150
gneiss rhyolite	pyrite, chalcopyrite, galéne, sphalérite, arsénopyrite, quartz, chlorite, séricite	Sb: 5000 Sn: 2000 Cd: 400	W: 300 Ag: 20 à 150 Au: 0,3 à 1

3. Les filons polymétalliques

3.1. Les filons polymétalliques a fort Pendage

On note dans la partie Nord du massif de l'Edough, le développement d'un remarquable réseau de filons polynmétalliques paralléles et en relais (Fig. 1 et Fig. 3).

Les principales caractéristiques de ces filons sont résumées dans le Tableau A.

Les effets du magmatisme miocéne (phénoménes hydrothermaux), sont responsables de la mise en place des mineralisations filoniennes (le stock métal proviendrait en grande partie des roches encaissantes, «lessivées» par les solutions hydrothermales).

3.2. Les filons «concordants»

En outre, des niveaux minéralisés stratoïdes sont é galement rencontrés dans le Nord du massif de l'Edough. Il s'agit de niveaux de quartz massifs minéralisés, concordants à sub-concordants avec la foliation des roches cristallophylliennes qui les encaissent.

Les principales caractéristiques de ces filons sont résumées dans le Tableau B.

Ces niveaux, seraient-ils des quartz d'exsudation minéralisés?, des filons hydrothermaux à faible pendage?, d'anciens niveaux gréseux minéralisés?, des niveaux résultant de sécrétions latérales?

N'ayant pas encore toutes les données, on se gardera pour le moment de toute interprétation hâtive.

4. Les indices d'antimoine

Les indices d'antimoine de l'Edough, sont situés pour la plupart aux extrémités Nord-Est et Sud-Ouest de ce massif. Parmi les indices les plus importants on distingue:

4.1. Les indices d'antimoine de Zlezelle, Kebira I, Tenfous

Les indices d'antimoine de Zlezelle, Kebiral I et Tenfous, sont en relation avec des fractures d'orientation moyenne Nord-Ouest et de pendage 50° à 90°, affectant une série de micaschistes et niveaux lenticulaires de marbres.

Les corps minéralisés se présentent sous forme de filons en chapelets, nids, stockwerks, d'extension horizontale 20 à 30 métres, de puissance pluridécimétrique et d'extension verticale inconnue.

L'essentiel de la minéralisation est constitué par des aiguilles (0.1 à 7 cm) de minéraux antimoniféres: stibine, berthiérite, fûlöppite. En outre, ces minéraux antimoniféres sont partiellement remplacés (corrodés) par la pyrite, pyrhotite, marcassite et arsénopyrite. Les autres minéraux rencontrés sont généralement sporadiques, il s'agit de: chalcopyrite, argents rouges, sphalérite, freïbergite, sphalérite, olivénite, thorianite.

Tableau B: Les filons concordants du Nord du massif de l'Edough: caractéristiques principales

FILON	Direction (D) pendage (α)	Extension horizontale (L) puissance (p)	Roche encaissante	Association minérale	Eléments en traces (en P. P. M.)
EDDIKRA	D: Est-Ouest α: 28 à 35° N	L: 300 m p: 50 cm	micaschistes gneiss	pyrite, arsénopyrite, chalcopyrite, galéne, marmatite, quartz α, muscovite, chlorite	Co: 80 à 1000 Ag: 25 à 50 Ni: 30 à 100 W: 30 à 400 Sb: 30 à 500 Mo: 50 Sn: 40 à 50 Au: 0.3 à 2
KOUDIAT BERGOUGA	D: Est-Ouest α: 15 à 30° N	L: 200 à 500 m p: 30 à 60 cm	micaschistes gneiss	pyrite, arsénopyrite, chalcopyrite, galéne, marmatite, quartz α, muscovite, chlorite	Sb: 300à1500 Au: 0.3 à 2 Sn: 80 Ag: 35 W: 300
SI LAKHDAR (SEMHOUT)	D: N 60 à N 80° E α: 30° N	L: 300 m p: 10 à 30 cm	amphibolite	pyrite, arsénopyrite, chalcopyrite, sphalérite	Ni: 40 à 2000 Mo: 30 Co: 20 à 2000 Ge: 60
	D: N 140 à N 160° E α: 38° E	L: 300 m p: 50 à 70 cm	amphibolite	pyrite, arsénopyrite, chalcopyrite, sphalérite	Ni: 40 à 2000 Mo: 30 Co: 20 à 2000 Ge: 60

Les éléments en trace décelés par analyse spectrale ont: Ag: 50 à 3000 P.P.M., Sn: – 10 à 3000 P.P.M., W: 60 P.P.M., Ge: 6 à 100 P.P.M., Au: 0.3 à 2 P.P.M.

La gangue est constituée essentiellement par du quartz et de la calcédoine. La mise en place de la minéralisation s'est faite par remplissage des fractures et remplacement métasomatique des roches encaissantes.

4.2. Les indices d'antimoine des Safsaf et Fernana el Kebira

Ces indices ne sont que des curiosités minéralogiques et ne présentent aucun interêt économique. Ils sont constitués par un réseau complexe de veinules de puissance millimétrique à centimétrique traversant les gneiss du socle métamorphique.

La minéralisation est constituée essentiellement par des aiguilles et des plages ovoîdes de stibine disséminées dans un quartz microcristallin.

4.3. Les indices d'antimoine de Semhout et Kebira II

Les indices d'antimoine de Semhout et Kebira II sont en relation avec des fractures affectant des amphibolites. La minéralisation est constituée essentiellement par des aiguilles de stibine disséminées dans une gangue constituée de quartz et calcédoine.

4.4. Conclusion

En plus du contrôle tectonique des minéralisations antimoniféres, on note également un contrôle lithologique. En effet, la minéralisation antimonifére, est trés bien développée dans les micaschistes et les marbres, faiblement développé dans les amphibolites, et trés rare dans les gneiss.

Par ailleurs, on note l'absence totale de mercure, qui normalement devrait être présent dans cette paragenése à Sb-Ag-As.

5. Les indices de scheelite associes aux rhyolites miocenes

Ces indices sont localisés essentiellent sur le flanc Sud du massif rhyolitique de Kef el Fedjel (Nord du massif de l'Edough). La minéralisation (scheelite, fluorite, topaze) se présente sous forme de veinules et miaroles millimétriques. Ces indices présentent beaucoup de similitudes avec les «topaz rhyolites» cénozoiques fréquentes dans le Sud-West des U.S.A.

Conclusion

Les minéralisations rencontrées dans le massif de l'Edough sont diverses et variées. Dans les parties centrales et Sud'Ouest du massif, on note surtout le développement de minéralisations «intimement» associées aux roches métamorphiques du socle: il s'agit de gisements stratiformes de fer ainsi que des gisements complexes «métamorphiques-pyrométasomatiques» à W (scheelite), As, Fe, Cu, Sn.

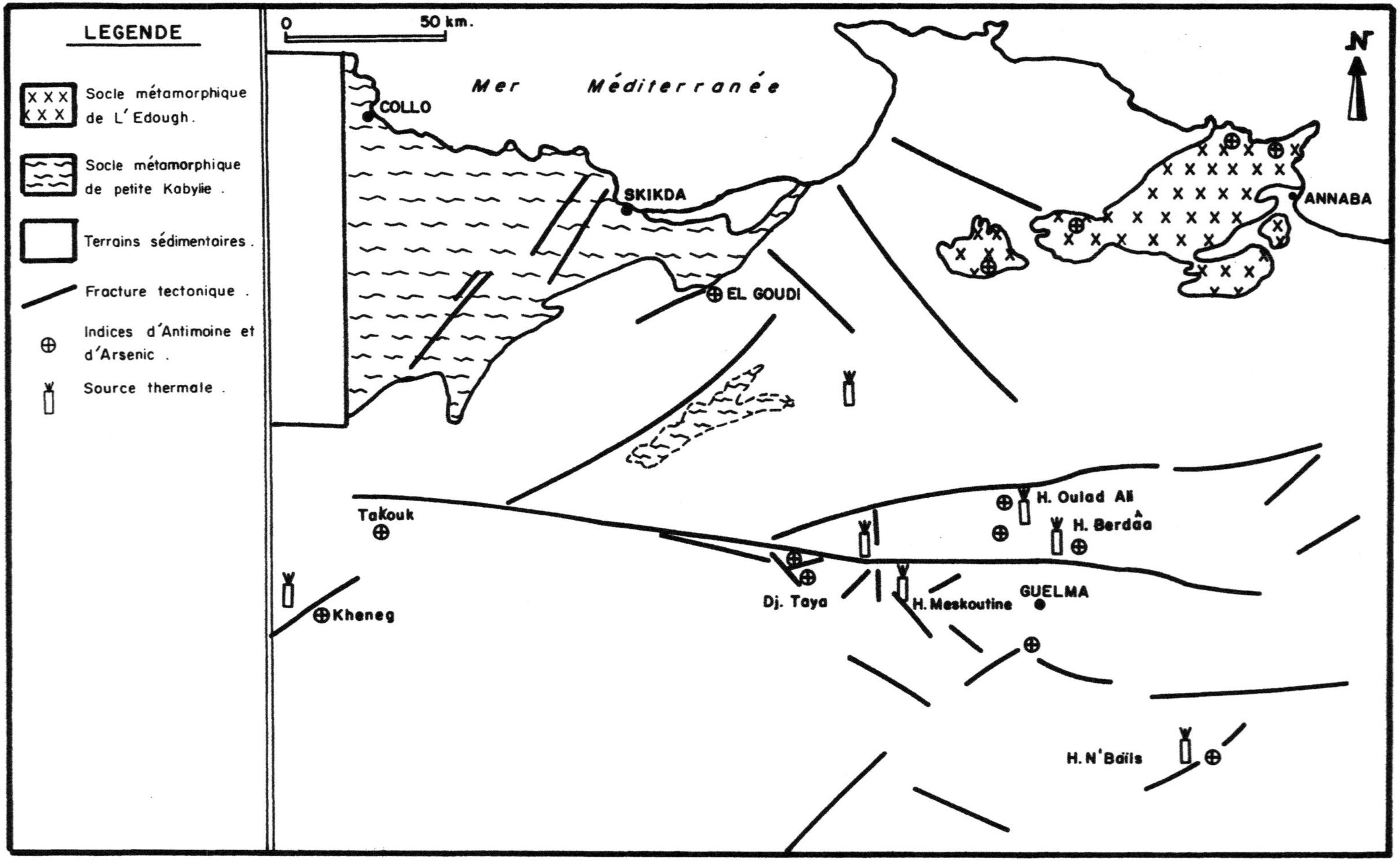

Fig. 7. Les Indices d'antimoine et d'arsenic du pourtour régionnal du massif de L'Edough. D'aprés Toubbal (1984), modifié

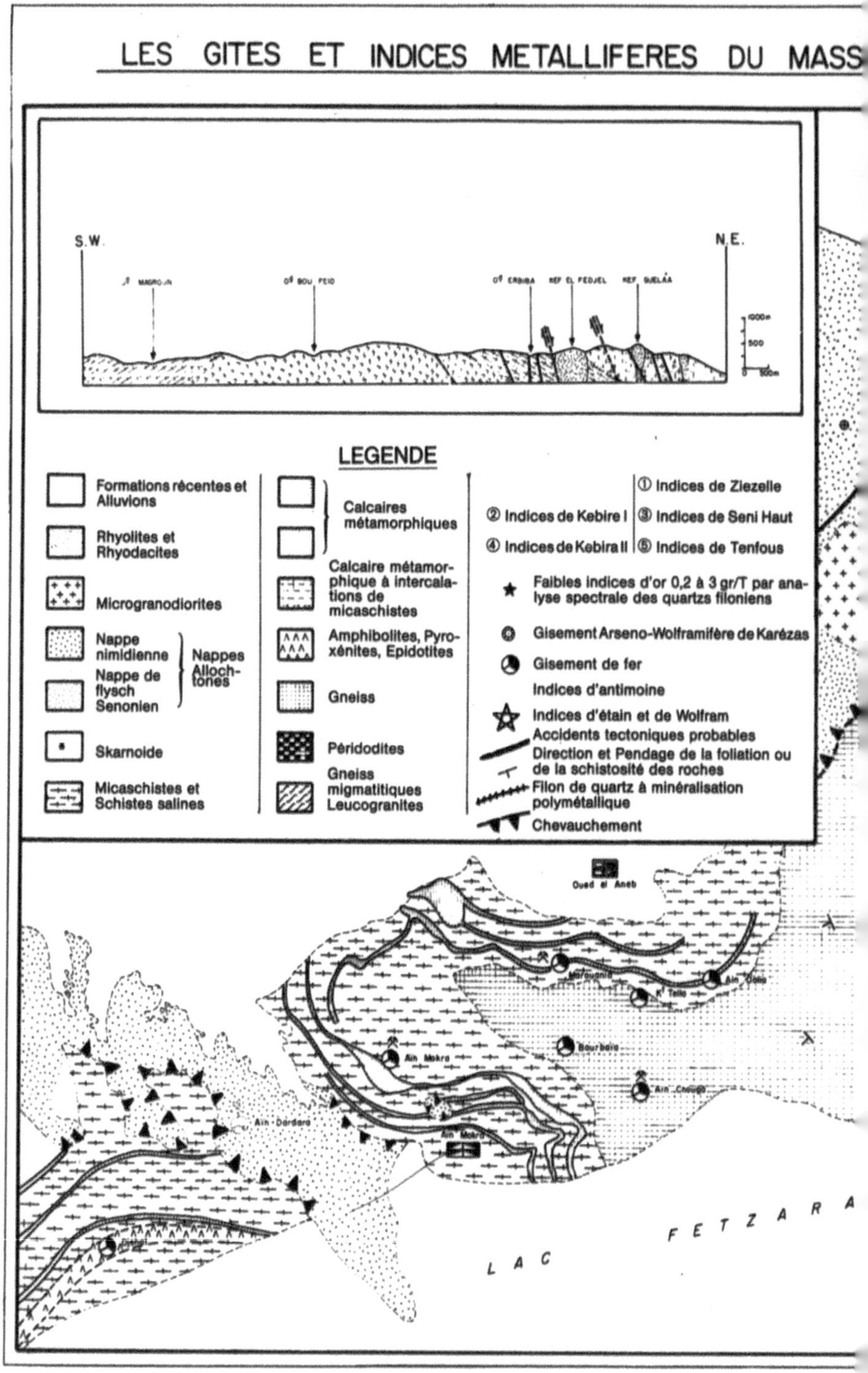

Fig. 8. Les Gites et indices Metalliferes

'EDOUGH DANS LEUR CONTEXTE GEOLOGIQUE

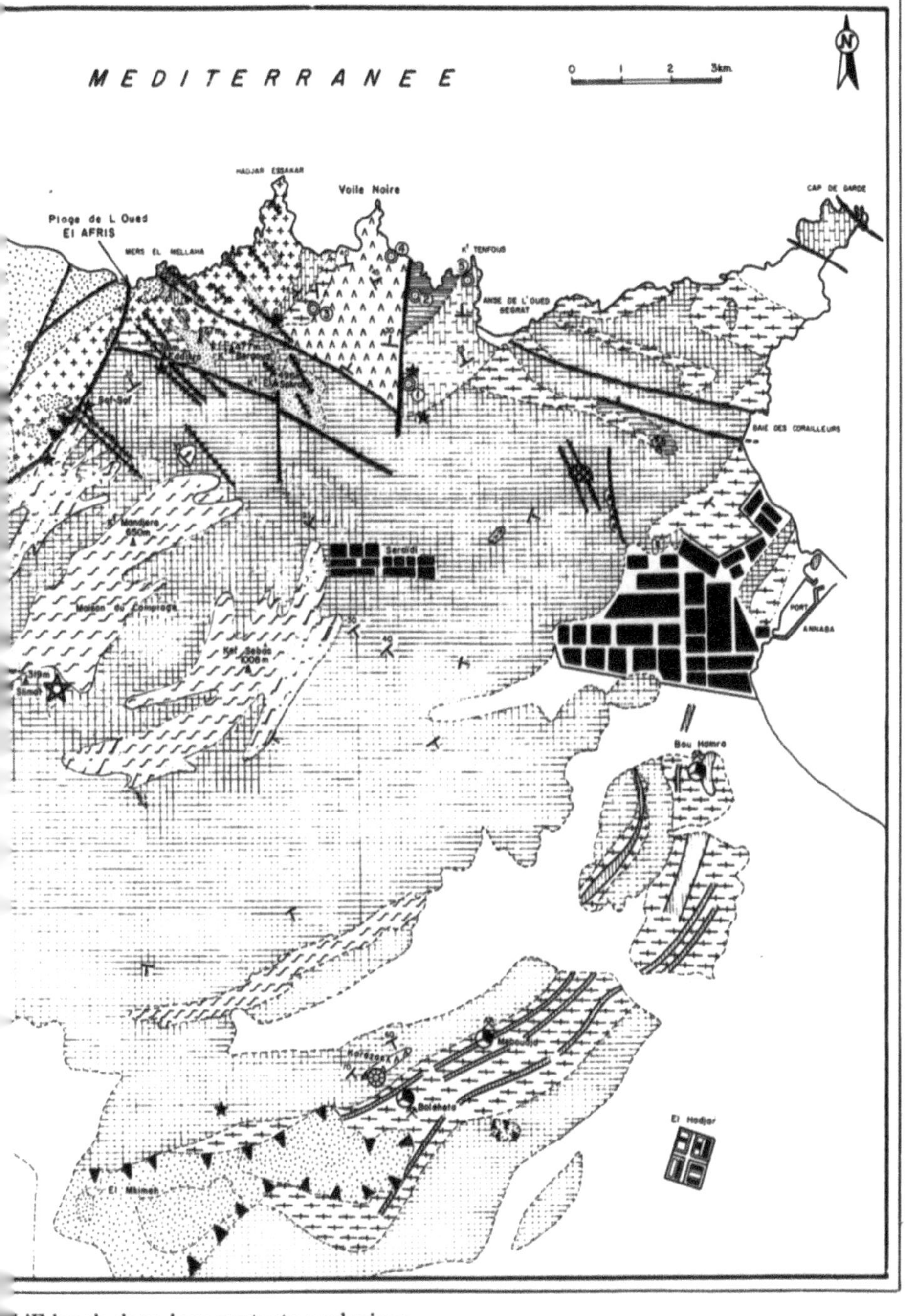

L'Edough dans leur contexte geologique.

Dans la partie Nord du massif par contre, on rencontre surtout des minéralisations hydrothermales à Cu, Pb, Zn, Bi, Sb, concentrées dans des fractures. Ces minéralisations sont essentiellement développées par le magmatisme miocéne qui aurait joué uniquement un rôle de moteur thermique ayant remobilisé les éléments en trace. En effet pour ces minéralisations, en plus du contrôle tectonique, on note également l'existence d'un contrôle lithologique.

En outre, on note la présence d'une permanence géochimique des éléments: Sb, As, Fe, F, Ag, Bi, Sn, W, non seulement à travers le massif de l'Edough, mais encore dans tout le pourtours régionnal de ce massif: El-Goudi, Takouk, Dj. Debbagh, Hammam N'baîls, Dj. Nador . . . Ces indices, sont pour la plupart situés à proximité de grandes fractures du socle ou de sources thermales (Fig. 7).

Ainsi, ce massif montre une bonne illustration des notions de permanence géochimique, héritage et remobilisations.

Bibliographie sommaire

AISSA, D. E. (1981): Les indices et gisements auriféres du Hoggar et d'Algérie Nord orientale. Congrés nat. des Sci. de la Terre. Tunis 1981.

AISSA, D. E. (1983): Les indices métalliféres du Nord du massif de l'Edough. 4° séminaire Nat. des Sci. de la Terre. Alger, Juin 1982.

AISSA, D. E. (1983): Les niveaux minéralisés stratoides du Nord du socle métamorphique de l'Edough (Annaba)-Publ. Serv. Géol. de l'Algérie-Bull. N° 50, pp. 7 à 25, 15 fig., 2 pl.

BURT, D. M., et al. (1982): Topaz rhyolites: distribution, origin and significance for exploration. – Econ. Geol. Vol. 77, pp. 1818–1836.

DUSSERT, M. (1912): Etude sur les gisements de fer de l'Algérie. – Ann. des Mines, Paris. – E.NA.R.E.M. (1968–70): Rapports sur les travaux de réévaluation des gisements de fer et du gisement de Wolfram de Karezas – rapport interne – inédit.

MARIGNAC, C. (1981): The polymetallic ore veins of Aïn Barbar (Algerie), as a consequence of alpine geothermal activity–in Mineral deposits of Alps–pp. 298–312, Ed. Springer.

VILLA, J. M. (1970): Le Dj. Edough: un massif cristallin externe du Nord-Est de la Berbérie. Bul. Soc. Géol. France (7) XII. N° 5, pp. 805–812.

Geotectonic Evolution and Metallogeny in Tunisia

By M. AMOURI*)

Zusammenfassung

Die Entwicklung des Landes wird seit der Trias bis ins Miozän an Hand paläofazieller Karten dargestellt (Fig. 2–11), wobei die den betreffenden Epochen zeitlich zugeordneten Lagerstätten verzeichnet sind. Die von älteren Autoren angenommene Verbindung zwischen Brüchen im Sockel, Triasdiapiren und Pb-Zn-Lagerstätten ist unwahrscheinlich. Die Mehrzahl der Erzvorkommen wird als primär syngenetisch und im Tertiär mobilisiert angesehen.

Summary

As the other Maghreb countries, Tunisia is included in the alpidic metallogenic province. By its geographic position, it occupies a privileged place in relation to the other mediterranean countries. Indeed, its situation is a hinge: on one hand, it is located between the saharian platform, in the South and the alpidic orogenic domain, in the North, and on the other, it is situated between the western part of the Maghreb and Sicilia which is the european extension of the Atlas. This position gives Tunisia particular geologic and metallogenic aspects.

Geologically, Tunisia is characterized by the absence of hercynian outcrops; the oldest being the permian in the south, at J. TEBAGA. The mesozoic and cenozoic cover, shoved off on the triassic level, is folded and faulted by tectonics which become more and more intensive to the North.

Tunisia shows an alpidic metallogeny with a Pb-Zn-F-Ba and Fe paragenesis. Ores are concentrated in the north and central parts of the country, whereas the eastern and southern parts do not contain metal deposits.

The genesis of the ore deposits was discussed by many authors. Except the phosphate series which are purely sedimentary, the genesis of the metallic ore de-

*) Département de Géologie, ENIS. 3038, Sfax, Tunisie.

posits is interpreted differently according to authors and schools. Some think that all ores result from the last alpidic orogenic event, called telethermal, others admit the synsedimentary origin, with diagenetic remobilization in some karstic and faulted environments.

The text shows the relation between the geotectonic and paleogeographic evolution and metallogeny since the triassic system. It is supported by a set of paleogeographic maps. Major ore deposits are represented by black symbols in their epoch of genesis.

Résumé

La Tunisie, pays du Maghreb, se situe sur la marge active de la plaque africaine. Son évolution géotectonique n'est connue qu'à partir du Trias. Le rapprochement Afrique-Laurasie, dès le Jurassique, a permi une tectonique souple puis cassante.

Cette tectonique a commencé discrètement à l'Aptien, puis au Crétacé supérieur et a engendré des structures, en haut-fond, favorables au piégeage des métaux lourds. Plus tard, au Miocène, la tectonique devient intense à la suite de la collision entre les deux plaques permettant ainsi l'expulsion des nappes telliennes puis numidiennes dans la partie septentrionale du pays. Au front de ces nappes se développent des bassins intramontagneux et une avant-fosse, le tout à remplissage molassique, affectés d'une tectonique post tangentielle. Sur les bordures de ces structures, de nombreux gîtes métallifères stratiformes se sont formés. Ces gîtes ont alimenté «per descensum» d'autres gîtes karstiques et/ou de cassures, situés dans des séries sous-jacentes, anté-néogènes.

Introduction

La Tunisie, comme les autres pays du Maghreb, fait partie de la province métallogénique alpine maghrébine (ROUTHIER, 1980). Par sa position géographique, elle occupe une place privilégiée par rapport aux autres pays méditerranéens. En effet, sa situation en charnière, entre d'une part, la plateforme saharienne, au Sud, et le domaine orogénique alpin, au Nord, et d'autre part, les pays occidentaux du Maghreb et la Sicile, prolongation européenne de l'Atlas, lui confère des aspects géologiques et métallogéniques bien particuliers.

Du point de vue géologique, la Tunisie est caractérisée par l'absence de tout affleurement hercynien; le plus ancien étant le Permien à Fusulines, d'extension limitée, dans le Jebel Tebaga de Medenine, au Sud. Elle est formée de terrains mésozoïque et cénozoïque se présentant en couverture décollée au niveau du Trias, surtout au Centre et au Nord. Ces terrains sont affectés d'une tectonique souple et cassante, devenant de plus en plus intense vers le Nord.

Du point de vue métallogénique, la Tunisie renferme des gisements de type alpin à paragenèses simples Pb-Zn, Fe, F-Ba et P, répartis essentiellement dans le Nord et le Centre. Les parties méridionales et orientales en sont dépourvues.

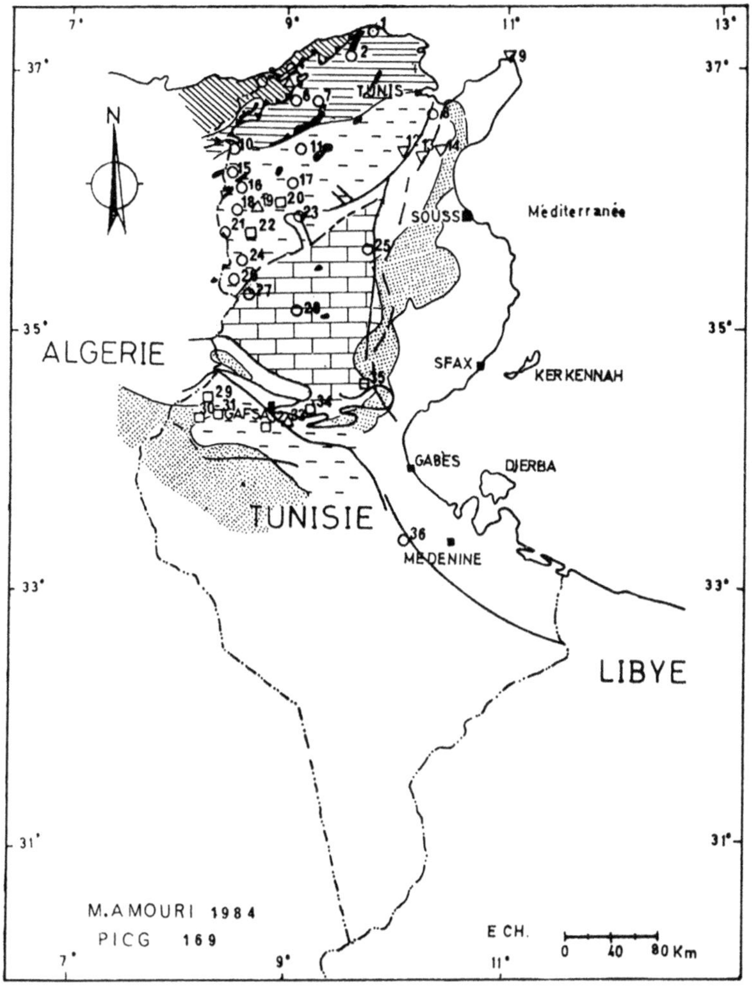

Fig. 1 - Esquisse structurale et localisation des principaux gisements.
Structural sketch and location of major ore deposits.

Flysch numidien *Numidian flysch*

Bassins miocènes *Miocene Troughs*

Diapirs triasiques *Triassic diapirs*

Bassins crétacés-tertiaires *Cretaceous-tertiary basins*

Ile de Kasserine *Kasserine Island*

Bassins intracratoniques *Intracratonic basins*

o Pb-Zn △ Fe ▽ F - Ba □ P

1-Bechateur;2-Jalta;3-AînAllega;4-Sidi Driss ;5-Tamera;6-J.Hallouf-Sidi Bou Aouane;7-El Haouaria;8-J.Ressas;9-Oued M'tak;10-O.Maden; 11-Fej El Adoum;12-Sidi Taya;13-Hammam Zriba-J.Guebli;14-H.Jedidi; 15-Sakiet;16-Garn Halfaya;17-Bougrine;18-Sidi Amor;19-Jerissa;20-Sra Ouartane;21-Boujabeur;22-Kalâat Khasba;23-Skarna;24-Lajered;25-Trozza; 26-EL Hamra;27-Chambi;28-Aîn Nouba;29-Moularès;30-Redeyef;31-Metlaoui; 32-Mdilla;33-J.Ank;34-El Ayacha;35-Mehri Zebbeus;36-Tebaga.

La mise en place des différents gîtes a été discutée par plusieurs auteurs. Hormis les gîtes de phosphate qui sont purement sédimentaires, la genèse des autres gisements métallifères fût discutée et interprétée différemment selon les époques et l'école des auteurs. Certains pensent que tous les gisements résultent de la dernière manifestation orogénique alpine, qualifiés de téléthermaux (SAINFELD, 1952; BOLZE, 1952). D'autres admettent l'origine synsédimentaire, avec remobilisation et piégeage dans des structures favorables tels les karsts et les cassures (FUCHS, 1973; NICOLINI, 1968 et d'autres).

Evolution Geotectonique

Dans le développement qui suit, et sur la base des nouvelles données relatives aux études géologiques et métallogéniques effectuées ces dernières années, nous essayons de montrer la liaison entre évolution géotectonique de la Tunisie et la mise en place des minéralisations. Cette évolution ne peut être examinée qu'à partir du Trias, du fait que le socle n'affleure nulle part; le seul affleurement permien du Sud est considéré comme faisant partie de la couverture mésozoïque; même les forages les plus profonds n'ont pas atteint le socle dans les parties centrale et septentrionale du pays.

Selon P. F. BUROLLET (1983) et J. MARIE et al. (1984), les pays atlasiques, dont fait partie la Tunisie, se développent au front de la plate-forme saharienne. Ce front constitue une zone intermédiaire entre le craton, au Sud, et les orogènes telliens, au Nord. La plate-forme possède une croûte continentale amincie ou étirée en direction du Nord. Elle a commencé à s'enfoncer par subsidence à partir du Carbonifère supérieur ou du Permien. Au Trias, elle était occupée par une mer à tendance évaporitique; le Jurassique est marin, souvent profond. Des contraintes en collision ou en décrochement, à partir du Crétacé, ont crée des ondulations assez vastes; les paléo-reliefs positifs ont réduit ou arrété la sédimentation; alors que ceux négatifs ont accéléré la subsidence. Ces contraintes se sont accentuées au Paléogène pour aboutir aux collisions et aux phases majeures des plissements néogènes.

Le schéma structural actuel (Fig. 1) montre du Nord au Sud les zones suivantes:

* flysch numidien, unité allochtone formée d'importantes masses gréseuses.
* bassins molassiques en auge du Miocène.
* bassins crétacés et tertiaires nord et sud.
* le môle de Kasserine.
* les bassins périatlasiques ou intracratoniques.
* l'axe nord-sud.
* la plate-forme saharienne.

Les gisements de phosphate se trouvent sur le pourtour de l'île de Kasserine, au niveau des bassins crétacé et tertiaire nord et sud; les gisements métalliques sont concentrés principalement dans les bassins crétacés-tertiaires nord et les bassins molassiques.

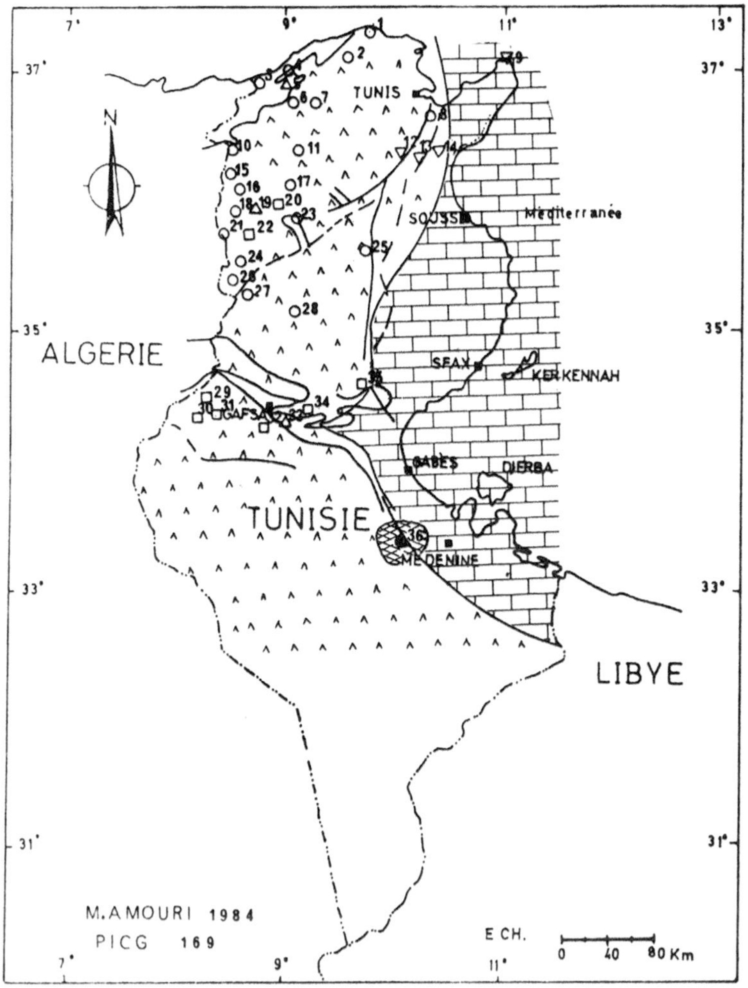

Fig.2 - Carte paléogéographique du Trias

Paleogeographic map for Trias

Faciès à évaporites
Evoporitic facies

Faciès carbonatés
Carbonated facies

Aires émergées
Emerged areas

Les symbôles noirs désignent les gisements formés pendant le Trias.

Black symbols show ore deposits formed during Trias

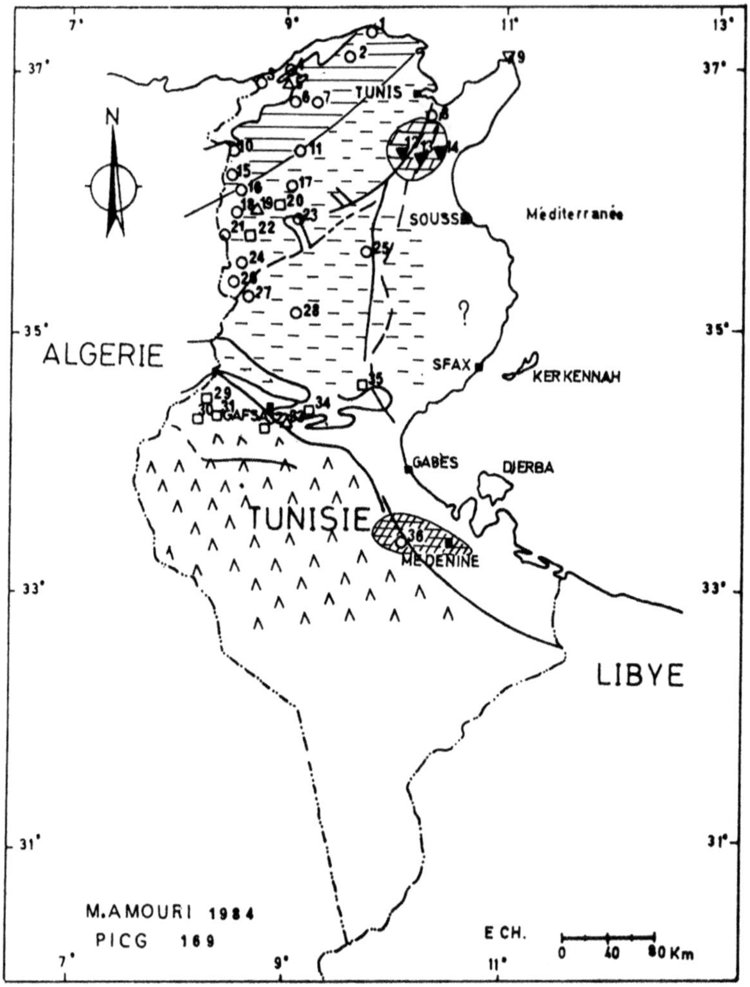

Fig. 3 - Carte paléogéographique du Jurassique inférieur

Paleogeographic map for lower Jurassic

Sillon tunisien
Tunisian trough

Aires emergées
Emerged areas

Faciès néritique
Neritic facies

Faciès à évaporites
Evaporitic facies

Facies récifal à para-récifal
Reefal to subreefal facies

Les symbôles noirs indiquent les gisements formés au Jurassique inférieur
Black symbols show ore deposits formed during lower Jurassic

Au Trias, la Tunisie est recouverte d'une mer, marge sud-ouest de la Tethys, peu profonde, à tendance évaporitique, responsable du dépôt de séries détritiques, carbonatées et évaporitiques. La nature de ces dépôts a facilité la mise en place de nombreuses intrusions diapiriques. A l'Est se développe un domaine carbonaté (BUROLLET, 1975), alors qu'au Sud, le môle de Tebaga reste émergé (Fig. 2).

Aucune forme de minéralisation ne se trouve ni dans le Trias ni dans les éléments de bréche au contact du Trias; ce qui exclu toute minéralisation ancienne, antérieure au Trias. La localisation de nombreux gîtes métalliques, à proximité des diapirs triasiques, a amené certains auteurs à penser au rôle métallisant du Trias (BOLZE, 1952; SAINFELD, 1952); or, on sait que les diapirs ont commencé à percer dès le Jurassique; leur affleurement, à l'Aptien, a été signalé par E. LAATAR (1980) dans la mine de Fej el Adoum. Par ailleurs, la liaison diapir-accident de socle n'est pas toujours claire. En effet, les failles du socle s'amortissent dans le niveau du décollement. Il est difficile d'expliquer, par conséquent, le cheminement des solutions minéralisantes, à partir du socle jusqu'aux sites de dépôt aptien, les plus proches, en traversant un niveau gypso-argileux décollé et une série fortement épaisse du Crétacé inférieur. D'ailleurs P. SAINFELD (1952) a émi quelques réserves à propos de la théorie téléthermale. Les travaux de V. PERTHUISOT et H. ROUVIER (1984) ont montré le rôle indirect joué par les diapirs pour la formation des concentrations métallifères; ce sont les structures positives engendrées par la montée diapirique qui en sont responsables.

Le Jurassique fait suite au Trias avec une sédimentation variée argilo-sableuse et carbonatée. Les domaines paléogéographiques commencent à peine à se différencier. Du Nord au Sud, on rencontre le sillon tunisien à sédimentation marneuse profonde, puis une zone à sédimentation marno-calcaire faisant transition avec le domaine carbonaté et évaporitique du Sud. Au Nord-Est, s'individualise un domaine récifal ou pararécifal surlequel se sont formées, à deux reprises, les minéralisations fluobarytiques et sulfurées de Zaghouan, au Sinémurien (Fig. 3) et au Portlandien (Fig. 4). Ces minéralisations, stratoïdes ou en amas, sont associées à des surfaces d'émersion intra et fini-Jurassique (FLORIDIA, 1973; FUCHS, 1973; THIEBEROZ, 1974; TOUHAMI, 1979). Ces minéralisations étaient considérées comme résultant de la dernière manifestation orogénique alpine (SAINFELD, 1952). R. TRUILLET (1981) considère les lacunes stratigraphiques comme dues à une ablation suivie d'une substitution tectonique de couverture. La minéralisation jalonne un grand contact mécanique qui tronque les séries sous-jacentes et sur-jacentes. La mise en place est le résultat d'un hydrothermalisme qui est encore actif. Malheureusement, aucune minéralisation actuelle n'a été trouvée au voisinage des sources actuelles thermales.

Le passage Jurassique-Crétacé est continu, sauf dans la zone fluorée où les lacunes persistent. A partier du Néocomien, les domaines paléogéographiques s'affirment davantage (Fig. 5). Du Nord au Sud, on rencontre le Sillon tunisien qui reste à sédimentation marneuse profonde, puis une zone de transition à sédimentation mixte gréso-argilo-carbonatée, future zone des dômes ou des diapirs; ensuite une plate-forme instable, à quelques ilôts émergés, est le siège d'opposition entre les apports clastiques sahariens au Sud-Ouest et les faciès de mer ouverte à la limite de la plate-forme stable orientale; ensuite, le sillon de Gafsa à sédimentation fortement

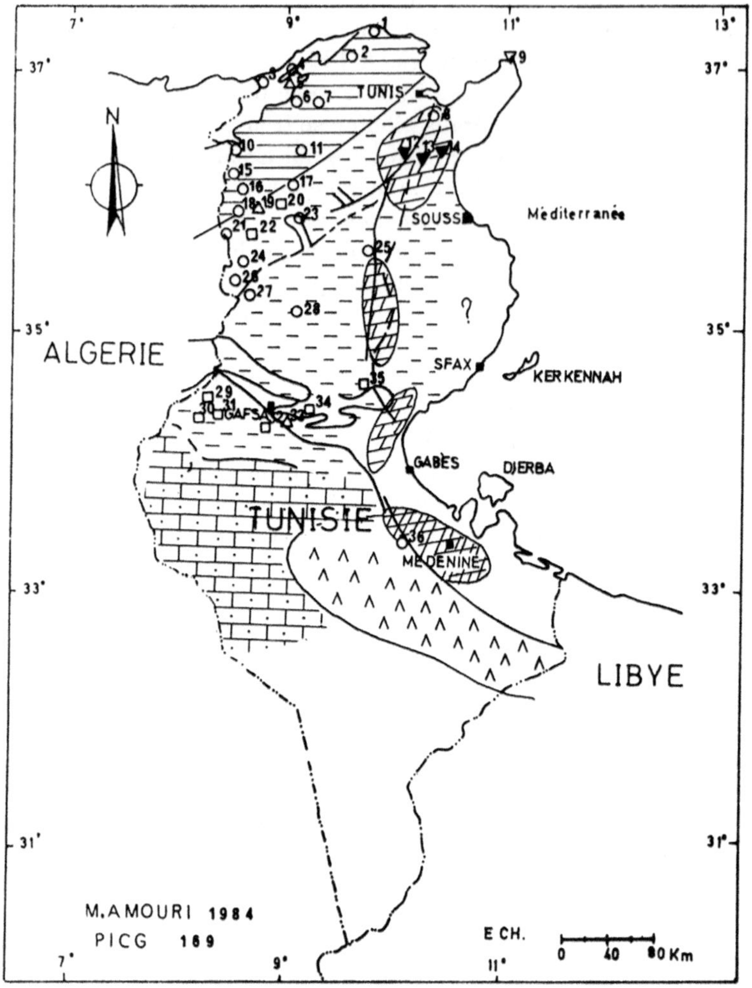

Fig. 4 - Carte paléogéographique du Jurassique supérieur
Paleogeogeographic map for upper Jurassic

Sillon tunisien *Tunisian trough*	Faciès sablo-carbonatés *Sandy-carbonated facies*
Faciès néritique *Neritic facies*	Aires émergées *Emerged areas*
Faciès récifal *Reefal facies*	Faciès à évaporites *Evaporitic facies*

Les symbôles noirs indiquent les gisements formés au Jurassique supérieur
Black symbols show ore deposits formed during upper Jurassic

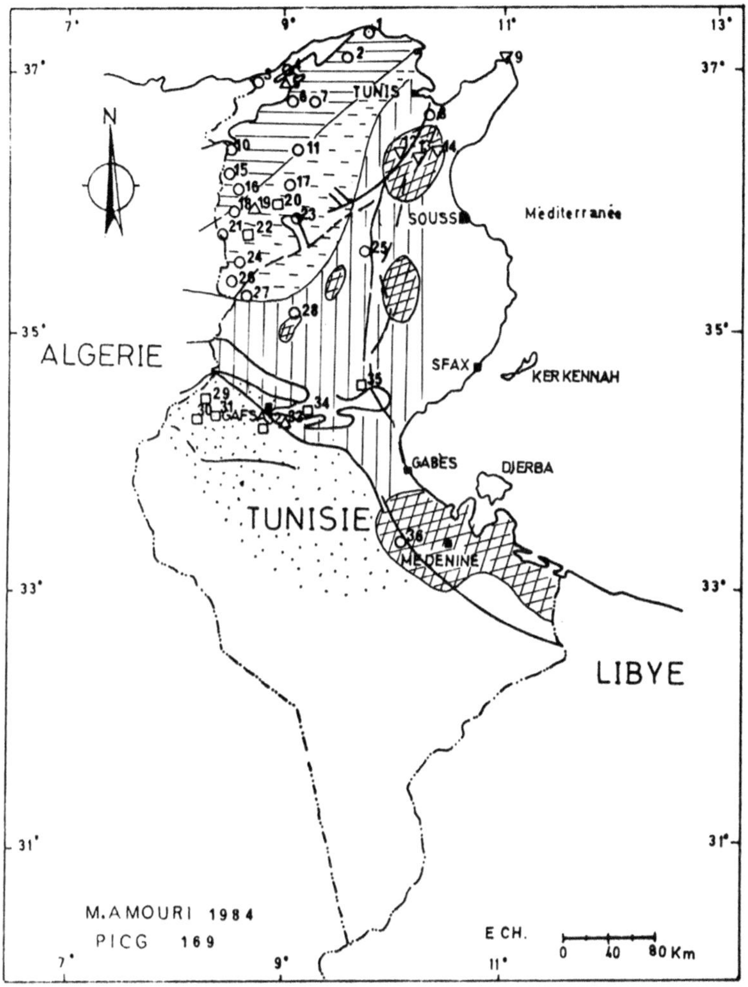

Fig. 5 - Carte paléogéographique du Crétacé inférieur
Paleogeographic map for lower cretacious

Sillon tunisien *Tunisian trough*	Aires émergées *Emerged areas*
Dépôts mixtes *Mixed deposits*	Sables grossiers *Coarse grained sand*
Dépôts argilo-sableux *Argilaceous sandstone deposits*	

subsidente; enfin, la plate-forme saharienne, stable, à sédimentation deltaïque avec un domaine émergé de Tebaga.

Pendant le Crétacé inférieur, aucune forme de minéralisation n'est connue. C'est à partir de l'Aptien que se forment la plupart des minéralisations plombo-zincifères et ferrifères du Centre tunisien. A la suite de mouvements compressifs, des percements diapiriques se font sentir, surtout dans la zone des dômes et la plate-forme centrale, entrainant ainsi l'installation de haut-fonds (Fig. 6) qui évolueront par la suite pour donner les anticlinaux actuels. Sur chaque haut-fond, se développe un faciès récifal à pararécifal passant latéralement à des marno-calcaires avec variation de puissance. Les faciès carbonatés sont porteurs des minéralisations Pb-Zn-Ba et Fe. Les exemples sont nombreux: El Hamra, Lajred, Aïn Nouba, Boujaber, Sidi Amor, Slata, Jerissa ... Le modèle génétique défini comme péridiapirique par V. Perthuisot et H. Rouvier (1984) est fondé sur la possibilité de transfert d'une partie des métaux dans la couverture sédimentaire vers le point haut constitué par la structure diapirique, grâce à des solutions particulièrement salées et chaudes. Les phénomènes diagénétiques comme la silicification et la dolomitisation peuvent jouer un rôle purificateur; elles permettent en effet de libérer les carbonates préenrichis de leur contenu métallifère. Les éléments ainsi libérés seront concentrés dans des structures piéges, comme les réseaux karstiques, par exemple. Les observations de terrain et les analyses minéralogiques et géochimiques isotopiques effectuées récemment confirment ces points de vue (Amouri, 1984). La source initiale des métaux doit être recherchée dans les séries jurassiques et crétacées, émergées, situées à l'Ouest, en Algérie. Le Maghreb se présentant comme un plan incliné vers l'Est (Vila, 1980), il est alors concevable que la remobilisation des concentrations hercyniennes ou alpines précoces puisse donner le stock métallifère nécessaire pour la formation de ces gîtes péridiapiriques.

Au Crétacé supérieur, les mêmes phénomènes se répètent en quelques endroits du Centre-ouest tunisien: J. Chambi, Bougrine, Fej el Adoum ... Les minéralisations stratiformes sont souvent remobilisées dans des réseaux de fractures plus ou moins ouvertes par dissolution karstiques (Laatar, 1980) (Fig. 7).

A la fin du Crétacé, s'individualise le môle résistant de Kasserine, large domaine émergé. Sur ses bordures s'installe une ligne de rivage fluctuante en fonction des transgressions et regressions.

Le début du Tertiaire est marqué par le dépôt des séries phosphatées tout autour du môle de Kasserine (Fig. 8). Les bassins de Gafsa, au Sud, et de Tajerouine, au Nord, sont les plus importants. Les couches phosphatées sont en nombre de huit. Elles ne sont pas toutes exploitables. La différence des teneurs en Uranium dans les phosphates entre les bassins sud et nord, le bassin sud étant le plus pauvre, résulte d'une différence du milieu de dépôt. Le bassin de Gafsa, au Sud, est intracratonique et partiellement fermé; alors que le bassin nord est ouvert, de type méditerranéen (Abdelhadi et al., 1981). Au sommet des séries phosphatées se dépose une couche de fer oolithique au J. Ank (Fig. 8); alors qu'a la base des mêmes séries se trouvent les minéralisations Pb-Zn du Sekarna, au Centre.

Au début du Tertiaire ou peut être avant, les terres émergées, à l'échelle du Maghreb, prennent de l'ampleur; la liaison terrestre devenant possible de la côte

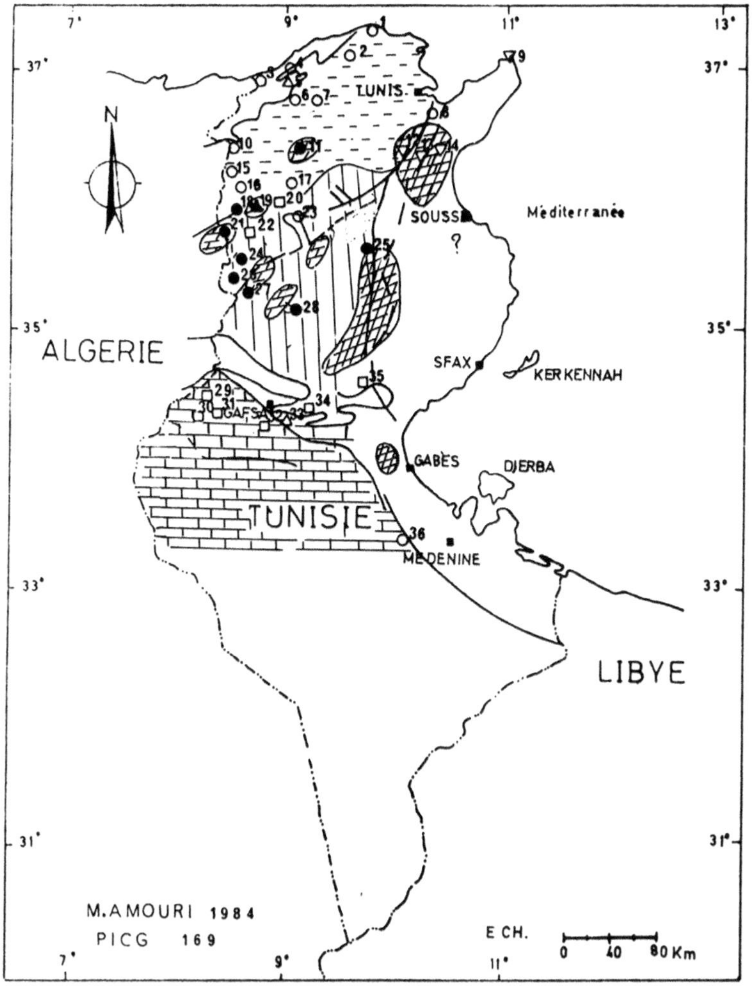

Fig 6 - Carte paléogéographique de l'Aptien
Paleogeographic map for Aptien

Sillon tunisien *Tunisian trough*	Aires émergées *Emerged areas*
1 Zone des hauts-fonds 2 *High ground zona*	Plate-forme carbonatée interne *Internal carbonated platform*

1. Sedimentation de la plate-forme externe
   Continious sedimentation of external platform
2. Faciès récifal subémergé
   Subermerged reefal facies

Les symbôles noirs indiquent les gisements formés à l'Aptien
Black symbols show ore deposits formed during Aptien

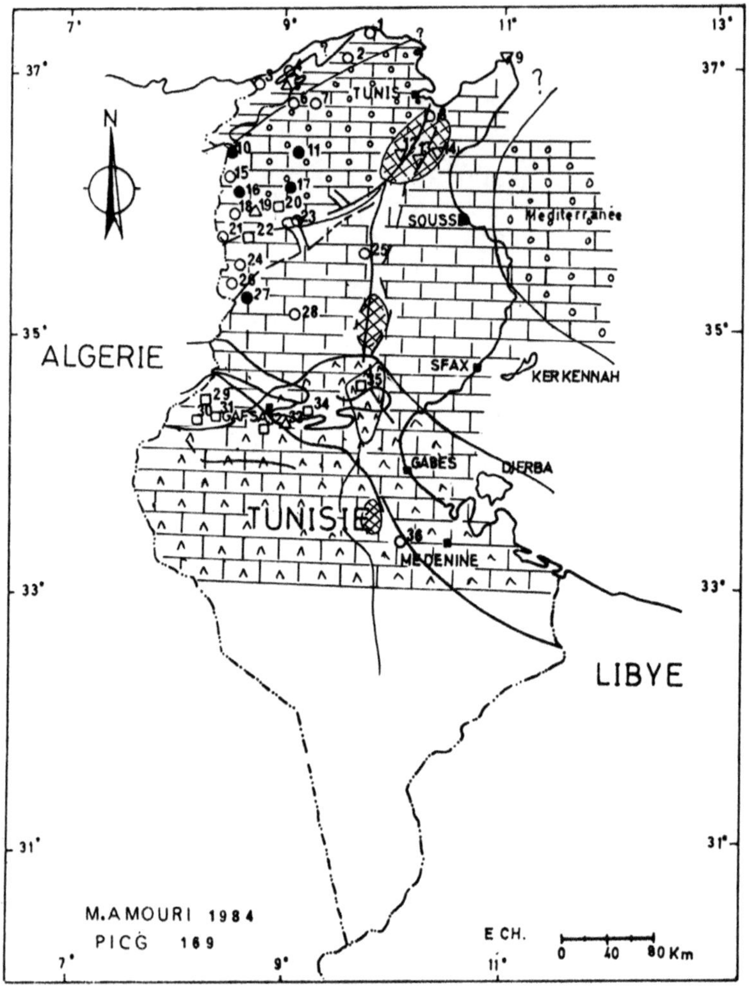

Fig.7 - Carte paléogéographique du Cénomanien
Paleogeographic map for Cenomanien

(symbol)	Sillon tunisien *Tunisian trough*	(symbol)	Aires émergées *Emerged areas*
(symbol)	Dépôts planctoniques *Planctonic deposits*	(symbol)	Evaporites *Evaporits*
(symbol)	Plate-forme carbonatée centrale *Central carbonated platform*	(symbol)	Plate-forme sud *South platform*

Les symbôles noirs indiquent les gisements formés au Cénomanien
Black symbols show ore deposits during Cenomanien

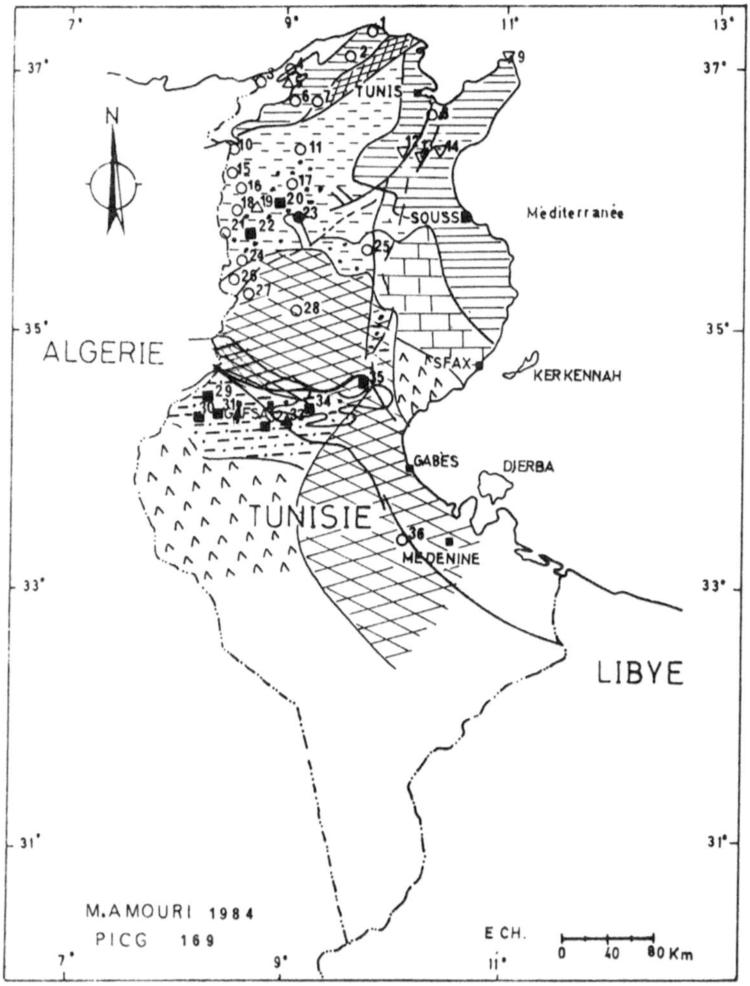

Fig. 8 - Carte paléogéographique de l'Eocène
Paleogeogaphic map for Eocene

Dépôts néritiques *Neritic deposits*	Plate-forme carbonatée *Carbonated platform*
Dépôts mixtes *Mixed deposits*	Dépôts phosphatés *Phosphated deposits*
Aires émergées *Emerged areas*	Evaporites *Evaporits*

Les symbôles noirs indiquent les gisements formés à l'Eocène
Black symbols show ore deposits during Eocene

atlantique à Bizerte; c'est ce qu'on appelle le paléo-Maghreb (VILA, 1980). Les mouvements en compression N-S s'accentuent aussi pendant cette même période, surtout au Priabonien, accompagnant des bouleversements paléogéographiques profonds de l'avant pays de la chaine alpine du Maghreb. Ces bouleversements consistent en l'accentuation des mouvements salifères et l'expulsion des premières nappes lors de l'ecrasement partiel de la marge nord-africaine.

De l'Oligocène au Burdigalien, la compression continue de manière inégale dans l'espace et dans le temps. La tectonique est plus précoce à l'Ouest du Maghreb qu'à l'Est, les déformations devenant de plus en plus intenses en allant vers l'Est (VILA, 1980).

A l'Oligocène, presque toute la Tunisie est émergée; on note toutefois un domaine sans dépôts, occupant tout le Centre et le Sud, alors qu'au Nord on enregistre d'importants dépôts gréseux (BUROLLET, 1975) (Fig. 9).

Pendant le Burdigalien et avant la mise en place des nappes numidiennes, une légère manifestation métallogénique se fait sentir au Nord, près de Béja et au Cap Bon (Fig. 10), dans des bassins internes, juste au front des unités oligocène-crétacées ou telliennes.

Du Burdigalien au début du Tortonien, les phénomènes compressifs demeurent discrets et peuvent être momentanément nuls.

Au Tortonien se déroule un nouvel épisode de collision dont les modalités sont différentes de celles de l'édifice priabonien. En effet, selon J. M. VILA (1980), toute la tectonique se fait à sec, au dessus d'un plan de sous charriage qui ne tient pas compte de la paléogéographie antérieure, c'est à dire de la nature et des caractéristiques mécaniques des séries accumulées jusqu'à la fin du Crétacé supérieur ou de l'Eocène. Ces caractéristiques n'influent que sur le style des déformations.

Pendant cette période et jusqu'au Pliocène, se sont développés plusieurs petits bassins internes (Fig. 11), à remplissage diversifié, dans la partie septentrionale du pays (VASSILEFF et al., 1979), au front des nappes numidiennes. Ces bassins sont tous minéralisés en Pb-Zn et quelques autres métaux accessoires; leur substratum est plissé et fracturé au même titre qu'eux. Ils sont orientés NE-SW avec un déversement vers le SE. Les minéralisations sont disséminées dans des séries laguno-continentales détritiques et/ou dans des calcaires lacustres. Ces remplissages des bassins peuvent être volcano-terrigènes dans quelques endroits, Sidi Driss par exemple. Ces bassins ont fait ou font actuellement l'objet de quelques exploitations.

A ces gîtes de bassins, qualifiés de néogènes (VASSILEFF et al., 1979), sont associés des gîtes encaissés dans les terrains anté-néogènes. Ces gîtes occupent les réseaux de fractures souvent élargies par dissolution karstique; l'exemple le plus typique est celui de J. Hallouf-Sidi Bou Aouane.

Un peu plus vers le Sud de ces bassins, deux avant-fosses se sont développées, longeant le cours inférieur et moyen de l'Oued Mejerda (VASSILEFF et al., 1979). Il s'agit d'un étroit chenal comblé d'une importante série molassique. Sur ses bordures se développent des séries à Pb-Zn (El Alia) et à Cu (Chouichia).

Du Tortonien supérieur au Quaternaire, correspond la période de genèse des reliefs actuels. En effet, l'édifice tortonien n'a subi que des modifications modérées qui correspondent à une fracturation et à un plissement le plus souvent à grand

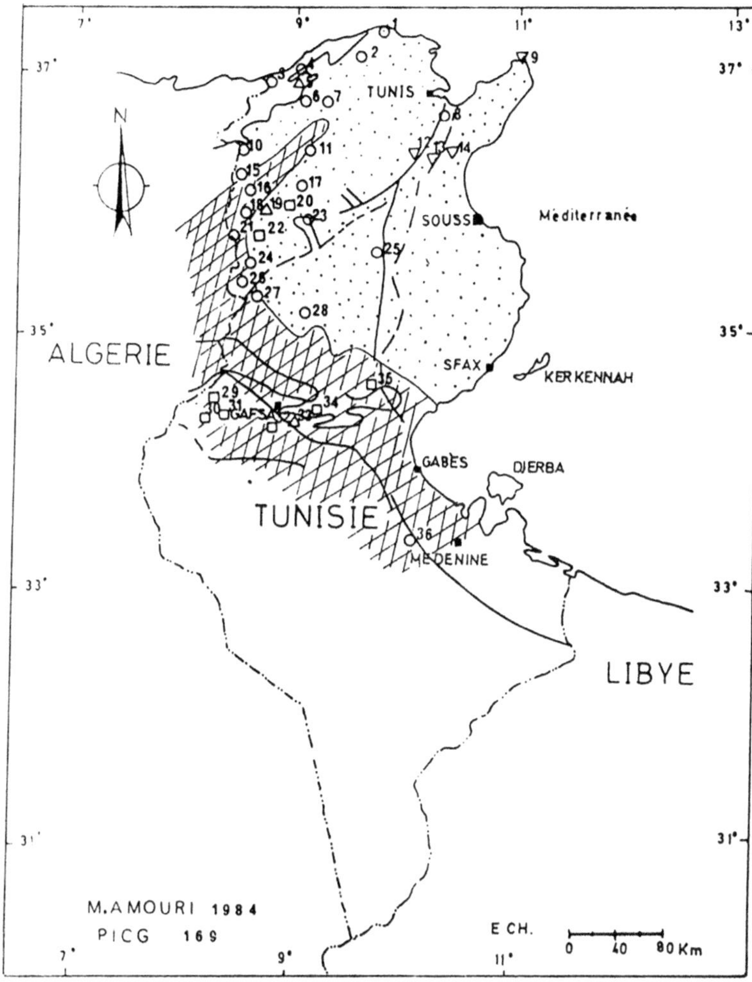

Fig. 9 - Carte paléogéographique de l'Oligocène
Paleogeographic map for Oligocene

Dépôts gréseux
Sandstone deposits

Aires émergées
Emerged areas

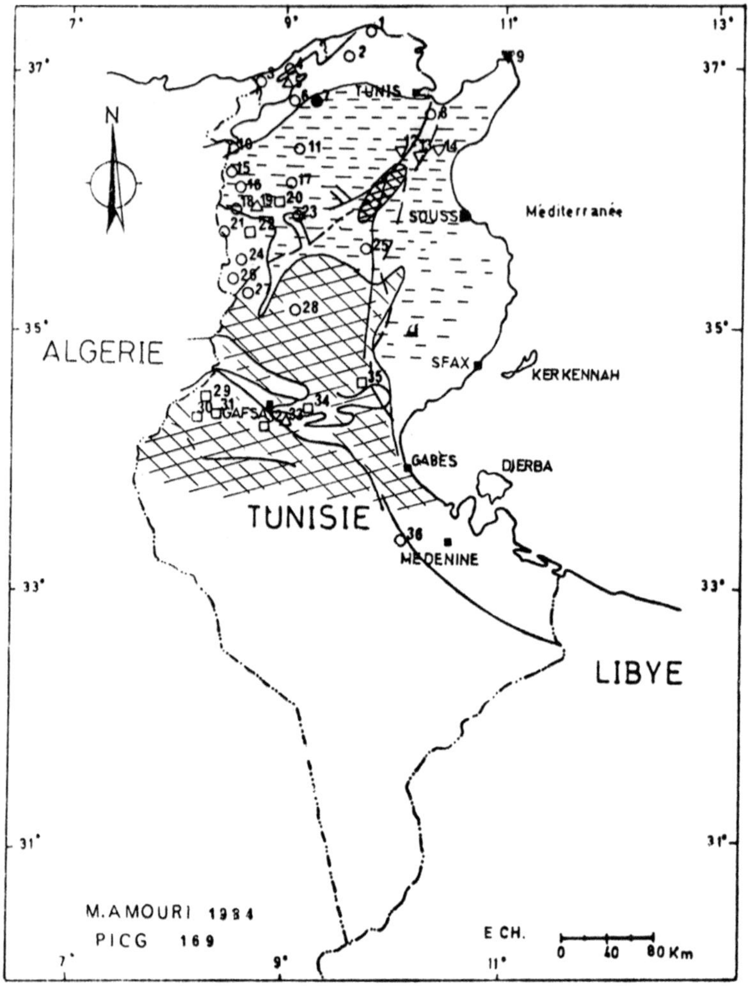

Fig. 10 - Carte paléogéographique du Miocène anté-nappes
Paleogeographic map for ante-nappe miocene

Dépôts néritiques
Neritic deposits

Aires émergées
Emerged areas

Les symbôles noirs indiquent les gisements formés au Miocène anté -nappes
Black symbols show ore deposits formed during ante-nappe Miocene.

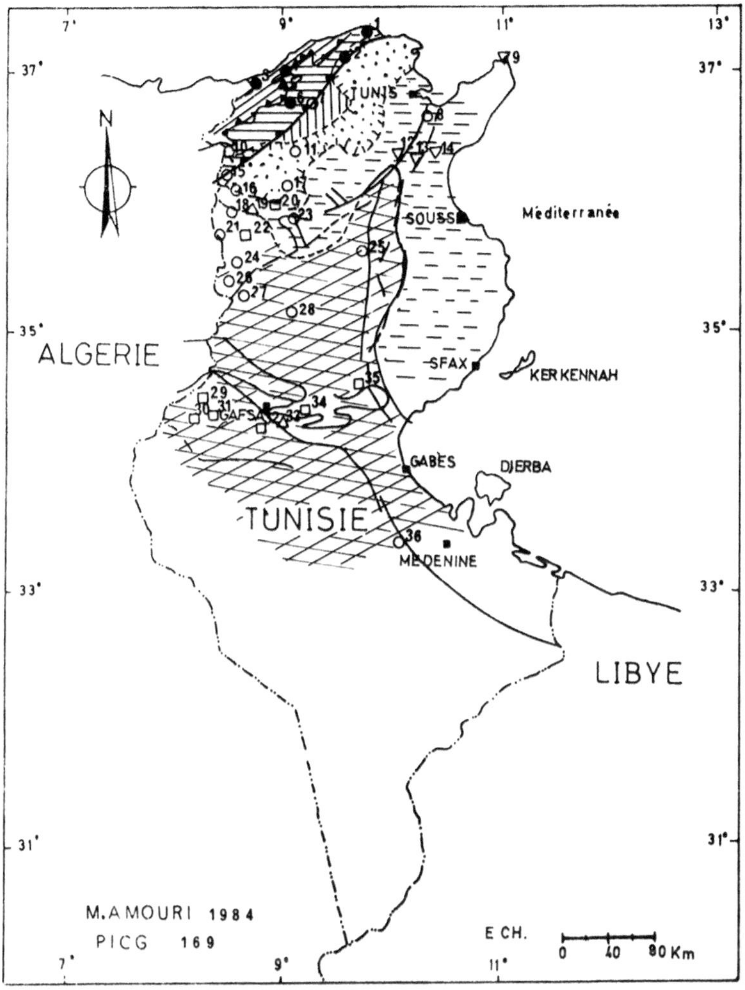

Fig. 11 - Carte paléogéographique du Miocène post-nappes
Paleogeographic map for post-nappes miocene

Unité numidienne *Numidian unit*	Avant-fosse *Fore-deep*
Unités telliennes *Tellian units*	Dépôts néritiques *Neritic deposits*
Paraautochtone *Paraautochtonous*	Aires émergées *Emerged areas*

Les symbôles noirs indiquent les gisements formés au Miocène post-nappes
Black symbols show ore deposits formed during post nappes Miocene

rayon de courbure des grandes unités charriées. On peut distinguer deux phases compressives successives précédant deux phases de relaxation des contraintes (ROUVIER, 1977). Les phases de relaxation correspondent à l'émission des roches effusives (volcanisme acide et basique des Nefza, Nord tunisien).

Enfin, l'accumulation des sédiments plio-quaternaires permet une forte subsidence en Tunisie septentrionale et orientale. On observe parfois un renversement spectaculaire des traits majeurs du relief dans la région de Bizerte (BUROLLET, 1975). Des massifs montagneux qui encadrent le sillon miocène post-nappe, s'effondrent rapidement au cours du Pliocène avec des basculements de plus de 2000 m entre le début du Pliocène et l'époque actuelle.

Conclusion

Les gisements tunisiens, à part les phosphates qui sont indiscutablement sédimentaires, présentent un caractère apparent épigénétique tardif, qualifié de «téléthermal». Cependant, les travaux récents de plusieurs auteurs ont démontré que ces gisements n'apparaissent pas comme issus de la dernière phase hydrothermale de l'orogenèse alpine; au contraire, ces gisements se sont formés à des époques différentes selon l'évolution géotectonique du pays. Cette évolution a permi l'installation de structures favorables au piégeage des métaux, des haut-fonds et des petits bassins à caractère réducteur notamment, durant le Crétacé et le Néogène post-nappes.

L'origine primaire des métaux, volcanique ou terrigène, n'a pu être démontrée. Une étude comparative régionale, à l'échelle du Maghreb serait-elle justifiée pour résoudre le problème; d'autant plus que la migration des métaux de l'Ouest vers l'Est est tout à fait plausible.

Les éléments métallifères ont passé par au moins une étape syndiagénétique, comme le témoignent les teneurs élevées des calcaires non transformés de Bougrine. La finesse des grains minéraux et leur disposition laminaire, dans des niveaux stratigraphiques voisinant la plupart des cas les gîtes métallifères, témoignent aussi de la présence des minéralisations primaires. Ces éléments sont libérés par dolomitisation et silicification et enfin concentrés dans des ouvertures de type karstique, par exemple, lors des mouvements tectoniques ultérieurs. C'est ce dernier phénomène qui donne l'aspect téléthermal.

Bibliographie

ABDELHADI, M., SASSI, S., et PREMOLI, C. (1981): Répartition de l'Uranium dans les phosphates tunisiens. In résumés. 1er Congrés national des Sciences de la Terre. Tunis. P. 60 bis.

AMOURI, M. (1984): Les minéralisations Pb-Zn liées à l'Aptien dans l'Atlas tunisien central. 5ème séminaire national des Sciences de la Terre. Alger.

BOLZE, J. (1952): Diapirisme et métallogénie en Tunisie. 19ème C. G. I. Alger. XII section. PP. 91–104.

BUROLLET, P. F. (1967): General Geology of Tunisia. In guidebook to the geology and history of Tunisia. Petrol. Explor. Soc. Libya. 9th ann. field conf. 1967. PP. 51–58.

BUROLLET, P. F. (1975): Géologie et sédimentologie de la Tunisie. Livret guide excursion N° 15. IX Congrès international de sédimentologie. Nice.
BUROLLET, P. F., et BUSSON, G. (1983): Plate-forme saharienne et Mésogée au cours du Crétacé. Notes et Mémoires. N° 18. CFP. Paris. PP. 17–26.
FLORIDIA, S. (1973): La province fluorée tunisienne. Aperçu géologique et métallogénique. Livre jubilaire M. SOLIGNAC. Annales des mines et de la géologie n° 26 Tunis. PP. 459–477.
FUCHS, Y. (1973): Sur les relations entre émersions et concentrations métallifères, quelques exemples tunisiens. Livre jubilaire M. SOLIGNAC. Annales des Mines et de la Géologie n° 26 Tunis. PP. 479–509.
LAATAR, E. (1980): Gisements de Plomb-zinc et diapirisme du Trias salifère en Tunisie septentrionale. Les concentrations péridiapiriques du district minier Nefate-Fedj El Adoum (région de Téboursouk). Thèse spécialité. Université P. et M. CURIE. Paris VI. 280 P.
MARIE, J., TROUVE, PH., DESFORGES, G., et DUPHAURE, PH. (1984): Nouveaux éléments de paléogéographie du Crétacé de Tunisie. Notes et mémoires n° 19 CFP. Paris. 37 P.
NICOLINI, P. (1968): Les gîtes Plombo-zincifères de Tunisie. Symposium sur les gisements Plombo-zincifères en Afrique Tunis 1966. Annales des mines et de la Géologie n° 23. Tunis. PP. 207–240.
PERTHUISOT, V., et ROUVIER, H. (1984): Halocinèse et métallogénie: les gisements de Pb-Zn péridiapiriques. Géochronique 1984, supplément au n° 10. Résumés 27e CGI. Moscou. PP. 42–43.
ROUTHIER, P. (1980): Où sont les métaux pour l'avenir? mémoire B. R. G. M. n° 105.
ROUVIER, H. (1977): Géologie de l'extréme – nord tunisien: tectoniques et paléogéographies superposées à l'extrémité orientale de la chaine nord-maghrébine. Thèse Es. Sciences Paris VI. 898 P.
SAINFELD, P. (1952): Les gîtes Plombo-zincifères en Tunisie. Annales des mines et de la géologie n° 9. Tunis.
THIEBEROZ, J. (1974): Hammam jedidi et Hammam Zriba (région de Zaghouan, Tunisie). Etude géologique et minière de deux gisements stratiformes dans le cadre de la province fluorée tunisienne. L'association des concentrations fluorées aux surfaces d'émersion. Thèse spécialité université Paris VI.
TOUHAMI, A. (1979): Contribution à l'étude géologique et métallogénique de la province fluorée tunisienne. (Tunisie orientale) rôle de l'altération superficielle dans la formation des concentrations fluorées. Thèse spécialité 321 P. Tunis.
TRUILLET, R. (1981): La structure de Hammam Zriba. Son histoire géologique et métallogénique. In résumés. 1er Congrès national des Sciences de la Terre Tunis. P. 51.
VASSILEFF, L., et STANICHEVA-VASSILEVA, G. (1979): Le Néogène en Tunisie septentrionale. Position géotechnique et métallogénie. I. Cadre géologique. Geologica Balcanica 9, 3. Sofia PP. 71–92.
VASSILEFF, G., et STANICHEVA-VASSILEVA, G. (1979): Le Néogène en Tunisie septentrionale. Position géotectonique et métallogénie. II. Tectonique du néogène. Geologica Balcanica 9, 4. Sofia 1979. PP. 59–76.
VILA, J. M. (1980): La chaine alpine d'Algerie orientale et des confins algéro-tunisiens. Thèse Es. Sciences. Université P. et M. CURIE. Paris VI. 639 P.

Comparative Metallogeny of Siderite Deposits

By W. Pohl*)

Zusammenfassung

Es wird eine Aufstellung der genetischen Lagerstättentypen gegeben, in denen das Mineral Siderit als Hauptbestandteil auftritt. Im Mittelpunkt der Diskussionen stehen die sogenannten metasomatischen Lagerstätten vom Typus Erzberg und Bilbao. Ihr tektonisch-paläogeographisches Milieu ist das der Vorländer oder Plattformen. Chemische Indikatorelemente für ihre Herkunft gibt es kaum; der REE-Gehalt ist im Falle des Erzberges niedrig. Es fehlen weltweite chemische Daten für die Erkenntnisse der Genese.

Summary

A comparison of metallogenetic features of siderite deposits must remain limited to general features, as a broad base of modern geochemical and complementary data is not yet available.

Nearly all larger siderite deposits have traditionally been thought to be of hydrothermal origin, including veins, volcano-sedimentary and "metasomatic" types. The latter are still an important source of iron ore in many countries, but their metallogeny remains quite enigmatic.

In metasomatic siderite deposits the iron appears to be foreign to the original sedimentary environment of the enclosing carbonates, and is typically thought to have been introduced into its present location a long time after deposition of the country rocks. Precambrian to Tertiary deposits of this type are known, and they occur in unmetamorphosed to amphibolite facies metamorphic rocks.

Their alpidic plate tectonic setting is not related to active subduction zones, but appears restricted to continental crust undergoing tectonic strain, and is

---

*) Institut für Geologie und Paläontologie, Technische Universität, Postfach 3329, D-33 Braunschweig, Germany.

variously accompanied by deep-seated basic intrusions or regional metamorphism; typically, however, the source of suspected hydrothermal solutions is unknown.

Non-hydrothermal genetic models for metasomatic siderite deposits invoke variously processes related to early or late diagenesis, karst-formation, or metamorphic mobilization affecting primary iron concentrations. Only broad comparative investigations of these ores and their country rocks with modern geoscientific methods will contribute towards a better understanding of their origin.

Introduction

At a worldwide scale actual mining of siderite deposits for iron ore is rather limited. In the Mediterranean-Alpine realm, however, siderite mines are more frequent as a locally or even nationally important source of iron. This economic interest, although of course secondary to oxidic iron ores, encourages continuing metallogenetic research.

The present review is conceived to summarize existing data on major siderite deposits and occurrences in the area covered by IGCP 169, to discuss their plate tectonic setting, and to encourage further geochemical and petrological work, supported by limited new data.

The Rôle of Siderite in Genetic Types of Iron ore Deposits

Iron ores occur in nearly all known geological environments, ranging from terrestrial bogs, rivers and lakes through shallow to deep seas. They are associated with many different types of magmas in plutonic and volcanic settings, forming within the magmas as an immiscible fluid or in their surrounding as a product of degassing and/or thermally introduced fluid convection. Many iron ore deposits, however, are not attributable to a clearly defined environment, and in such cases genetic explanations often range from telethermal "magmatic" through metamorphogenic to diagenetic fluids or even solutions derived from weathering rocks. Many siderite deposits belong to this enigmatic group, and command therefore a special scientific interest.

Because of the relatively narrow physico-chemical stability field of siderite in natural environments (Garrels & Christ, 1965; Seguin, 1978; Frost, 1979) economic iron carbonate deposits are less frequent than oxidic ores (see Table 1). Siderite is practically absent from magmatic and contact-metasomatic iron ores, as its low dissociation temperature (~ 450° C) precludes formation from melts except under very high pressures (Wyllie, 1966).

In nearly all types of sedimentary iron ore deposits siderite is ubiquitous, either as a primary or as a diagenetic constituent. In Precambrian banded iron formations (BIF), in continental-sedimentary settings, and even in marine sedimentary deposits ("Minette-type" in the widest sense) siderite is, however, hardly ever considered to be economic iron ore.

Nevertheless it should be noted here that siderite is a geologically important constituant in Paleozoic (Algeria, CSSR, France, Lybia, Maroc, Spain) and in Mesozoic (Lorraine/France, Germany) marine sedimentary iron ore deposits. Although karst-formation is often invoked for the origin of siderite deposits in carbonates there is only one larger undisputed example: The NE-Bavarian Cretaceous district (PFEUFFER, 1983; GUDDEN, 1984). There, fluvial transport of iron into non-marine karst-lakes and its deposition under oxidic and locally slightly reducing conditions has been demonstrated convincingly (GUDDEN, 1984; RUPPERT, 1984).

Most siderite deposits of present or former economic significance have traditionally been called "hydrothermal". Those in direct relationship and proximal to submarine volcanism (the volcano-sedimentary "Lahn-Dill type") can be compared with recent Fe-carbonate precipitation near degassing volcanoes (for example Santorini/Greece: PUCHELT et al., 1973). Rather distal situations (Vareš/Yugoslavia: JANKOVIĆ, 1967) of submarine exhalations related to volcanism are not infrequent. The same applies to siderite veins ("Siegerland-type"), which are of very local economic significance only. The epigenetic character of these veins, the often pronounced structural control, and geochemical data (Siegerland: STAHL, 1971) clearly indicate the hydrothermal origin. The source of the warm solutions, however, obviously need not be a magmatic body; metamorphic or deeply circulating formation waters could also be invoked.

The most disputed, economically important, and still enigmatic siderite deposits are those of the "hydrothermal"-metasomatic type. They comprise typically irregularly shaped ore-bodies within marine platform carbonates with conspicuous replacement textures, but also associated veins, stratiform lenses and sideritized breccias. Of the important examples cited in Table 1, however, probably not one single deposit has remained undoubted as being of hydrothermal origin.

Within this group, two sub-types are often differentiated (BOTTKE, 1981): The Erzberg-type (siderite only being mined), and the Bilbao-type (mainly oxidic iron ore mined, although most certainly derived by weathering of Fe-carbonates).

Metasomatic siderite deposits occur in quite different geological settings: non-metamorphic folded platform carbonates (Ouenza/Algeria), non-metamorphic faulted marine carbonates (Hüggel/NW-Germany), epimetamorphic carbonates (Erzberg/Austria) and amphibolite-fazies metamorphic carbonates (Batère/France). The sideritized carbonates may be dolomite or limestone, and include reef, lagoon and basinal fazies. Typically, the siderite bodies do not reflect any lateral or vertical facies zoning, which is so characteristic for nearly all sedimentary iron ores (JAMES, 1954; BORCHERT, 1960). Many are surrounded by irregular shells of ankeritic carbonates, but direct siderite-limestone or -dolomite contacts are frequented.

Hydrothermal-Metasomatic Siderite Deposits?

Genetic models

Traditionally, warm ascending iron-rich solutions were thought to have replaced limestone and/or dolomite with siderite; that this may have happened before,

Table 1. Rôle of siderite in genetic types of iron ore deposits (data from Walther & Zitzmann, 1977)

Genetic type of iron ore deposit	Occurrence siderite versus oxidic ores	Economic rôle of sid.	Important examples of siderite deposits
Magmatic	absent	–	–
Contact Metasomatic	very rare	–	–
Hydrothermal veins	equivalent	small, only local	○ Rudňani/CSSR, ○ Siegerland/FRG, △ Cercal do Alentejo/Port., ○ Madaras/Rom.
Metasomatic*)	predominant	locally important	● Ouenza/Alg., ● Eisenerz/Austria, ● Kremikovci/Bulg., ○ Batère/France, ○ Hüggel/FRG, ● Rudabánya/Hung., ● Shamsabad/Iran, ○ Delnita/Rom., ● Bilbao/Spain, □ Sierra Menera/Spain, ● Djerissa/Tun., ● Deveci/Turkey, □ Bakal/USSR, ● Ljubija/Jug.
Volcano-sedimentary	less frequent	only local	● Nisna Slaná/CSSR, ○ Ghelar/Rom., ○ Teliuc/Rom., ● Tajmište/Yug., ● Vareš/Jug.
Marine-sedimentary	less frequent	rather small	□ Gara Djebilet/Alg., ● Barrandian/CSSR, ● Normandie/France, □ Lorraine/France, □ Gifhorn/FRG, □ Salzgitter/FRG, □ Shatti Valley/Libya, ○ Ait Amar/Maroc., ○ Leon/Spain
Continental-sedimentary	rarely significant	unusual	● Ruhr/FRG, □ Lisakovsk/USSR
Karst	rarely significant	unusual	● NE-Bavaria/FRG
BIF	often present but secondary	–	–

Size of deposits: △ < 10 Mt ○ 10–100 Mt ● 100–1000 Mt □ > 1000 Mt of ore in virgin deposit

*) The existence of this type of genesis is now often disputed

during or after lithification of the country rocks has already been stated by REDLICH (1904). The source of such solutions may conceivably be quite variable:

- fluids derived from cooling deep intrusions of acidic, intermediate or basic magmas;
- degassing of subvolcanic magmatic bodies;
- intrusive magmatic bodies creating hydrothermal convection cells;
- ascending metamorphogenic fluids;
- ascending deeply circulating formation waters.

Rarely can the mineralizing solutions unequivocally be attributed to one of these classes. The only deposits of the metasomatic type with a plausible connection to a magmatic body appear to be the ones in Zechstein limestones about 5 kms above the top of the basic Cretaceous Bramsche intrusion (STADLER, 1971; WALTHER, 1983). Ordinarily, however, such links are quite cryptic and disputable.

Apart from the prevailing model of a hydrothermal-metasomatic origin (DAVYDENKO, 1972; WALTHER & ZITZMANN, 1977; BOTTKE, 1981; PETRASCHECK, 1982), various other processes have been invoked for the origin of different deposits of this type:

- volcano-sedimentary
  (Erzberg: lately again BERAN & THALMANN, 1978)
- marine-sedimentary, remobilization by subsurface karst formation
  (Sierra Menera/Spain: FERNANDEZ-NIETO et al., 1981)
- karst trapping continent-derived Fe
  (Djerissa/Tunisia: MAHJOUBI & SAMAMA, 1980)
- descending formation waters
  (Glamorgan/U.K.: SLATER & HIGHLEY, 1976)
- laterally migrating formation waters.

Later diagenetic or metamorphic remobilization are often an important aspect of some of these models. These recent trends of genetic thinking about metasomatic siderite deposits have been discussed succinctly and put into a regional metallogenetic perspective for Central and SE-Europe by W. E. PETRASCHECK (1982).

Geochemical investigations

Associated trace elements including REE, fluid inclusion chemistry (and thermometry) and stable isotope ratios are increasingly used as genetic indicators for rocks and mineral deposits. The regional and methodic coverage of metasomatic siderite deposits, however, is too limited to allow general conclusions.

Published analyses of the exploited ore (WALTHER & ZITZMANN, 1977) provide a very limited spectrum of elements only, which is little promising for genetic interpretation. On this base, a diagram linking the size of deposits or districts with %P is presented in Fig. 1. From this one might tentatively conclude, that deposits traditionally classed as hydrothermal have P-contents typically below 0.1%, while marine sedimentary and the NE-Bavarian karst ores range from 0.3–1.4%P. Volcano-sedimentary deposits vary from very low to very high P-contents.

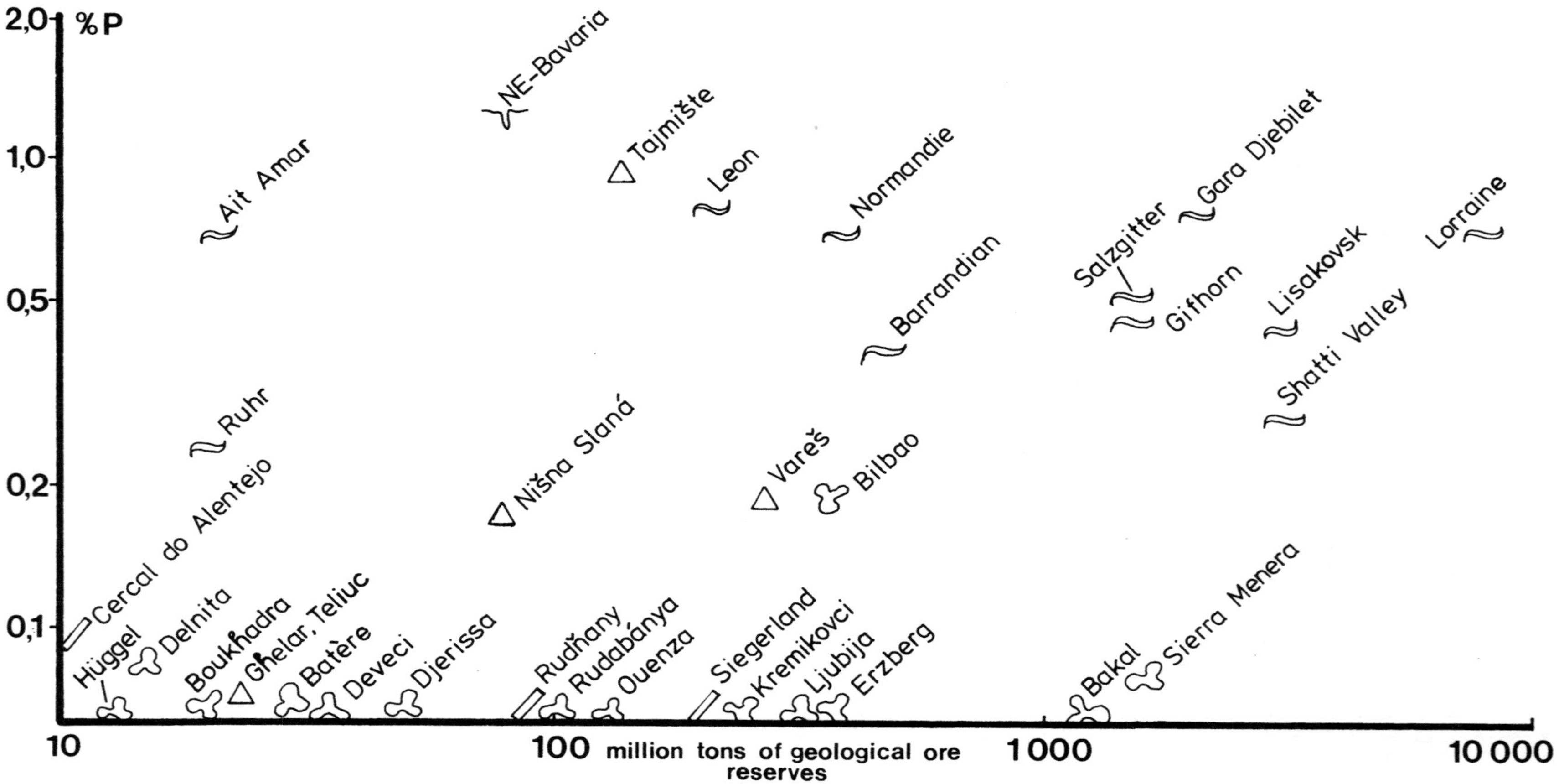

Fig. 1. Average P-contents versus size of major siderite deposits (data from Walther & Zitzmann, 1977).

The siderite ores are usually monomineralic, and only traces of sulfides (pyrite, Cu, Pb, Zn, As, Sb, Hg) and baryte ± quartz are found in small, secondary veinlets. Mg, Mn, and a number of trace elements (Co, Sc, Ni etc.) in siderite have been determined by DOLEZEL & SCHROLL (1979) to characterize siderites of various sedimentary and hydrothermal environments. The results are not clearly conclusive, as stated by the authors themselves (ibid.).

Other methods mentioned above have only occasionally been applied to siderites, and accordingly the data base concerning REE-spectra, fluid inclusions and stable isotopes is too sketchy to allow meaningful comparison. Tentatively the REE-contents of two siderite samples from Erzberg/Austria have been determined (see Fig. 2). They are characterized by a low total REE-content, lower light REE (probably due to the restricted substitution of LREE for Fe), negative Ce and Eu-anomalies (the first either because of a marine environment, or the passage of the hydrothermal solutions through an oxidic rock mass; the second not surprisingly indicating low Eh at deposition—MÖLLER et al., 1976; MORTEANI et al., 1982). By itself, and compared with the data from Santorini (PUCHELT et al., 1973), these results remain inconclusive, especially without comparative analyses of the country rocks. They do not exclude a hydrothermal origin, however.

The Alpidic Plate Tectonic Setting of "Hydrothermal"-Metasomatic Siderite Deposits

Obviously it is of great interest to examine the plate tectonic setting of metasomatic siderite deposits. As all models of Paleozoic and Precambrian plate tectonics in the project area are still highly speculative (for ex. East Alpine Paleozoic: POHL, 1984), we exclude deposits of pre-Mesozoic age from the following discussion.

Unfortunately the age attribution of some of the remaining eleven deposits (see Fig. 3) is based on quite circumstantial arguments (for ex. Erzberg/Austria: the "geologically" indicated Middle-Upper Cretaceous age may only reflect metamorphic mobilization—see W. E. PETRASCHECK, 1982, for discussion). The age of mineralization is so difficult to establish, because the sideritized carbonatas as a rule are older or even much older than the event (orogeny, faulting, intrusion etc.) which is interpreted to be responsible for the rise of hydrothermal solutions. As at Ouenza/Algeria for example (THIBIEROZ & MADRE, 1976) it is, however, characteristic for most of these deposits, that the iron is foreign to the original sedimentary environment of the country rocks. In view of this situation, the traditional geologically indicated age attribution of the mineralization in the various deposits has been adopted for the purpose of this paper (as depicted in Fig. 3).

In a very simplifying approach, three tectonic environments are differentiated to examine the alpidic plate tectonic setting of "hydrothermal"-metasomatic siderite deposits (see Table 2):

- a folded foreland of Mesozoic platform carbonates,
- a zone of continental collision and/or crustal stacking, and a

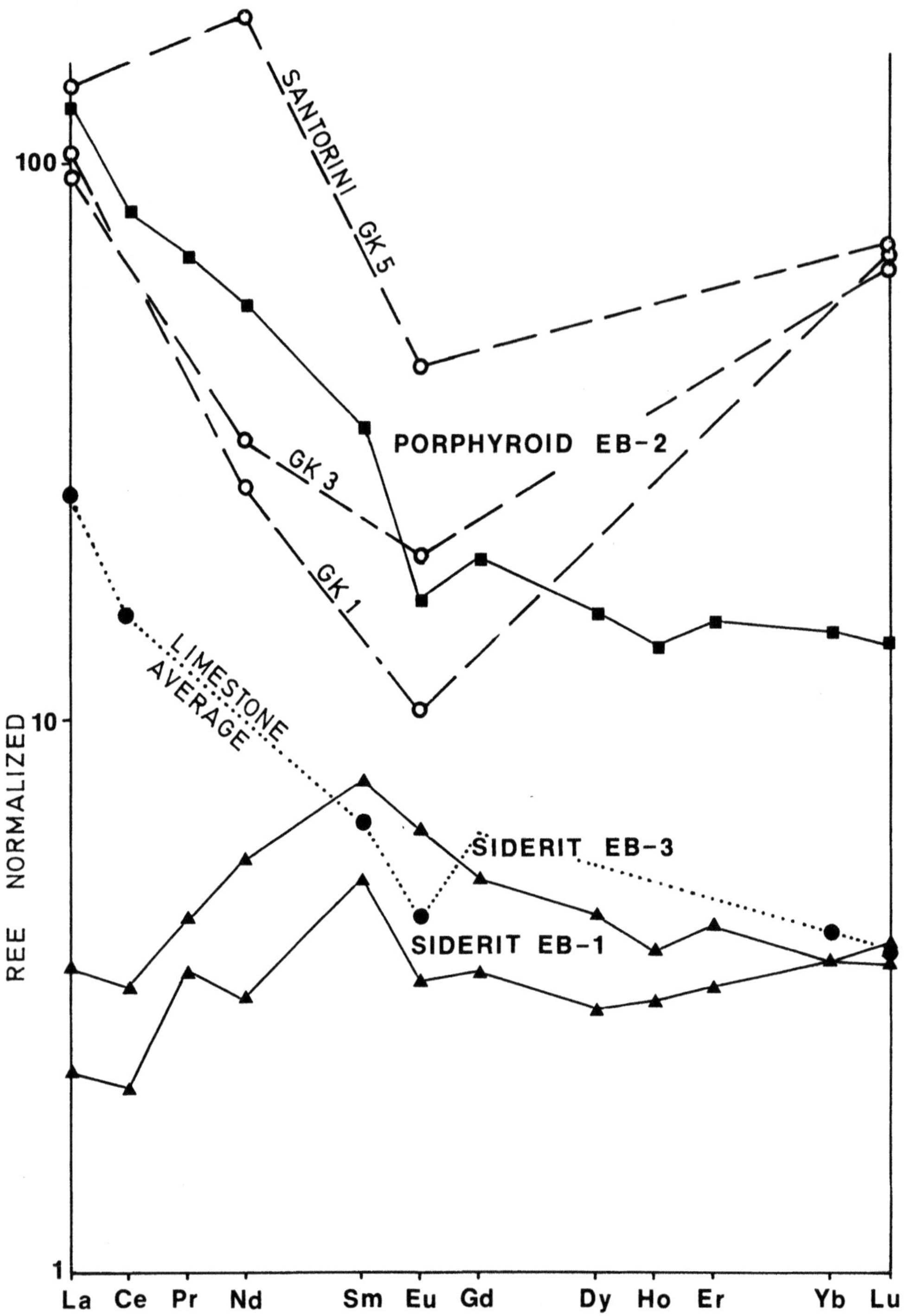

Fig. 2. Chondrite-normalised REE patterns in siderite and quartz-porphyry (Erzberg/Austria, analysed by Geomet Services, England 1983) compared with recent Fe-sediment from Santorini (from PUCHELT et al., 1973) and the limestone average (RONOV et al., 1974).

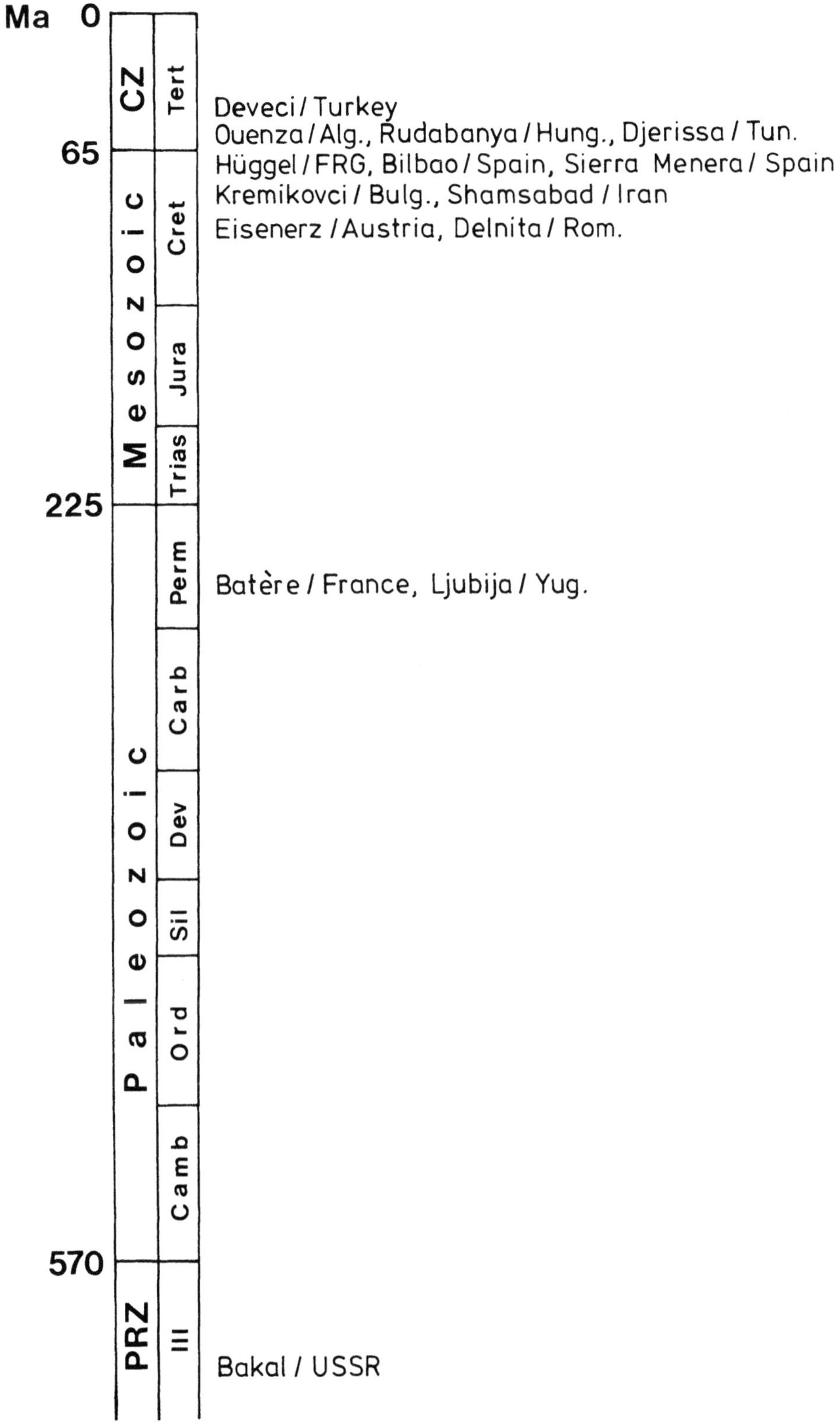

Fig. 3. Traditional age attribution of the formation of "hydrothermal"-metasomatic siderite deposits.

Table 2. Alpidic plate tectonic setting of "hydrothermal metasomatic siderite deposits (accepting the traditional age attribution of their formation)

Folded foreland (Carbonate platform)	Zone of continental collision or crustal stacking	Blockfaulted foreland (Carbonate platform)
Ouenza/Alg. (+ Ba, Cu, Zn, Pb, As)	Rudabanya/Hung. (+ Ba, Cu)	Hüggel/FRG (+ Cu, Ba, Pb, Zn)
Djerissa/Tun. (+ Cu, Pb, Zn, As)	Deveci/Turkey (+ Pb, As)	Sierra Menera/Spain (+ Ba)
Bilbao/Spain (+ Cu)	Kremikovci/Bulg. (+ Ba, Pb, Zn, Cu, Ag)	
	Eisenerz/Austria (+ Cu, Hg, Sb)	
	Shamsabad/Iran (+ Ba, Cu, Pb, Zn)	
	Delnita/Rom.	

– blockfaulted foreland, whose general nature during the supposed time of mineralization is a shallow marine (carbonate) platform, the sideritized rocks, however, being of Paleozoic age.

Far reaching conclusions on this base are certainly not justified. Keeping the above outlined limitations in mind one may state that the characteristic environment of metasomatic siderite deposits are not active subduction zones, but rather ± ordinary continental crust under tectonic stress possibly accompanied by magmatic or metamorphic heat domes.

Conclusions

Siderite occurs in a wide range of geological environments, but former or present economic ore bodies are restricted to volcano-sedimentary, hydrothermal vein and "hydrothermal"-metasomatic types of deposits, with very few exceptions only. In many respects similar to the latter, the clearly karst-related NE-Bavarian siderite-goethite district commands a special interest for comparative purposes.

The precise origin of the "hydrothermal"-metasomatic siderites in carbonates remains enigmatic. Deposits of this type have been formed from the Precambrian to Tertiary times, and they occur in non-metamorphosed to amphibolite fazies metamorphic rocks. The time of sideritization is typically thought to be much later than the original deposition of the carbonates, and the iron is obviously foreign to the original sedimentary environment. If of hydrothermal origin, many of these deposits must be called "telethermal"—a good example being the orebodies in Zechstein carbonates some 5 kms above the top of the Cretaceous Bramsche massiv/NW-Germany.

Of course, the sources of the iron are not necessarily fluids directly derived from cooling magmas, but may well be diagenetic, or deeply circulating formation waters, or metamorphic fluids. In addition, many different models based on non-hydrothermal sources have been proposed for various deposits of this type, but these are at present at least as vague in explaining all features observed as are the traditional hydrothermal-metasomatic ones.

Geochemical data on these ores are still insufficient to allow unequivocal genetic conclusions, but more modern, comparative research on their trace element contents, REE-spectra, fluid inclusions, and stable isotope ratios including country rocks is expected to enhance their understanding considerably. Additional genetic criteria may be derived from Mössbauer-effect studies (Amigo & Fortune, 1981).

The plate tectonic setting of those of the metasomatic siderite deposits thought to have been formed during the Alpidic orogenic cycle is apparently not related to active subduction zones. They occur in the zone of continental collision and crustal stacking as well as in the folded and unfolded foreland. Their environment may be classified as ± ordinary continental crust, although under active stress, which is probably accompanied by magmatic or metamorphic heat domes producing large fluid convection cells.

Acknowledgements

I wish to thank the Austrian IGCP-committee for travel grants extended to me over nearly 10 years for participation in IGCP-projects 3, 164 and 169. This was, of course, only possible through Prof. DDr. W. E. Petrascheck's never-ceasing active help and interest, and to him I offer my most sincere thanks.

References

Amigo, U. M., & Fortuné, J. P.: Mineralogical application of the Mössbauer effect to the study of siderites from Batère (France); N. Jb. Miner. Mh., 1981, *5,* 237–244, Stuttgart 1981.

Beran, A., & Thalmann, F.: Der Bergbau Radmer-Buchegg – ein Beitrag zur Genese alpiner Sideritlagerstätten. Tschermaks Min. Petr. Mitt. *25,* 287–303, 1978.

Borchert, H.: Genesis of marine sedimentary iron ores. Trans. Inst. Min. Metall. *69,* 261–279, 1959/60.

Bottke, H.: Lagerstättenkunde des Eisens. Glückauf V. Essen 1981.

Davydenko, Y. A.: Relationship of the siderite ore of Bakal in the Urals to bedded rocks. Doklady Akad. Nauk. SSSR *206,* 160–163 (USA) 1972.

Dolezel, P., & Schroll, E.: Beitrag zur Geochemie der Siderite in den Ostalpen. Verh. Geol. B.-A. Wien, Vol. 1978, *3,* 293–299, 1979.

Fernandez-Nieto, C., Fernandez-Rubio, R., Gutierrez Elorza, M., & Arrese Serrano, F.: Papel de la karstificación en la génesis de los yacimientos de hierro de Sierra Menera (Ternel y Guadalajara). Buletin Geol. Min. *92–2,* 127–140, 1981.

Frost, B. R.: Mineral equilibria involving mixed volatiles in a C-O-H-fluid phase: The stabilities of graphite and siderite. Am. J. Sci. *9, 279,* 1033–1059, 1979.

GARRELS, R. M., & CHRIST, C. L.: Solutions, minerals and equilibria. Harper & Row, New York 1965.

GUDDEN, H.: Zur Entstehung der nordost-bayrischen Kreide-Eisenerzlagerstätten. Geol. Jb. *66 D*, 3–49, Hannover 1984.

JAMES, H. L.: Sedimentary facies of iron-formation. Economic Geology *49*, 235–293, 1954.

JANKOVIĆ, S.: Wirtschaftsgeologie der Erze. Springer-Verlag, Wien 1967.

MAHJOUBI, H., & SAMAMA, J. C.: Modèle de concentration supergène d'amas sidéritiques: le cas du Jebel Jerissa, Tunisie. Congr. Géol. Int. 26/1980, *3*, 963, 1980.

MÖLLER, P., PAREKH, P. P., & SIMON, P.: Seltene Erden als geochemische Indikationen für die Genese von Fluorit und Calzit auf Gang- und Kluftlagerstätten im Weserbergland (Nordwestdeutschland) und benachbarten Gebieten. Geol. Jb. *D 20*, 77–112, 1976.

MORTEANI, G., MÖLLER, P., & SCHLEY, F.: The Rare Earth Element Contents and the Origin of the Sparry Magnesite Mineralizations of Tux-Lanersbach, Entachen Alm, Spiessnägel, and Hochfilzen, Austria, and the Lacustrine Magnesite Deposits of Diani-Kozani, Greece, and Bela Stena, Yugoslavia. Econ. Geol. *77*, 617–631, 1982.

PETRASCHECK, W. E.: Introduction. Pp 1–12. *In:* Mineral Deposits of Europe, Volume 2: Southeast Europe. IMM, London 1982.

PFEUFFER, J.: Zur Genese der Eisenerzlagerstätten von Auerbach–Sulzbach–Rosenberg–Amberg (Oberpfalz). Geol. Jb. *D 64*, 3–69, Hannover 1983.

POHL, W.: Metallogenetic evolution of the East Alpine Paleozoic basement. Geol. Rdsch. *73*, 1, 131–147, 1984.

PUCHELT, H., SCHOCK, H. H., & SCHROLL, E.: Rezente marine Eisenerze auf Santorin, Griechenland. Geol. Rundschau *62*, 786–803, 1973.

REDLICH, K.: Über das Alter und die Entstehung einiger Erz- und Magnesitlagerstätten der steirischen Alpen. Jb. Geol. R.–A. *53* (1903), 285–294, Wien 1904.

RUPPERT, H.: Physiko-chemische Betrachtungen zur Entstehung der Kreide-Eisenerz-Lagerstätten in Nordost-Bayern. Geol. Jb. *D 66*, 51–75, Hannover 1984.

SEGUIN, M. K.: Applications géologiques des resultats expérimentaux du système Fe-C-O-S (+ H_2O). Chem. Geol. *21*, 3–4, 267–290, 1978.

SLATER, D., & HIGHLEY, D. E.: The iron ore deposits in the United Kingdom of Great Britain and Northern Ireland. Pp 393–409. *In:* WALTHER & ZITZMANN (eds.), Iron ore deposits of Europe, Vol. I, Hannover 1977.

STADLER, G.: Die Vererzung im Bereich des Bramscher Massivs und seiner Umgebung. Fortschr. Geol. Rheinld. Westf. *18*, 439–500, Krefeld 1971.

STAHL, W.: Isotopen-Analysen an Carbonaten und Kohlendioxid-Proben aus dem Einflußbereich und der weiteren Umgebung des Bramscher Intrusivs und an hydrothermalen Carbonaten aus dem Siegerland. Fortschr. Geol. Rheinld. Westf. *18*, 429–438, Krefeld 1971.

THIBIEROZ, J., & MADRE, M.: Le gisement de sidérite du Djebel Ouenza (Algérie) est controlée par un golfe de la mer aptienne. Bull. Soc. Hist. nat. Afr. Nord Alger *67*, 3 & 4, 125–149, 1976.

WALTHER, H. W., & ZITZMANN, A.: The iron ore deposits of Europe and adjacent areas. 2 Vols., Schweitzerbarth V., Hannover 1977.

WALTHER, H. W.: The Alpidic Metallogenetic Epoch in Central Europe north of the Alps. Pp 313–328. *In:* SCHNEIDER, H. J. (ed.): Mineral Deposits of the Alps and of the Alpine Epoch in Europe, Berlin-Heidelberg 1983.

WYLLIE, P. J.: Experimental studies of Carbonatite problems: The origin and differentation of Carbonatite magmas. Pp 311–352. *In:* TUTTLE, O. F., & GITTINS, J. (eds.): Carbonatites, Interscience P., New York 1966.

Coordination of diagnostic features in ore occurrences of base metals in dolomites and limestones (Final report of IGCP-Project 6)

By L. KOSTELKA*), I. CERNY**), and E. SCHROLL***)

Zusammenfassung

Die triassischen Blei-Zinkerz-Lagerstätten im mediterranen Raum können in zwei Gruppen eingeteilt werden: in endogene mit der höheren Bildungsenergie und in exogene mit der niedrigeren Bildungsenergie. Die zweite Gruppe überwiegt im betrachteten Bereich; ihre Lagerstätten sind an die Randzonen von transgredierenden Flachmeeren gegen Hochgebiete gebunden (Fig. 1) und zumeist in Kalk-Dolomitserien auftretend. Form und Gehalt der Erzkörper variieren sehr. Die Herkunft der Metalle ist meist diskutabel; submarine Tiefenbrüche, Durchpausen älterer Lagerstätten des Untergrundes oder Einschwemmung von Verwitterungsprodukten der Randgebiete kommen in Frage. – Die Isotopenverhältnisse werden ausführlich diskutiert. Die Blei-Isotopen der alpinen Lagerstätten weisen auf älteres Gesteinsblei, doch fehlt noch eine regionale vergleichende geochemische Bearbeitung der Lagerstätten des gesamten Mediterranraumes. Die Mannigfaltigkeit der Erscheinungen läßt trotz – oder wegen – der verschiedenartigen beschriebenen Fälle noch keine eindeutigen Merkmale für eine genetische Diagnose ableiten, aus der generelle Leitlinien für die Prospektion gefolgert werden können.

1. Introduction

The geographical area of research work stimulated during the period of the project has been limited to ore occurrences of the Mediterranean orogen and the bordering platforms of the European and African continents including Asia Minor. The

*) Prof. Dr. Ludwig KOSTELKA, A-9020 Klagenfurt, Tristangasse 34.

**) Dr. Immo CERNY, A-9530 Bad Bleiberg, Bleiberger Bergwerksunion.

***) Prof. Dr. Erich SCHROLL, A-1031 Wien, P.O.B. 8, Geotechnisches Institut der BVFA Arsenal.

observations are restricted mainly to ore occurrences of base metals hosted in Mesozoic, especially Triassic dolomites and limestones. However, extensive investigations by AMSTUTZ and his team (Heidelberg) about Pb-Zn mineralizations in carbonate rocks in South America, Peru and Argentina have been included in this report, too (AMSTUTZ, 1978; DAHLHEIMER, 1980; LEVIN et al., 1973; HIRDES et al., 1978; SAMANIOGO, 1982; SUREDA et al., 1984).

Comprehensive works on Eastern Alpine deposits were composed by ASSARETO et al. (1977 a, b, c), BRIGO et al. (1977), BRIGO et al. (1978), SCHULZ et al. (1977), OMENETTO (1978) and others. From local and special studies the papers of OMENETTO et al. (1977), BRUSCA (1978, HAGENGUTH et al. (1982), HAGENGUTH (1983) may be mentioned.

Beyond the region of the Eastern Alps research work about deposits in general should be mentioned, such as about strata-bound mineralizations in the Little Carpathians (ILAVSKY, 1979; KANTOR, 1977), about the polymetallic deposits of the Western Balkan by MINČEVA-STEFANOVA (1978), about the Swiss part of the Western Alps (SCHMIDT et al., 1983), and about strata-bound deposits of the Iberian Peninsula by FONTBOTÉ et al. (1983) and CASTROVIEJO (1982).

In addition to these there are comprehensive publications from Algeria and Tunisia (FUCHS, 1978; VASILEFF et al., 1977). In the Mediterranean area however a systematic investigation and listing of strata-bound deposits in carbonate rocks has still to be carried out, as for instance in Libya and Turkey, or have been set in motion in Greece. About Triassic ore mineralizations in connection with acid effusive rocks in the Dobrogea (Romania) is reported by VLAD (1978).

Investigations of lead-zinc occurrences hosted in Mesozoic rocks and covering the Hercynian Platform in France and Germany, are especially valuable for the comparison with the ore mineralizations of the Mediterranean geosyncline (AMSTUTZ et al., in press).

There are numerous publications about the mineral deposits in Upper Silesia available too, which give reason for new genetical interpretations (DZULINSKI et al., 1980; NIEC, 1980; PAWLOSKA et al., 1980; GRUSZINSKY, 1982).

Mainly as a result of organization problems methodical geochemical investigations were more or less confined to the Eastern Alps and for obvious reasons the large deposits of Bleiberg-Kreuth (Austria) and Mežica (Yugoslavia/Slovenia) were investigated most carefully.

Further contributions and special studies can be subdivided into the fields of general geological observations, lithogenesis and dolomitization, ore structures, mineralogy and geochemistry including isotope geochemistry.

Although considerable progress was achieved by the studies and correlated discussions within the working groups of the project, it must be confessed that the target, as outlined in the title, could not be attained. The complexity of the factors forming a single deposit and the varying significance of the heterogeneous features of the deposits prevented the set up of a system of features which could enable a geologist to discover a new deposit with a reasonable risk by following the schedule. This aim was too ambitious and a lucky chance will furthermore remain a most welcome compilation to careful geological studies while prospecting for a deposit.

2. Further contributions

2.1. General geological contributions

For the formation of hydrothermal ore mineralizations the following facts of a model are accepted:

a) a source, resp. source rock supplying the metals,

b) a hydrous medium for leaching and transport,

c) geological traps combined with geochemical barriers.

The leaching process is primarely controlled by the temperature and the chemical state of the leaching medium, especially solved gases or/and the salt content, mainly NaCl being in question.

Discussing the superpositions for the lead-zinc ore mineralization hosted in carbonate rocks, it is evident that at least two genetical groups can be distinguished:

a) ore deposits connected and formed by sedimentary or diagenetic processes,

b) ore deposits more or less directly related to volcanic or magmatic events.

The two types of deposits are showing significant differences with respect to the energy level which can be observed in connection with the ore occurrences. The lead-zinc deposits which are related to endogen processes have been formed at a much higher energy level, than the deposits produced by exogen events. The term energy is considered for the temperature level of the leaching process depending on energetically controled geotectonic events. Of course deposits exist which show features pointing to both types. It seems to be significant that the ambiguous observations correspond to the position of these deposits in the whole system between endogen and exogen dominated by the ranges of distribution. That apparently is the reason for the nearly endless discussions about syngenesis and epigenesis.

The different geoenergetic conditions in platforms, along their coastal margins and in areas of developing geosynclines, subduction and rifting zones, cause the conditions for different initial processes and the development of multiple ore mineralizations, ranging from the sedimentary preenrichment on platforms up to black smokers in zones of rifting.

A significant contribution to the understanding have been given by members of the working group, such as FUCHS (1981a) who set up a genetical system of base metal mineralizations hosted in carbonatic sediments and related to regional geotectonical events.

The region of the project had been directed originally to the area of the Mesozoic Mediterranean orogen including the bordering epicontinental sedimentary cover of adjacent platforms. All the mentioned geotectonic and geoenergetic situations are found in this part of the earth crust (Fig. 1). However, the most of the contributions to the project are related to the European platform and the Eastern Alps. Therefore the emphasis has been on the exogen type of ore mineralizations.

In the geotectonic stage of transgression epicontinental platforms are covered by the weathering products of older crystalline complexes forming horsts. If the climatic conditions exclude lateritization processes, the exogenic processes at first

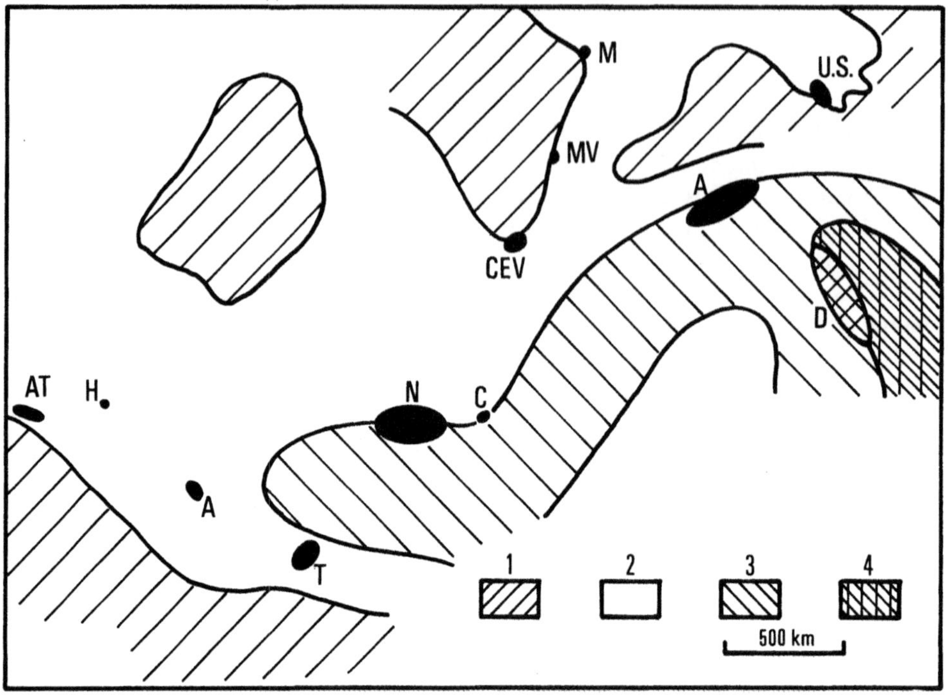

Fig. 1. Lead-zinc ore mineralizations of Triassic age in the Western Mediterranean system (after a schematic representation according to FONTBOTÉ et al., 1983)

Abbreviations:
1 = Horsts
2 = Transgressed platforms
3 = Thinned and faulted continental crust
4 = Oceanic crust of the Thetis

Ore districts:

AT, H	Atlas Mountains, Morocco	N	Betic Cordillera
A	Atlas Mountains, Algier	CEV	Cevennes
T	Atlas Mountains, Tunis	US	Upper Silesia
ALP	Eastern and Southern Limestone Alps	MV	Morvan
D	Dinarides	M	Mechernich-Maubach

supply clastic sediments. Chemical sediments follow in form of evaporitic intercalations and carbonatic rock series. The preenrichment of base metals can became apparented as well in the clastic sediments as in the superimposed carbonatic rock series or in intercalated pelitic schists. Generally geochemical anomalies can be found in the lowest carbonatic rock strata of mostly dolomitic composition. In many cases a widespread halo of metals can be observed distinctly above the Clark values, but richer metal concentrations are rather exceptional. The source in the metal content of the weathered crystalline complexes, such as apical parts of granitic intrusions or preexisting ore deposits (KÖPPEL et al., 1983). Fig. 1 clearly shows the connection of horsts with richer ore concentrations (FONTBOTÉ et al., 1983). Analogous to the Permian copper shale the distribution od lead and zinc is conditioned by the morphology; lead prefers swells and zinc basins (CERNY, 1977). In the carbonatic environment lead and zinc dominates and can be separated completely from copper and silver. For instance DROVENIK et al. (1983) reported that galena was purified by remobilization from some trace elements.

The further concentration of lead and zinc is determined by the geological development of the platform.

If the sediment cover is thin enough, the decomposition of biogenic substances into coal, oil and gas, salt diapirism and the activity of formation waters must be taken into consideration for late diagenetic and epigenetic metal concentrations. Younger magmatic activity in the crystalline basement is mostly an additional factor to exogenic heat (WALTHER, 1983). Tectonic events producing deep faults are the precondition for geothermal systems. Remobilization of metals from the basement were described by SCHNEIDERHÖHN (1953) at first. He called it "tracing of deposits" (Durchpausen). For instance the remobilization of fahlore from Old Paläozoic deposits into the superimposed mesozoic carbonate rocks was mentioned by GSTREIN (1983).

The migration of the elements happens in different ways (ROUTHIER, 1980). There is no doubt, that the evaporitic environment causes the migration of base metals during sedimentation and diagenesis (SCHNEIDER, 1972, 1978, 1980a, b, 1981a, b; FUCHS, 1976; LAGNY, 1980 etc.). Therefore the identification of brine relics in carbonate rocks and in the ore bodies could be used as a geochemical prospecting tool (see below in the following chapter Geochemistry).

The conditions for the formation of lead zinc deposits in the geosynclinal stage are similar to these in the overflooded platforms. Islands originated from the crystalline and Palaeozoic basement, coast-lines and reefs with lagoons are possible. An additional factor is the geosynclinal volcanism. The contribution of the Triassic volcanism to the formation of lead zinc deposits in the Eastern and Southern Alps seems to be still an unsolved problem. The mostly potassium-rich acid volcanites could be preferably the source of lead, barium, thallium or fluorine, but the lead isotope composition of the ore lead doesn't confirm this hypothesis. However, later tectonic processes forming the Alpine orogen have prevented oil formation and activities of formation waters. Only in the Eastern part of the Northern Limestone Alps indications for ore mineralizations being probably related to formation waters, could be found.

According to the platetectonics, the Eastern part of Mediterranean beginning with the Dinarides was influenced by the opening of the Tethys. The lead zinc deposits are often directly related to volcanogenic, resp. magmatogenic processes (ŠTRUCL, 1981). In the North of the Dinarides the first endogenic lead-zinc occurrences have been mined at Zavrh/Slovenia (DROVENIK et al., 1980). The ore occured in silicified and recristallized Ladinian reef limestone connected with hydrothermally altered keratophyric tuff. The high-hydrothermal volcano-sedimentary ores are similar to some ore occurrences of Middle Triassic age in Črna Gora (Montenegro). Other ore deposits are the siderite layers of Vareš combined with base metal mineralizations in the same area, Supleja Stijena, Podrinje etc. Ore structures and textures, mineralogy and geochemistry are significantly different from the exogenic type.

While the exogen type of lead zinc deposits usually occurs in well defined parts of more or less thick sequences of carbonatic rock series, the lithological situation of endogen deposits is mostly an extraordinary one. The alteration phenomena on the surface are often remarkable. Therefore this type is easier to be prospected. Endogen deposits are very often characterized by pretty high grade ores. Generally it can be observed that the limits of variablity of ore deposits related to plutonic or volcanogenic events are somehow smaller than patterns of the exogenic group. Naturally there are also deposits which can be classified as an intermediate type, showing features of both kinds of deposits, exogenic and endogenic ones. Submarine volcanogenic sources can be associated with sedimentary ore depositions in further distance. The same situation applies to hot solutions emanating from feeder faults, for instance very well described by MINČEVA-STEFANOVA (1978) for the ore deposits of the type Sedmočislenici (Stara Planina, Bulgaria).

Considering the suppositons for the formation of lead-zinc deposits in carbonate rocks we have to distinguish between regional and local factors:

The basic regional factor is the presence of the metals, the basic local factor the existence of suitable geological traps to collect the metal compounds.

The sources of the metals resp. their origin can be:

a) Magmatogenic lead and zinc originated during some phases of the geosynclinal stage. Mostly tuffites are the evidence for such events which often are not detectable at all. In that case sea water was very likely the medium for transport.

b) Feeders which have received their metal compounds from the crust mostly by deep reaching tectonic faults or lineaments.

c) Mobilization and re-deposition of lead and zinc pounced from elder deposits in an overlying younger sequence of strata.

d) Decomposition of elder geological units which contained Pb-Zn (or other metals) at least in parts and redeposition of these minerals.

The three suppositons a)—c) may have been efficacious and two or even all three of these phenomena may have been active with various intensity in the same deposit.

Besides the sources and the traps the processes connected with the lithification, resp. diagenesis, generally are of major importance for the formation of an exogen deposit. Locally concentration of metals may have been produced by karst-phenomena.

It is a consequence of these manifold genetical models and the often afterwards overlapping evolutionary components, that the selection of prospecting strategy is much more difficult than in case of deposits, the origin of which is directly detectable from endogen events.

Very likely this system of exogen and endogen deposits may occur in different areas and under varying conditions even in pre-Triassic sequences, but it certainly will be more difficult to distinguish these two types due to alterations originating from their even more complicated history.

Finally we can characterize these two types and can expect in the areas of their distribution that methods selected and adapted to the observed features will lead to optimal results. This may facilitate the set up of a prospecting strategy but not more. The real work still has to be done by geological, geochemical and geophysical methods.

2.2. Lithogenesis and dolomitization

The importance of facies bounderies for ore mineralization was reported by CERNY (1982). The microfacies is as revealing as macro-facial rock structures. In this respect the sedimentological interpretations of facial changes within the carbonate sequence is of vital importance, especially in the paleogeographic connection with reefs as proved by BECHSTÄDT (1875, 1979 a, b) in the Bleiberg-Kreuth deposit.

The importance of dolomitization for ore concentrations has already been mentioned to on several occasions.

A geochemical model for the processes of dolomitization was set up by MÖLLER et al. (1976). An investigation of the process of dolomitization by microprobe analysis was carried out by MINČEVA-STEFANOVA (1981).

2.3. Ore structures

The genetic interpretation of ore structures of this type of deposits was summarized by SCHULZ (1976). It was completed by FONTBONTÉ and AMSTUTZ (1980/1981) with respect to diagenetically crystallized rhythmites.

In the project the genetic conception of Bleiberg-Kreuth was put into question as a result of outdated suggestions and short-time observations question by reason of the comparability of some types of ore structures occuring in the Pb-Zn ore mineralization of Upper Silesia (DZULINSKY et al., 1977 a, b, 1978).

2.4. Mineralogy

Among contributions of ore mineralogy especially the results by MINČEVA-STEFANOVA (1979, 1980) should be emphasized. AMOURI et al. (1978) and DEVIGNE (1980) drew attention to the possible importance of sulphur-bacterial action in the formation of whisker-shaped galena. In the course of other investigations, such micro-crystallite formations were also found in the Eastern Alps (SCHROLL, 1980).

In-situ measurements with secondary mass spectrometry (SIMS) made for the first time in Vienna by PIMMINGER (1983) and investigations with the scanning electronic microscope (SCHROLL, 1981) showed that the insertion of important trace elements in sphalerites (such as germanium, gallium, thallium and arsenic) down to a range which cannot be resolved any more (ca. 1 μ^2), must be present in the form of micromineral phases or domaines, so that a substitution in the sphalerit lattice can be excluded.

2.5. Geochemistry

The geochemical knowledge about carbonate rock hosted lead-zinc ores was presented by SCHROLL (1978) at the beginning of the project. The geochemical characterization of the Bleiberg type in comparison with other carbonate hosted lead-zinc mineralizations has again been discussed by SCHROLL (1983) at the end of the project.

Geochemical investigations of the sedimentary rocks were connected with profiles of mineralized Triassic rock series (KRANZ, 1973, 1976; CERNY, 1977, 1978; HAGENGUTH, 1983). In green marls of lower Ladinian age tuffitic products could be identified (HAGENGUTH, 1983). Also in Raibl volcanogenic intercalations have been studied by JOBSTRAIBIZER, 1981).

The content of chlorine in carbonate rocks and in ore bodies was recently investigated by WOLTER et al. (1983) in the Bleiberg-Kreuth deposit. He reports on indications of halos with considerable brine concentrations in, as well as around the sulphidic ore bodies. This phenomenon can be interpreted as remnant of the transporting solutions. Moreover there are investigations going on whether this phenomenon could be used as a prospecting tool.

The distribution of fluorine in host rock sequences and especially its conspicuous linkage with comparatively few and thin layers indicate that the substance for the primary syndiagenetic formation of fluorine cannot originate from the host rock.

The lateral and vertical occurrence of fluorine anomalies and their spatial connexion with Pb-Zn mineralizations may be used as a proximity indicator which may be valuable for paleogeographic reconstructions. This was proved by the Berlin working team for numerous large deposits of the Eastern and Southern Alps.

The syndiagenetic origin of the fluorite is also proved by its characteristic content of the elements of the rare earth (REE) group, which is roughly the same as in the carbonate sediment. The subsequent diagenetic remobilization of fluorites can be shown by the fractionating in the REE content (SCHNEIDER et al., 1975, 1977; HEIN et al., 1983). MÖLLER et al. (1983) could interpret the distribution of gallium and germanium in sphalerites on the base of a physicochemical model. The results give evidence that the metals of Bleiberg type deposits have been leached at a low temperature range. Besides this, new data on the trace element distribution of galena and sphalerite are available but have only partly been published.

2.6. Isotope geochemistry

2.6.1. Carbon and oxygen isotopes

The investigation of carbon and oxygen isotopes of the carbonatic rocks and the gangue carbonate minerals informs us about the origin of the water participating in the ore forming processes and the origin of the carbon dioxide. Diagenetic processes influence the isotopic composition of oxygen preferably depending on the water-wallrock ratio. Hydrothermal ore fluids or meteoric waters can be recorded.

The distribution of the isotopes of both elements in the carbonates of the Bleiberg type show that diagenetic processes are operating up to late diagenesis (SCHROLL et al., 1978; KAPPEL et al., 1982; DOLENEC et al., 1983).

2.6.2. Sulphur isotopes

Systematical investigations of the sulphur isotope composition of sulphate and sulphide minerals support the evidence that in the Bleiberg type the sulphate is derived from the sea water sulphate and biogenic processes have influenced the sulphide sulphur during the early diagenesis. The remobilization of the sulphidic ores has mostly taken place in an open system (SCHROLL et al., 1983). Other investigations were carried out by DROVENIK et al. (1976, 1980) and PAK et al. (1980), SCHROLL et al. (1983).

The sulphur isotope composition reported from sulphidic ore minerals of the Bleiberg type alone however doesn't allow general statements about the admixture or the full evidence of volcanogenic sulphur. Ore mineralizations of endogenic origin show sulphur isotope distributions, which are distinctly different (SCHROLL et al., 1983). But a range of δ^{34}S-values around zero was found too. In the case of a high temperature source and the existence of feeder faults transitions of δ^{34}S-values to biogenic sulphur can be observed in some distance.

2.6.3. Lead isotopes

Extensive work for a better understandig of the origin of lead ore resulted from investigations of the lead isotopes in the area of the Eastern and Southern Alps. Abundant data of lead isotope compositions are available (KÖPPEL et al., 1976, 1979; KÖPPEL, 1983). Part of these data have to be published in detail. Not only the isotopic composition of the ore lead, but also of rock lead was studied. The resulting data indicate that in deposits of the type Bleiberg the ore lead doesn't correspond with the rock lead. This applies also for some investigated potassium-rich acid volcanites of Triassic age. The ore lead is isotopically very homogenous and must be introduced several times during a short period of time. The ore lead shows a model age of about little more than 300 my and it can be derived from ore lead and rock lead of the Old Paleozoic basement, originating from the Precambrian African platform. Ore lead from lead-zinc deposits of Morocco and Algier is comparable. Lead isotopes from the Hercynean platform and the Penninic area of the Alps are significantly different. This is true in the Alps for most of the lead ores of the Anisian and other Triassic mineralizations deviating in their chemical and/or mineralogical characteristics.

Lead isotope data are reported from the strata-bound polymetallic deposits of the Western Stara Planina Mountains (Bulgaria) (AMOV et al., 1979) and from Upper Silesia (PAWLOWSKA et al., 1980).

Endogen and exogen ore mineralizations can't be distinguished by the aid of lead isotopes alone. Well mixed homogeneous leads are found in both types. Mixing of the magmatic melt with pelitic substances is operating in the same manner. However, radiogenic ore lead is limited to the area of the Alps indicating uneconomic small ore occurrences (KÖPPEL et al., 1983).

2.7. Geochemical characterization

The geochemical characterization (SCHROLL, 1978, 1984) can be of practical importance as it turned out that relations to metal concentration seem to exist, but these relations have to be worked out for certain parts of the Earth crust which to their applicability in specific regions and at different times. The period of time necessary for carrying out our project (6) of the Correlation Programme was too short in relation to the amount of the problems as a whole. Standardized comparative investigations covering the whole Mediterranean area or other parts of the crust could not be achieved. On the basis of systematic geochemical and isotope-chemical data characteristic parameters for the Triassic lead-zinc ores of the crustal area of the Eastern and Southern Alps could be deduced.

Detailed geochemical investigations in the Eastern Alps suggest that in this part of the Earth crust larger Triassic deposits with high metal concentrations ($x \cdot 10^6$ t) are bound to a certain concentration of trace elements in lead-zinc ore, to a certain pattern of distribution of the sulphur isotopes in the sulphides and sulphates and, above all, to a well defined, very homogenous distribution of lead isotopes. These results were for the first time applied with good results in a regional study about prospecting of lead-zinc mineralizations in the Austrian part of the Karawanken mountains (Cerny et al., 1981).

It is necessary to continue this work in this direction, as the importance of this problem for the exploration of base metals is obvious, especially for the developing countries, and it seems likely that with modern methods and concepts the manifold problems can be solved.

3. Conclusions

Although it was not possible to find general "diagnostic features" for the mineralization of the base metals in carbonate rocks on a global scale, it turned out that in principle there do exist two types of deposits: one group which is related to exogen processes and a second group depending directly to endogen events. In the Mediterranean these two types do occur in seperate areas which are marked by their specific geological position. The deposits related to endogen processes are not at all occuring in carbonate rocks only so. They are characteristic for the active orogenic belt in the Dinarides. The deposits depending on exogen processes occur in Central Europe in the pre-Alpine and Alpine geosyncline in North Africa and on the epicontinental platforms as f. g. in Upper Silesia.

The methods of prospecting must be appropriate to the two differing genetical models which are owing to respective features of the deposits.

The mineralizations in carbonate rocks, from late pre-Cambrian onwards, are showing very similar phenomena even if they are different in origin and geologic evolution. Synchronous metallic deposits in carbonate rocks are confined to shallow water sediments with delta-reef- and lagoonal deposits.

On purpose we suggest the term "synchronous" instead of syngenetic or sedimentary, because it is only possible in very rare cases to verify that the ore has been deposited on the bottom of the sea, which would be the supposition when speaking of "sedimentary". Synchronous—as we understand the terminus—may also include the processes of lithification—or the events during diagenesis.

Acknowledgement

Finally we thank all officers and official bodies having supported the IGCP-Project 6, which—although not successful with regard to the title—could achieve considerable progress concerning the global occurrence of lead-zinc deposits and especially those in the Mediterranean.

References

(selected papers related to the project 73/0/6)

AMOURI, M., DEVIGNE, J. P., and FUCHS, Y. (1978): Paleobacteria and lead-zinc mineralization at Bled Zelfane (Central Tunisia). Miner. Deposita. (Berlin), Vol. 13, 183–189.

AMOV, B. G., MINČEVA-STEFANOVA, J., and BALDJIEVA, TS. T. (1979): New data on the lead isotopic composition of galenas from the stratabound polymetallic deposits in the Western Stara Planina Mountain, Bulgaria (in Russian). Geochimija (USSR), *4,* 535–554.

AMSTUTZ, G. C., and FONTNONTÉ, L. (1985): Correlative observations on the lead-zinc-(barite-fluorite) occurrences in the Germanic Triassic lithofacies of Central Europe. Geol. Jb. (Hannover) (in press).

AMSTUTZ, G. C. (1978): Zu einer Metallogenie der Zentralen Anden von Peru (about a metallogeny of the Central Andes of Peru). Forsch. Geol. Paläont., 44/45, 151–158.

ASSERETO, R., JADOUL, F., and OMENETTO, P. (1977 a): Stratigraphia et metallogenesi del settore occidentale del distretto a Pb, Zn, fluorite e barite di Gorno (Alpi Bergamasche). Riv. Ital. Paleont. (Milano), Vol. 83, *3,* 395–532.

ASSERETO, R., BRUSCA, C., GAETANI, M., et al. (1977 b): The Pb-Zn mineralization in the Triassic of the Dolomites; geological history and genetic interpretations. L'Industria Mineralia (Roma), No. 231, 1–32.

ASSERETO, R., BRUSCA, C., GAETANI, M.,and JADOUL, F. (1977 c): The Pb-Zn mineralization in the Triassic of the Dolomites; geological and genetic interpretations. L'Industria Mineralia (Milano), 1–34.

BECHSTÄDT, TH. (1975): Zyklische Sedimentation in erzführendem Wettersteinkalk in Bleiberg-Kreuth (Kärnten, Österreich) (Cyclic sedimentation in the ore bearing Wettersteinlimestone of Bleiberg-Kreuth, Carinthia, Austria). N. Jb. Geol. Paläont. Abh. (Stuttgart), Vol. 149/1, 73–95.

BECHSTÄDT, TH. (1978): Faziesanalyse permischer und triadischer Sedimente des Drauzuges als Hinweis auf eine großräumige Lateralverschiebung des Ostalpins. Jb. Geol. B.-A. (Wien), Vol. 121/1, 1–121.

BECHSTÄDT, TH. (1979): The lead-zinc deposit of Bleiberg-Kreuth (Carinthia, Austria): Palinspastic situation, paleography and ore mineralization. Verh. Geol. B.-A. (Wien), 1978/3, 221–235 (1979), Proceed. 3rd ISMIDA (leoben 1977, 47–61.

BRIGO, L., KOSTELKA, L., OMENETTO, P., et al. (1977): Comparative reflections on four Alpine Pb-Zn deposits. *In:* Time- and Strata-bound Ore Deposits, KLEMM, D. D. et al., (ed.), Springer-Verlag Berlin, 273–293.

BRIGO, L., and OMENETTO, P. (1978): The lead and zinc ores of the Raibl (Cave del Predil) zone in northern Italy. *In:* Ergebnisse der Österr. Projekte des Internationalen Geologischen Korrelationsprogrammes (IDCP) bis 1976. (ZAPFE, H., ed.), Österr. Akad. Wiss. Erdwissenschaftl. Komm., Schriftenreihe Vol. 3, 103–110.

BRUSCA, C. (1978): Il giacimento di Salafossa nel quadro della geologia del Trias delle Dolomiti. Bollettino Assoc. Mineralia Subalpina, Torino, Vol. 16/1, 21–35.

Castroviejo, R. (1982): Strata-bound ores of Spain: Newly discovered features and their interpretations. *In:* Ore Genesis. The State of Art, Amstutz, G. C., et al. (ed.), Springer-Verlag Berlin-Heidelberg-New York, 565–572.

Cerny, I. (1977): Zur Fazies und Blei/Zink-Verteilung im „Anis" der Karawanken. Carinthia II (Klagenfurt), 167/87, 59–78.

Cerny, I. (1978): Geochemie „anisischer" Sedimentgesteine in den Nordkarawanken. Carinthia II (Klagenfurt), 168/88, 55–70.

Cerny, I. (1982): Sulfidvererzung an der Wende Karn/Nor in Karbonatgesteinen der Draukalkalpen. Mitt. Ges. Geol. Bergbaustud. Österr. Wien, *28*, 109–120.

Cerny, I., Scherer, J., and Schroll, E. (1982): Blei-Zink-Verteilungsmodell in stilliegenden Blei-Zink-Erzrevieren der Karawanken. Archiv f. Lagerstättenforschung GBA (Wien), Vol. 2, 15–22.

Cerny, I. (1983): Pb-Zn Erzmineralisation in Dolomitgesteinen der Draukalkalpen (Kärnten, Österreich). *In:* Ore Mobilization in the Alps and SE-Europe (Petrascheck, W. E., ed.), Schriftenreihe der Erdwissenschaftlichen Komm., Österr. Akad. Wiss., Springer-Verlag Wien-New York, Vol. 6, 39–52.

Dalheimer, M. (1980): Konkordante und diskordante Erzführung am Domo de Yauli, Zentral Peru. Geow. Lateinamerika Koll. Heidelberg, 17–18.

Devigne, J. P. (1980): Action paléo-bacterienne dans les mineralisation plombiferes stratiformes. XXVI Int. Geol. Congr. Abstr. Paris, Resumes, No. 26, Vol. 3, 927.

Dolenec, T., Kušej and Pezdič, J. (1980): The isotopic composition of oxygen and carbon in lead and zinc deposits of Northern Karavanke. *In:* Mineral Deposits of the Alps (Schneider, H. J., ed.), Springer-Verlag Berlin-Heidelberg-New York-Tokyo, 176–188.

Drovenik, M., Štručl, I., and Pezdič, J. (1980): Izotopska sestava žvepla v svincevocinkovih nahajališč severnik Karavank. Rud.-Metal Zb. (Ljubljana), Vol. 27, No. 2–3, 179–197, No. 4, 413–436.

Drovenik, M., Duhovnik, J., and Pezdič, J. (1976): The sulfur isotope composition of sulfides from ore deposits in Slovenia (in Slovenian) Rud. Met. Zb. (Ljubljana) Vol. 24, No. 2–3, 195–246.

Drovenik, M., Plenicar, M., and Drovenik, F. (1980): Nastanek rudisc v SR Sloveniji. Geologija, Razprave in Porocila (Ljubljana), Vol. 23/I, 1–157.

Drovenik, M. (1983): Mobilization of ore and gangue minerals in some slovenian mineral deposits. *In:* Ore Mobilization in the Alps and SE-Europe (Petrascheck, W. E., ed.), Schriftenreihe der Erdwissenschaftlichen Komm., Österr. Akad. Wiss., Springer-Verlag Wien-New York, Vol. 6, 75–81.

Dzulynski, S. (1977 a): Comments on the genesis of the Eastern Alpine Zn-Pb deposits. Miner. Deposita, *12*, 219–233.

Dzulynski, S., and Sass-Gustkiewicz, M. (1977 b): Comments on the genesis of the Eastern-Alpine Zn-Pb deposits. Miner. Deposita (Berlin), Vol. 12, No. 2, 219–233.

Dzulynski, S., and Sass-Gustkiewicz, M. (1978): Ore breccias in the Triassic rocks of the Cracow-Silesian region (Poland). *In:* Ergebnisse der Österr. Projekte des Internationalen Geologischen Korelationsprogrammes (IGCP) bis 1976. (Zapfe, H., ed.), Österr. Akad. Wiss., Erdwissenschaftl. Komm., Schriftenreihe Vol. 3, 125–130.

Dzulynski, S., and Sass-Gustkiewicz, M. (1980): Dominant ore-forming processes in the Crakow-Silesian region. *In:* Proceed. of the 5th IAGOD Symp. (Snowbird), Schweizerbart Stuttgart (Ridge, J. D., ed.), Vol. I, 415–430.

Fontbonté, L., and Amstutz, G. C. (1980): New observations on diagenetic crystallization rhythmites in the carbonate facies of the Triassic of the Alpujarrides (Betic Cordillera,

Southern Spain). I. Symp. Diagenesis. Barcelona 1980. Rev. Inst. Inv. Geol. Barcelona 1980, Vol. 34, 293–310.

FONTBONTÉ, L., and AMSTUTZ, G. C. (1982): Observations on ore rhythmites of the Trzebionka Mine, Upper Silesia-Cracow, Poland. *In:* Ore Genesis. The State of the Art (AMSTUTZ, G. C., et al., eds.), Springer-Verlag Berlin-Heidelberg-New York, 83–91.

FONTBONTÉ, L., and AMSTUTZ, G. C. (1983): Facies and sequence analysis of diagenetic crystallization rhythmites in strata-bound Pb-Zn-(Bba-F) deposits in the Triassic of Central and Southern Europe. *In:* Mineral Deposits of the Alps (SCHNEIDER, H. J., ed.), Springer-Verlag Berlin-Heidelberg-New York-Tokyo, 347–358.

FUCHS, Y. (1976): A propos de quelque exemples de relation entre altérations et concentrations métallifères (Pb, Zn, Cu, U, Ba, Sr, F, ...) dans le domaine épicontinental. Bull. Soc. Géol. Fr., XVIII, *1,* 91–94.

FUCHS, Y. (1978 a): Über die Erzlagerstätten im Tertiär von Tunesien und Constantinois (Ost-Algerien). Verh. Geol. B.-A., Vienna, *3,* 311–319, 1978/3, Proceed. 3rd ISMIDA (Leoben 1977), 127–145.

FUCHS, Y. (1978 b): Sur un exemple de relation entre une minéralisation barytique et un milieu à évaporites; le gîte de Pessens (Aveyron) Réunion R.C.P. 459, avril 1978, Paris. *In:* Sc. de la Terre, Nancy, *22, 2,* 127–146.

FUCHS, Y. (1980 a, b): Sur l'importance des eaux connées pour la formation de certains gisement en milieux évaporitiques. Bull. Centr. Rech. Explor. Prod. Elf Aquitaine, *4, 1,* 433–443.

FUCHS, Y. (1980 b): Sulfates et sulfures – Gîtes en milieu carbonaté. Quelque exemples abstr., XXVI Intern. Geol. Congress, Paris, Sect. XIII.

FUCHS, Y. (1981 a): Les minéralisation des milieux carbonatés de la platforme épicontinentale. Mem. Soc. Géol. Ital., *22,* 35–51.

FUCHS, Y. (1981 b): On some aspects of the relationship between mineralization and evaporitic facies, abstr. Fourth Int. Symp. Min. Alps (ISMIDA), Berchtesgaden, 71.

GRUSZINSKY, H. (1982): The genesis of the zinc-lead ore deposits of Upper Silesia. *In:* Ore Genesis. The State of Art (AMSTUTZ, G. C., et al., eds.), Springer-Verlag Berlin-Heidelberg-New York, p. 92–95, 1982.

GSTREIN, P. (1983): Über mögliche Umlagerungen von Fahlerz devonischer Schwazer Dolomit wie auch in der angrenzenden Schwazer Trias. *In:* Ore Mobilization in the Alps and SE-Europe (PETRASCHECK, W. E., ed.), Schriftenreihe der Erdwissenschaftlichen Komm. Österr. Akad. Wiss., Springer-Verlag Wien-New York, Vol. 6, 65–74.

HAGENGUTH, G. (1983): Geochemische und fazielle Untersuchungen an den Maxerbänken im Pb-Zn-Bergbau Bleiberg-Kreuth/Kärnten. Doctor thesis Univ. Wien, 1–120.

HAGENGUTH, E., BOBER, E., GÖTZINGER, M. A., and LEIN, S. (1982): Beiträge zur Geologie, Mineralogie und Geochemie der Pb-Zn-Vererzungen Annaberg und Schwarzenberg (Niederösterreich). Jahrb. Geol. B.-A. Wien, *125,* 1/2, 155–218.

JOBSTRAIBIZER, P. (1981): Mineralogia, chimismo e transformazioni diagenetiche nelle tufiti mediotriassico delle serie calcareo-tufacea del giacimento di Raibl (Cave di Predil-Friuli). Mem. Sci. Geol. Ital. (Padova), Vol. 34, 355–386.

HEIN, U. F., and SCHNEIDER, H. J. (1983): Fluorine anomalies accompanying the Alpine Pb-Zn deposits compared to the geochemistry of their fluorites. *In:* Mineral Deposits of the Alps (SCHNEIDER, H. J., ed.), Springer-Verlag Berlin-Heidelberg-New York-Tokyo, 198–212.

HIRDES, W., and AMSTUTZ, G. C. (1978): Geologische und metallogenetische Untersuchungen im nördlichen Atacocha-Distrikt, Zentralperu, am Beispiel der Blei-Zink-Grube Machcán. Geol. Jb. D., Vol. 28, 55.

ILAVSKY, J. (1979): Stratiforme Erze in den Westkarpaten und ihre gegenseitigen Korrelationen. Verh. Geol. B.-A., Wien, 1978/3, 393–402 (1979), Proceed. 3rd ISMIDA (Leoben 1977), 219–228.

LEVIN, P., and AMSTUTZ, G. C. (1973): Neue Untersuchungen über schichtgebundene Lagerstätten im zentralen Ost-Peru. Forsch. Geol. Paläont., Münster, *31/32,* 211–237.

KANTOR, J. (1977): Pb-Zn-ores of the Westcarpathian Triassic and the distribution of their sulphur isotopes. *In:* Time- and Stratabound Ore Deposits (KLEMM, D. D., et al., eds.), Springer-Verlag Berlin.

KAPPEL, F., and SCHROLL, E. (1982): Ablauf und Bildungstemperatur der Blei-Zink-Vererzung Bleiberg-Kreuth/Kärnten. Carinthia II (Klagenfurt), Vol. 127/2, 49–62.

KÖPPEL, V., and KOSTELKA, L. (1976): The isotopic composition of ore-lead from Bleiberg (abstr.). Symposium Delft, Amsterdam, 1976.

KÖPPEL, V., and SCHROLL, E. (1979): Bleiisotopenzusammensetzung von Bleierzen aus dem Mesozoikum der Ostalpen. *In:* Third international symposium on the mineral deposits of the Alps (HOLZER, H. F., et al., eds.), Austria, Geol. Bundesanst., Verh., No. 3, 403–409.

KÖPPEL, V. (1983): Summary of lead-isotopes data from ore deposits of the Eastern and Southern Alps. Some metallogenetic and geotectonic implications. *In:* Mineral Deposits of the Alps (SCHNEIDER, H. J., ed.), Springer-Verlag Berlin-Heidelberg-New York-Tokyo, 162–168.

KRANZ, J. R. (1973): Die Strontium-Verteilung in den Arlbergschichten (Oberes Ladin) des Klostertales (Vorarlberg), Nördliche Kalkalpen. N. Jb. Geol. Paläont. Mh., Stuttgart, *3,* 170–187.

KRANZ, J. R. (1976): Stratiforme und diskordante Zink-Blei-Anomalien im erzhöffigen Oberen Wettersteinkalk (alpine Mitteltrias). Mineral. Deposita, Berlin, Vol. 11, 6–23.

LAGNY, P. (1980): Les gîsements stratiformes associés aux évaporites. Position dans le temps et place dans l'espace des bassins sédimentaire évaporitiques. Bull. Centr. Rech. Expl. Elf Aqu., Pau, *4, 1,* 445–478.

MINČEVA-STEFANOVA, J. (1978): Geological position and mineralogy of the polymetallic deposits in the Western Balkan Mountains confined to Triassic sediments. *In:* Ergebnisse der Österr. Projekte des Internationalen Geologischen Korelationsprogrammes (IGCP) bis 1976. (ZAPFE, H., ed.), Österr. Akad. Wiss. Erdwissenschaftl. Komm., Schriftenreihe Vol. 3, 111–123.

MINČEVA-STEFANOVA, J. (1978): On the causes for localization of stratabound-type ores in the Iskâr gorge (in Russian): Compt. rend. Acad. bul. Sci., T. 31, No. 10, 1329–1331.

MINČEVA-STEFANOVA, J. (1979): Paragenetische und geochemische Bedeutung der Mineralien von den Bravoit-, Cobaltin- und Linneit-Gruppen in den Polymetallvererzungen des „Stratabound"-Typs in den Balkaniden. Verh. Geol. B.-A. (Wien) 1978/3, p. 419–435, Proceed. 3rd ISMIDA (Leoben 1977), p. 245–261.

MINČEVA-STEFANOVA, J., with BONEV, I., and PUNEV, L. (1980): Pearceite with an intermediate unit cell–first find in nature (in Bulgarian). Geochem. mineral. petrol., Sofia, Vol. 11, 13–34.

MINČEVA-STEFANOVA, J. (1981): Electron-microprobe investigations of the dolomitization and its relationship with the sulphide mineralizations in the polymetallic deposits confined to the Triassic calcareous sediments in the Western Balkan, abstr. Proceed. 4th ISMIDA, Berchtesgaden, Oct. 4–10, 45, 1981.

MÖLLER, P., RAJAGOPALAN, G., and GERMANN, K. (1976): A geochemical model for dolomitization based on material balance; Part II. Trace element distribution during dolomitization. Geol. Jb. D. 20, Hannover, 57–76.

MÖLLER, P., DULSKI, P., and SCHNEIDER, H. J. (1983): Interpretations of the Ga and Ge distribution in sphalerite. *In:* Mineral Deposits of the Alps (SCHNEIDER, H. J., ed.) Springer-Verlag Berlin-Heidelberg-New York-Tokyo, 213–222.

NIEC, M. (1980): Model of formation of Cracovian-Silesian type zinc-lead deposit of Bolesan near Olkuz (Poland). *In:* Proceed. of the 5th IAGOD Symp. (Snowbird), Schweizerbart Stuttgart (RIDGE, J. D., ed.), Vol. I, 431–444.

OMENETTO, P., and VAILATI, G. (1977): Ricerche geominerarie nel settore centrale del distretto a Pb, Zn, fluorite e barite di Gorno (Lombardia). L'Ind. Miner. (Milano), 25–44.

OMENETTO, P. (1979): Significant ore fabric relationships in the lead, zinc, fluorite and barite deposits of the Triassic province (Italian Southern Alps). Ann. Soc. Geol. Belgique, Vol. 102, 519–529.

PAK, E., and SCHROLL, E. (1980): Schwefelisotopenzusammensetzung von Baryten aus den Ost- und Südalpen (Sulphur isotope composition of barites of the Eastern and Southern Alps). Tschermaks Min. Petr. Mitt., Vol. 27, 79–91.

PAWLOWSKA, J., and WEDOW, H. (1980): Strata-bound zinc-lead deposits of the Upper Silesian region, Poland—a review of some recent research. *In:* Proceed. of the 5th IAGOD Symp. (Snowbird), Schweizerbart Stuttgart (RIDGE, J. D., ed.), Vol. I, 467–486.

PETRASCHECK, W. E. (1977): Metallogenese von Laurion-Granitischer Herd oder Paleokarst? – Annales Géologiques des Pajies Helleniques, Tom. Bd. 46, Athen.

PETRASCHECK, W. E. (1967): Die Blei-Zinklagerstätten in Kalken des westlichen Taurus, M.T.A. Bull. 68. Ankara.

PIMMINGER, M. (1983): In-situ Spuren- und Isotopenanalyse mit Sekundärionen-Massenspektromonie (SIMS). Doctor thesis Techn. Univ. Wien, 1–137.

SAMANIEGO, A. (1982): Correlation of strata-bound mineral deposits in the Early Cretaceous Santa Metallotect of North and Central Peru. *In:* Ore Genesis. The State of Art (AMSTUTZ, G. C., et al., eds.), Springer-Verlag Berlin-Heidelberg-New York, 508–527.

SCHMIDT, S. TH., and AMSTUTZ, G. C. (1983): Mineralogical investigations of strata-bound Pb-Zn-Ba occurrences in the Triassic in Central Switzerland. *In:* Mineral Deposits of the Alps (SCHNEIDER, H. J., ed.), Springer-Verlag Berlin-Heidelberg-New York-Tokyo, 117–127.

SCHNEIDER, H. J. (1972): The problem of the ore-mobilization and -transport in the lead-zinc-ore deposits of the East-Alpine Middle Triassic Beds. Proc. 2nd Internat. Symposium on the Mineral Deposits of the Alps, Bled 1971, Geologija, Razprave in Porocila (Ljubljana), Vol. 15, 217–219.

SCHNEIDER, H. J., MÖLLER, P., and PAREKH, P. P. (1975): Rare Earth element distribution in fluorites and carbonate sediments of the East-Alpine Mid-Triassic sequences in the Nördliche Kalkalpen. Miner. Deposita (Berlin), Vol. 10, No. 4, 330–344.

SCHNEIDER, H. J., MÖLLER, P., PAREKH, P. P. et al. (1977): Fluorine contents in carbonate sequences and rare earth distribution in fluorites of Pb-Zn deposits in East-Alpine Mid-Triassic. Miner. Deposita (Berlin), Vol. 12, 22–36.

SCHROLL, E. (1978): Zur Korrelation geochemischer Charakteristika der Blei-Zink-Lagerstätte Bleiberg-Kreuth mit anderen schichtgebundenen Vererzungen in Karbonatgesteinen (Bleiberg-Kreuth). *In:* Ergebnisse der österr. Projekte des Internationalen Geologischen Korrelationsprogrammes (IGCP) bis 1976 (ZAPFE, H., ed.), Österr. Akad. Wiss., Erdwissenschaftliche Komm., Schriftenr., Vol. 3, 131–158.

SCHROLL, E., and EICHER, U. (1978): Sauerstoff- und Kohlenstoffisotopen-Bestimmungen von einigen Kalksteinen und Kalziten aus der Blei-Zink-Lagerstätte Bleiberg-Kreuth/Kärnten. Anz. Öst. Akad. Wiss., math.-naturw. Kl., Nr. 2, 1–3.

SCHROLL, E. (1981): REM-Untersuchungen an Schalenblenden: Ein Beitrag zur As- und Tl-Führung von Sphaleriten. Fortschritte der Mineralogie, Vol. 59/1, 178–179.

SCHROLL, E., SCHULZ, O., and PAK, E. (1983): Sulphur isotope distribution in the Pb-Zn-deposit Bleiberg/Carinthia). Miner. Deposita, Vol. 18, 17–25.

SCHROLL, E. (1983): The geochemical characterization of the Bleiberg type in comparison with other carbonate hosted lead-zinc mineralizations. *In:* Mineral Deposits of the Alps (SCHNEIDER, H. J., ed.), Springer-Verlag Berlin-Heidelberg-New York-Tokyo, 189–197.

SCHROLL, E. (1984): Geochemical indicator parameters of lead-zinc ore deposits in carbonate rocks. *In:* WAUSCHKUHN et al. (eds.), Syngenesis and Epigenesis in the Formation of Mineral Deposits, Springer-Verlag Berlin-Heidelberg, 294–305.

SCHULZ, O. (1976): Typical and non-typical sedimentary ore fabrics: Handbooks of Stratabound and Stratiform Ore Deposits (WOLF, K. H., ed.), Elsevier, Amsterdam, Vol. 3, 295–338.

SCHULZ, O., and SCHROLL, E. (1977): Die Pb-Zn-Lagerstätte Bleiberg-Kreuth; Stand der geowissenschaftlichen Forschung 1976, Projekt 2437, 2776 S. Verh. Geol. B.-A., Austria, Vol. 1977, No. 3, 375–386.

ŠTRUCL, I. (1981): Die schichtgebundenen Blei-Zink-Lagerstätten Jugoslawiens. Mitt. Österr. Geol. Ges., Vol. 74/75, 307–322.

SUREDA, R. J., and AMSTUTZ, G. C. (1981): Neue Untersuchungen über die schichtgebundenen Pb-Zn-Lagerstätten in der Sierra de Aguilar, Provinz Jujuy, Argentinien. Zbl. Geol. Paläont. I, 494–504.

VASILEFF, L., and POPOV, A. (1977): Formations minerales et zones metallogeniques de la Tunesie. Geol. Balcanica (Sofia), Vol. 7/4, 43–56.

VLAD, S. (1978): Metallogeneza Triasica di zona Tulcea (Dobrogea di Nord). St. Cerc. geol. geofiz. geogr. Geologie (Bucaresti), Vol. 23, 249–258.

WOLTER, R., and SCHNEIDER, H. J. (1983): Saline relics of formation water in the Wettersteinkalk and their genetical connection with the Pb-Zn mineralization. *In:* Mineral Deposits of the Alps (SCHNEIDER, H. J., ed.), Springer-Verlag Berlin-Heidelberg-New York-Tokyo, 223–230.

Other references:

ROUTHIER, P. (1980): Où sont les mètaux pour l'avenir? BRGM, Orleans.

SCHNEIDERHÖHN, M. (1953): Fortschritte in der Erkenntnis sekundär-hydrothermaler und regenierter Lagerstätten. N. Jb. Min. Mh. 9/10, 223-237.

GPSR Compliance

The European Union's (EU) General Product Safety Regulation (GPSR) is a set of rules that requires consumer products to be safe and our obligations to ensure this.

If you have any concerns about our products, you can contact us on ProductSafety@springernature.com

In case Publisher is established outside the EU, the EU authorized representative is:

Springer Nature Customer Service Center GmbH
Europaplatz 3
69115 Heidelberg, Germany

Batch number: 10726298

Printed by Printforce, the Netherlands